LIST OF ELEMENTS WITH THEIR SYMBOLS AND ATOMIC MASSES

ELEMENT	SYMBOL	ATOMIC NUMBER	ATOMIC MASS[a] (amu)
Actinium	Ac	89	(227)
Aluminum	Al	13	26.9815
Americium	Am	95	(243)
Antimony	Sb	51	121.75
Argon	Ar	18	39.948
Arsenic	As	33	74.9216
Astatine	At	85	(210)
Barium	Ba	56	137.34
Berkelium	Bk	97	(247)
Beryllium	Be	4	9.01218
Bismuth	Bi	83	208.9806
Boron	B	5	10.81
Bromine	Br	35	79.904
Cadmium	Cd	48	112.40
Calcium	Ca	20	40.08
Californium	Cf	98	(251)
Carbon	C	6	12.01115
Cerium	Ce	58	140.12
Cesium	Cs	55	132.9055
Chlorine	Cl	17	35.453
Chromium	Cr	24	51.996
Cobalt	Co	27	58.9332
Copper	Cu	29	63.546
Curium	Cm	96	(247)
Dysprosium	Dy	66	162.50
Einsteinium	Es	99	(254)
Erbium	Er	68	167.26
Europium	Eu	63	151.96
Fermium	Fm	100	(257)
Fluorine	F	9	18.9984
Francium	Fr	87	(223)
Gadolinium	Gd	64	157.25
Gallium	Ga	31	69.72
Germanium	Ge	32	72.59
Gold	Au	79	196.9665
Hafnium	Hf	72	178.49
Hahnium[b]	Ha	105	(262)
Helium	He	2	4.00260
Holmium	Ho	67	164.9303
Hydrogen	H	1	1.0080
Indium	In	49	114.82
Iodine	I	53	126.9045
Iridium	Ir	77	192.22
Iron	Fe	26	55.847
Krypton	Kr	36	83.80
Lanthanum	La	57	138.9055
Lawrencium	Lr	103	(257)
Lead	Pb	82	207.2
Lithium	Li	3	6.941
Lutetium	Lu	71	174.97
Magnesium	Mg	12	24.305
Manganese	Mn	25	54.9380
Mendelevium	Md	101	(256)
Mercury	Hg	80	200.59
Molybdenum	Mo	42	95.94
Neodymium	Nd	60	144.24
Neon	Ne	10	20.179
Neptunium	Np	93	237.0482
Nickel	Ni	28	58.70
Niobium	Nb	41	92.9064
Nitrogen	N	7	14.0067
Nobelium	No	102	(255)
Osmium	Os	76	190.2
Oxygen	O	8	15.9994
Palladium	Pd	46	106.4
Phosphorus	P	15	30.9738
Platinum	Pt	78	195.09
Plutonium	Pu	94	(244)
Polonium	Po	84	(209)
Potassium	K	19	39.102
Praseodymium	Pr	59	140.9077
Promethium	Pm	61	(145)
Protactinium	Pa	91	231.0359
Radium	Ra	88	226.0254
Radon	Rn	86	(222)
Rhenium	Re	75	186.207
Rhodium	Rh	45	102.9055
Rubidium	Rb	37	85.4678
Ruthenium	Ru	44	101.07
Rutherfordium[b]	Rf	104	(261)
Samarium	Sm	62	150.4
Scandium	Sc	21	44.9559
Selenium	Se	34	78.96
Silicon	Si	14	28.086
Silver	Ag	47	107.868
Sodium	Na	11	22.9898
Strontium	Sr	38	87.62
Sulfur	S	16	32.06
Tantalum	Ta	73	180.9479
Technetium	Tc	43	98.9062
Tellurium	Te	52	127.60
Terbium	Tb	65	158.9254
Thallium	Tl	81	204.37
Thorium	Th	90	232.0381
Thulium	Tm	69	168.9342
Tin	Sn	50	118.69
Titanium	Ti	22	47.90
Tungsten	W	74	183.85[c]
Unnilennium	Une	109	(266)
Unnilhexium	Unh	106	(263)
Unniloctium	Uno	108	(265)
Unnilseptium	Uns	107	(262)
Uranium	U	92	238.0
Vanadium	V	23	50.9
Xenon	Xe	54	131.3
Ytterbium	Yb	70	173.0
Yttrium	Y	39	88.9
Zinc	Zn	30	65.3
Zirconium	Zr	40	91.2

[a] Based on the assigned relative atomic mass of ^{12}C = exactly 12; parentheses denote the mass number of the isotope with the longest half-life.
[b] Name and symbol not officially approved.

5TH EDITION

IN PREPARATION FOR COLLEGE CHEMISTRY

5TH EDITION

IN PREPARATION FOR COLLEGE CHEMISTRY

G. WILLIAM DAUB · WILLIAM S. SEESE
HARVEY MUDD COLLEGE   CASPER COLLEGE

Pearson Education
Upper Saddle River, NJ 07458

Library of Congress Cataloging-in-Publication Data

Seese, William S.
In preparation for college chemistry / William S. Seese,
G. William Daub.—5th ed.
p.   cm.
Includes index.
1. Chemistry.   I. Daube, G. William.   II. Title.
QD33.S393   1993
540—dc20   89-8395
                                                   CIP

Editor in Chief: Tim Bozik
Cover designer: Karen Salzbach
Design Director: Jayne Conte
Pre-press buyer: Paula Massenero
Manufacturing buyer: Lori Bulwin
Photo editor: Lori Morris-Nantz
Production editor: Jennifer Fischer

Cover photo: Phototake, Inc.

Photo credits: p. 16, Jeff Gnass/Stock Market • p. 38, Stephen Frink/
Stock Market • p. 54, United Press International • p. 62, University
of Chicago • p. 76, The Bettmann Archive • p. 92, Fundamental
Photographs • p. 116, Lairia Druskis • p. 128, Intel • p. 147, NASA •
p. 168, Paul Silverman/Fundamental Photographs • p. 187, Dr. E. R.
Degginger • p. 190, Fundamental Photographs • p. 192, Katherine P.
Daub • p. 197, Shirley Zeilberg • p. 213, R. Folwell/Photo
Researchers • p. 234, Rick Altman/Stock Market • p. 255, AP/Wide
World Photos • p. 272, Michael Weisbrut/Stock Boston •

A Pearson Education Company
Upper Saddle River, NJ 07458

Printed in the United States of America
10 9 8 7 6

ISBN 0-13-120627-3

Prentice-Hall International (UK) Limited,London
Prentice-Hall of Australia Pty. Limited, Sydney
Prentice-Hall Canada Inc., Toronto
Prentice-Hall Hispanoamericana, S.A., Mexico
Prentice-Hall of India Private Limited, New Delhi
Prentice-Hall of Japan, Inc., Tokyo
Pearson Education Asia Pte. Ltd., Singapore
Editora Prentice-Hall do Brasil, Ltda., Rio de Janeiro

Contents

Chapter 7   nomenclature of inorganic compounds   133

Chapter 8 chemical calculations 152

Chapter 9 chemical equations 172

Chapter 10   ionic equations   202

Chapter 11   stoichiometry   217

Preface

To the Student: We hope that you will read this preface. It will assist you in your study of chemistry. This book is written *for you.* The purpose of this book is simple: to help you prepare to take college chemistry. This text concentrates on the *fundamentals* that provide a foundation for future studies in chemistry. When you master these fundamentals, the more complex material becomes easier to learn. We have assumed that you have had little science background, let alone chemistry, and that your mathematics needs review. Therefore, we have included cartoons and analogies to help you understand some of the principles of chemistry and a chapter on basic mathematics (Chapter 1) to help you with your mathematics review.

In this text there are seven features that help you learn chemistry. They are:

1. *countdown,* a review of previous material

2. *tasks* and *objectives, keyed* to the examples, exercises, and problems (except general problems)

3. *problem solving,* using the *factor-unit* method

4. *examples,* followed by a solution and a comparable *exercise* for you to work

5. *problems,* at the *end* of the *chapter* similar to examples and exercises

6. *quiz,* at the end of each chapter to help you review the material in that chapter

7. discussion of a *compound* or *element* familiar to you.

See if you can identify these features in your text.

The *countdown* includes five questions from previous chapters, which act as a review or basis for the material to be introduced in the new chapter. The answers to these questions are in parentheses. If you need help in answering the question, you will find a section number in parentheses immediately following the question. You should refer to that section for further help. This is a new feature in this edition.

The *tasks* are things you must know to accomplish the objectives. The *objectives* give you specific pieces of information and tell you exactly what we want you to do with that information in order to master the material in the chapter. After each objective a section or example, exercise, and problem are listed. To accomplish the objective you should be able to do the exercise and problem. All problems except the general problems are *keyed* to the objectives. New terms for each chapter and the section in which they are introduced are listed in the first objective. These terms appear in bold print in the appropriate chapter section and in the running glossary in the margin adjacent to the section. The definitions for these terms also appear alphabetically in the Glossary at the end of the text. Also in the margins you may find short *Study Hints.* These Study Hints give you tips on how to remember or think through a solution.

To help you learn *to think* about chemistry and science in general and *to solve* problems, we have included a number of features. The *factor-unit method* (dimensional analysis), a general **problem solving** method, is introduced in Chapter 2. This is a general method for solving problems and can be applied to word problems you may encounter in any science. Various **examples** follow the new material presented in the chapters. Accompanying the example is a complete solution to the problem. Following the solution is an **exercise** for you to do. The answers to these exercises are found at the end of the chapter.

At the *end* of each *chapter* are a number of **problems,** and a **quiz** to help you check your understanding. We have included just a sufficient number of problems to cover the basics, so do **ALL** of them. The answers and selected solutions to these problems are found at the end of the book (Appendix 5). At the end of the problems is a special group of problems, called general problems. *General problems,* which often require you to use material from several sections of a chapter and also from previous chapters, challenge your overall understanding. They should be left until last. The chapter quiz is designed to help you test yourself on the material in the chapter. It is important that you use the quiz as a learning tool, to make sure that you are understanding the material. You should take the quiz after you have studied the material and worked all the problems in the chapter. The answers to the quiz questions are in Appendix 5.

At the very end of each chapter is a new feature in this edition titled **Compound** or **Element.** You will probably be familiar with the compound or element. In this short essay, we try to show you how chemicals play a role in your life. You may find some very interesting and useful facts in these short essays.

A few words of warning in studying chemistry or any subject:

1. DON'T read this book in bed. We don't think it will put you to sleep, but you will learn a lot more if you sit up and use paper and pencil to work the examples as you go. Your book will also be in better shape at the end of the semester for whatever use you may have for it.

2. DON'T try to learn it all the night before the exam. The easiest way is to do a little bit at a time. You should spend about two hours outside of class studying for every one hour spent in class. This then would mean about six hours study time per week outside of class. This study time doesn't count if you spend it talking to a friend, even if it is in the library! It counts only if you are actually studying. The slow-but-steady approach *does* pay off and you will *feel* better about school.

No one person writes a textbook. Many people advise the authors and many people have contributed to this textbook. We appreciate their support and input. Among those we specifically would like to thank are Dr. Daniel T. Haworth, Marquette University, Milwaukee, Wisconsin and the many reviewers of the text. We would especially like to thank our wives, Sandra Anne Hollenberg and Ann Reeves Seese, for their advice, suggestions, and support.

You are the final judge of any textbook. Therefore, we would greatly appreciate your writing to us and letting us know how you did in the course, what you thought of the book, what you had trouble with, and what you found helpful. With your help, we hope to keep this text a text for students who succeed in chemistry.

G. William Daub
Department of Chemistry
Harvey Mudd College
Claremont, California 91711

William S. Seese
Emeritus, Department of Chemistry
Casper College
Casper, Wyoming 82601

1

BASIC MATHEMATICS

COUNTDOWN   Perform the indicated mathematical operations. You may use a calculator if you wish. See Appendix 2 (Your Calculator), if you need help in using it.

5 Add:

$$6.21$$
$$25.11$$
$$\underline{3.87}$$ (35.19)

4 Subtract:

$$12.21$$
$$\underline{-3.92}$$ (8.29)

3 Multiply:

$$2.50 \times 11.0 =$$ (27.5)

2 Multiply:

$$2.24 \times 0.875 \times 3.25 =$$ (6.37)

1 Divide:

$$\frac{616}{14} =$$ (44)

TASKS   **1** Know the rules given in Section 1–1 for identifying significant digits.

2 Know the rules given in Section 1–2 for rounding off the nonsignificant digits.

OBJECTIVES

1 Give the distinguishing characteristicis of each of the following terms:
   (a) significant digits (Section 1–1)
   (b) exponent (Section 1–3)
   (c) exponential notation (Section 1–3)
   (d) scientific notation (Section 1–4)

2 Given various numbers, identify the number of significant digits in the number (Example 1–1, Exercise 1–1, Problem 1).

3 Given various numbers, add, subtract, multiply, and divide these numbers, and round off your answers correctly to the proper number of significant digits (Examples 1–2 and 1–3, Exercises 1–2 and 1–3, Problems 2 and 3).

4 Given a number, express that number in exponential notation as requested (Examples 1–4 and 1–5, Exercises 1–4 and 1–5, Problems 4 and 5).

5 Given numbers in exponential notation, add, subtract, multiply, or divide these numbers (Examples 1–6 and 1–7, Exercises 1–6 and 1–7, Problems 6 and 7).

6 Given a number in exponential notation, determine the square root of the number or raise the number to a requested power (Examples 1–8 and 1–9, Exercises 1–8 and 1–9, Problems 8 and 9).

7 Given a number, express that number in scientific notation to three significant digits (Example 1–10, Exercise 1–10, Problem 10).

8 Given a linear equation, solve the equation for the unknown (Example 1–11, Exercise 1–11, Problem 11).

Why do I have to take chemistry? Thousands of students have asked this question for nearly 100 years. The standard answer has been that chemistry will be useful in your profession and in the modern technological world in which you live. But the real answer for many students is that it is required for a degree in engineering, nursing, home economics, forestry, the agricultural sciences, and other sciences. Now, chemistry can be more to you than a requirement; in fact, some people even find it so fascinating that they devote a lifetime to it as a teacher, a researcher, or a technical sales representative.

Chemistry is everywhere because chemicals are everywhere. In the morning when you get up, you wash your face and hands with a chemical (soap) and brush your teeth with toothpaste (calcium carbonate or baking soda, soap, sodium fluoride, and other ingredients). You may wear a wool (natural organic fiber) or nylon (synthetic organic fiber) sweater. Even your jeans are chemicals. They are made of cotton, a natural organic fiber; or of cotton blended with a synthetic fiber).

Before we can begin a preparation for college chemistry, we need to review some basic mathematics. In this chapter we consider significant digits (figures), exponents, and the solving of linear equations.

1–1 Significant Digits

significant digits (figures) Digits in a measurement that are known to be precise, along with a final digit about which there is some uncertainty.

The **significant digits** (figures) are the digits in a measurement that are known to be precise, along with a final digit about which there is some uncertainty. We know that there are exactly 100 cents in a dollar, so this is an exact number. Exact numbers are precisely known and can have as many significant digits as needed. The exchange rate of the U.S. dollar to the British pound is about $1.50 to the pound. This is a variable number, for it changes from day to day. The last digit recorded (0) is uncertain (varies), and in some cases even the next to last digit (5) may vary.

To determine the number of significant digits, we follow certain rules.

1. *Nonzero digits:* 1, 2, 3, 4, 5, 6, 7, 8, 9 are always significant.

2.1	two significant digits
21.7	three significant digits
21.75	four significant digits

2. *Leading zeros:* zeros that appear at the start of a number are *never* significant because they act only to fix the position of the decimal point in a number less than 1.

0.675	three significant digits
0.0675	three significant digits
0.00675	three significant digits

3. *Confined zeros:* zeros that appear between nonzero numbers are *always* significant.

208	three significant digits
2008	four significant digits
20,008	five significant digits

4. *Trailing zeros:* zeros at the end of a number are significant *only* if the number (a) contains a decimal point *or* (b) contains an overbar.

125.0	four significant digits
125.00	five significant digits
125$\overline{0}$	four significant digits
12,5$\overline{00}$	five significant digits
1250	three significant digits (The number does not contain a decimal point and the zero does not have an overbar).

EXAMPLE 1–1

Determine the number of significant digits in the following numbers.

ANSWERS (RULES)

(a) 12.5	3(1)
(b) 908	3(1,3)

ANSWERS (RULES)

(c)	6.50	3(1,4)
(d)	0.056	2(1,2)
(e)	35$\bar{0}$	3(1,4)
(f)	6.02	3(1,3)
(g)	7065	4(1,3)
(h)	0.604	3(1,2,3)
(i)	20.05	4(1,3)
(j)	12,$\bar{0}$00	3(1,4)

Exercise 1–1

Determine the number of significant digits in the following numbers.[1]

(a) 111   (b) 101$\underline{1}$ (c)   7.50
(d) 0.00520   (e) 58$\bar{0}$ (f) 62.080

1–2 Mathematical Operations Involving Significant Digits[2]

We will now use these significant digits to do simple calculations.

Addition and Subtraction

In addition and subtraction, *the answer must not contain a smaller place* (i.e., decimal units, tenths, etc.) *than the number with the smallest place*. The sum of **25.1** + 22.11 is 47.21, but the answer must be expressed to only the tenths decimal place because the tenths decimal place is the smallest place in the number 25.1; hence, the answer is 47.2. The reason becomes obvious if you note that the hundredths decimal place is not measured in the number 25.1 and, thus, could vary widely. The difference of 4.732 − **3.62** is 1.112, but the answer must be expressed to only the hundredths decimal place, because the hundredths decimal place is the smallest place in the number 3.62; hence, the answer is 1.11.

---

[1] Answers to all exercises are at the end of the chapter.

[2] In chemistry you will be solving many mathematical problems. We suggest that you buy an inexpensive calculator, but before you do this, ask your instructor what model he or she recommends. The calculator should have the following functions: +, −, ×, ÷, and $\sqrt{}$. For Chapter 14 you will need the function log x. In college chemistry you will need not only the function log x, but also the function 10^x. See Appendix 2 (Your Calculator) if you need help in using your calculator. Use your calculator throughout this book.

Multiplication and Division

In multiplication and division, *the answer must not contain any more significant digits than the **least** number of significant digits in the numbers used in the multiplication or division.* The product of 22.23 × 2.15 is 47.7945, but the answer must be expressed to only three significant digits since 2.15 has only three significant digits; hence, the answer is 47.8. The quotient of $\dfrac{22.23}{2.15}$ is 10.339535, but the answer again must be expressed to only three significant digits since 2.15 has only three significant digits; hence, the answer is 10.3.

Rounding Off

The next problem facing us is how to round off the *nonsignificant* digits and arrive at the desired number of significant digits. The following rules apply to rounding off nonsignificant digits.

1. If the first nonsignificant digit is *less* than 5, drop it and leave the last significant digit the same. Hence, in the preceding example, 10.3395 is equal to 10.3 to three significant digits.

2. If the first nonsignificant digit is *more* than 5 or *is* 5 followed by *numbers other than zeros*, drop the nonsignificant digit(s) and increase the last significant digit by one. Thus, 47.7945 and 47.752 are each equal to 47.8 to three significant digits.

3. If the first nonsignificant digit is 5 and is followed by *zeros*, drop the 5 and:

    (a) increase the last significant digit by *one if it is odd*; or

    (b) leave the last significant digit *the same if it is even.* Thus, 47.150 is equal to 47.2 to three significant digits and 47.450 is equal to 47.4 to three significant digits.

4. Nonsignificant digits to the *left of the decimal point* are not discarded, but are replaced by zeros. Thus, 1263 becomes 1260 and not 126 when rounded off to three significant digits. Similarly, 13,269 is equal to 13,300 and not 133 to three significant digits. Note that these new zeros are not significant digits [see rule 4 in Section 1–1].

EXAMPLE 1–2

Round off the following numbers to three significant digits.

ANSWERS (RULE)

(a)	261.3	261 (1)
(b)	453.6	454 (2)
(c)	474.5	474 (3b)
(d)	687.54	688 (2)

ANSWERS (RULE)

(e) 687.50	688 (3a)
(f) 3572	3570 (1, 4)
(g) 688.50	688 (3b)
(h) 12.750	12.8 (3a)
(i) 0.027650	0.0276 (3b)
(j) 93,271,853	93,300,000 (2, 4)
(k) 0.027654	0.0277 (2)
(l) 0.027750	0.0278 (3a)

Exercise 1–2

Round off the following numbers to three significant digits.

(a) 262.4   (b) 434.8   (c) 61.250
(d) 61.253   (e) 61.350   (f) 629.6

Now let us consider some mathematical operations applying the rules governing addition and subtraction, multiplication and division, and rounding off.

EXAMPLE 1–3

Perform the indicated mathematical operations and express your answer to the proper number of significant digits.

(a) 0.647 + 0.03 + 0.31

SOLUTION The smallest place is the hundredths decimal place in the numbers 0.03 and 0.31; express the answer to the hundredths decimal place.

$$
\begin{array}{l}
0.647 \\
0.03 \quad \longleftarrow \text{numbers with the smallest places} \\
\underline{0.31} \quad \swarrow \\
0.987 \qquad 0.99 \qquad \textit{Answer}
\end{array}
$$

Rounded off to the hundredths decimal place, 0.987 is 0.99.

(b) 24.78 − 0.065

SOLUTION The smallest place is the hundredths decimal place in the number 24.78; express the answer to the hundredths decimal place.

$$
\begin{array}{l}
24.78 \quad \longleftarrow \text{number with the smallest place} \\
\underline{-0.665} \\
24.715 \qquad 24.72 \qquad \textit{Answer}
\end{array}
$$

Rounded off to the hundredths decimal place, 24.715 is 24.72.

(c) 753×13

SOLUTION   The number with the least number of significant digits is 13, which has two, so the answer must be expressed to two significant digits.

$$753 \times 13 = 9789 \qquad 9800 \qquad Answer$$

Rounded off to two significant digits, 9789 is 9800. The trailing zeros in the number 9800 are not significant, as there is no decimal point in the number and there is no bar (‾) above the zeros.

(d) $\dfrac{181.8}{75}$

SOLUTION   the number with the least number of significant digits is 75, which has two; hence, the answer must be expressed to two significant digits.

$$\frac{181.8}{75} = 2.424, \text{ from your calculator}$$

Rounded off to two significant digits, 2.424 is 2.4.   *Answer*

(e) $\dfrac{9.74 \times 0.12}{1.28}$

SOLUTION   The number with the least number of significant digits is 0.12, which has two, so the answer must be expressed to two significant digits.

$$\frac{9.74 \times 0.12}{1.28} = 0.913125, \text{ from your calculator}$$

$$0.91 \qquad Answer$$

In performing this operation on your calculator, do the entire operation in *one* step. You can either divide and multiply, or multiply and then divide. Both ways give the same answer on your calculator, 0.913125. Do **not** carry out the operation in *two* steps. Do not perform the multiplication or division, *remove* the answer from your calculator, and then put this answer back in your calculator and do the next step. If you do the operation this wrong way, you may make a mistake in writing down the answer from one of the steps.

Exercise 1–3

Perform the indicated mathematical operations and express your answer to the proper number of significant digits.

(a) $1.25 + 6.732 + 0.843$

(b) $7.86 + 4.30 - 6.45$

(c) 0.073×453.3

(d) $\dfrac{173.2}{25.3}$

(e) $\dfrac{6.45 \times 0.25}{1.61}$

1-3 Exponents

In scientific work it is often necessary to use large numbers or extremely small numbers. As a practical example, the world population is estimated at 5,600,000,000 people and the U.S. population at 258,000,000 people. The U.S. population represents a mere 0.046 factor of the world population. This section presents a method that allows such numbers to be written in a more condensed form.

exponent A number or symbol written as a superscript above another number or symbol, the base, denoting the number of times the base is to be multiplied by itself.

An **exponent** is a number or symbol written as a superscript above another number or symbol, the *base*, denoting the number of times the base is to be multiplied by itself. The number of times a base is repeated as a factor is called the "power of the base." For example, in the symbol x^n, n is the exponent and x is the base. We read "x^n" as "x raised to the nth power," which is equal to

$$\underbrace{x \cdot x \cdot x \ldots x}_{n \text{ factors}}$$

We read "10^6" as "10 raised to the sixth power," and it equals

$$\underbrace{10 \cdot 10 \cdot 10 \cdot 10 \cdot 10 \cdot 10}_{6 \text{ factors}} \text{ or } 1,000,000$$

As review, let us consider some powers of 10 as given in Table 1–1. We can express 1000 as $10 \cdot 10 \cdot 10$, or 10^3; $\frac{1}{1000}$ can be expressed as $\frac{1}{10} \cdot \frac{1}{10} \cdot \frac{1}{10}$ or $1/10^3$—hence, 10^{-3}.

exponential notation A form for expressing a number using a product of two numbers; one of the numbers is a decimal and the other is a power of 10.

Now let us consider expressing numbers in exponential notation. **Exponential notation** is a form for expressing a number using a product of two numbers; one of the numbers is a decimal and the other is a power of 10. For example, 24.1×10^4 is in exponential notation with 24.1 being the decimal and 10^4 being the power of 10.

TABLE 1–1 POWERS OF 10

$$1000 = 10^3$$
$$100 = 10^2$$
$$10 = 10^1$$
$$1 = 10^{0a}$$
$$\tfrac{1}{10} = 0.1 = 10^{-1b}$$
$$\tfrac{1}{100} = 0.01 = 10^{-2b}$$
$$\tfrac{1}{1000} = 0.001 = 10^{-3b}$$

[a] Any nonzero number raised to the zero power is equal to 1, such as $x^0 = 1$, or $10^0 = 1$.

[b] Any number written with a base and a negative exponent is the inverse of another number using the *same base*, and the corresponding *positive* exponent. For example x^{-n} is the inverse of x^n (since $x^{-n} = 1/x^n$), and 10^{-6} is the inverse of 10^6 (since $10^{-6} = 1/10^6$).

To express a number in exponential notation, you may find the following guidelines helpful.

1. A positive exponent means a number *larger* than 1. A negative exponent means a number *smaller* than 1.

2. The exponent is equal to the *number of places* you move the decimal point.

3. Changing a number by shifting the decimal point to the *left* of its original position results in a *positive* exponent. Changing a number by shifting the decimal point to the *right* of its original position results in a *negative* exponent.

EXAMPLE 1–4

Express the estimated world population of 5,600,000,000 people or 5,600,000,000 in exponential notation as 5.6×10^n.

SOLUTION   Moving the decimal point to the left to obtain 5.6 as the decimal means that the exponent (n) must be positive. To obtain 5.6 we must move the decimal point *nine* places to the left of its original position. Hence, the answer is 5.6×10^9 people.   *Answer*

EXAMPLE 1–5

Express the factor of the U.S. population to that of the world population, 0.046, in exponential notation as 4.6×10^n.

SOLUTION   Moving the decimal point to the right to obtain 4.6 as the decimal means that the exponent (n) must be negative. To obtain 4.6 we must move the decimal point two places to the right of its original position. Hence, the answer is 4.6×10^{-2}.   *Answer*

Exercise 1–4

Express 7,280,000 in exponential notation as 7.28×10^n.

Exercise 1–5

Express 0.000543 in exponential notation as 5.43×10^n.

Addition and Subtraction of Exponential Numbers

To add or subtract exponential numbers, we must express *each* quantity to the *same power of 10*. Add or subtract the decimal parts in the usual manner and record the powers of 10.

EXAMPLE 1–6

Carry out the operations indicated on the following exponential numbers

(a) $3.40 \times 10^3 + 2.10 \times 10^3$

SOLUTION   Both numbers have the same power of 10 (10^3); hence, they can be added.

$$3.40 \times 10^3$$
$$\underline{2.10 \times 10^3}$$
$$5.50 \times 10^3 \quad \textit{Answer}$$

(b) $4.20 \times 10^{-3} + 1.2 \times 10^{-4}$

SOLUTION   To add these numbers, first convert them to the *same power of 10*. The number 1.2×10^{-4} converts to 0.12×10^{-3}, following the guidelines given earlier. We can now add these numbers.

$$4.20 \times 10^{-3}$$
$$\underline{0.12 \times 10^{-3}}$$
$$4.32 \times 10^{-3} \quad \textit{Answer}$$

Exercise 1–6

Carry out the operations indicated on the following exponential numbers.

(a) $7.35 \times 10^5 - 2.45 \times 10^5$ (b) $2.7 \times 10^4 + 3.54 \times 10^5$

(c) $5.41 \times 10^{-4} - 2.3 \times 10^{-5}$

Multiplication and Division of Exponential Numbers

For multiplying or dividing exponential numbers, the only requirement is that the numbers are expressed to the same *base*, which is 10 in exponential notation. In multiplication, multiply the decimal parts in the usual manner and ***add*** algebraically the *exponents* of the two numbers to give the new exponent. In division, divide the decimal parts of the two numbers in the usual way and ***subtract*** algebraically the *exponents* of the two numbers to give the new exponent.

EXAMPLE 1–7 ————

Carry out the operations indicated on the following exponential numbers.

(a) $1.70 \times 10^6 \times 2.40 \times 10^3$

SOLUTION   Multiply the decimals and then add the exponents algebraically, as

$$(1.70 \times 2.40)(10^6 \times 10^3) = 4.08 \times 10^{(6+3)}$$
$$= 4.08 \times 10^9 \quad \textit{Answer}$$

(b) $1.70 \times 10^6 \times 2.40 \times 10^{-3}$

SOLUTION

$$(1.70 \times 2.40)(10^6 \times 10^{-3}) = 4.08 \times 10^{6-3}$$
$$= 4.08 \times 10^3 \quad \textit{Answer}$$

Note that the exponents are *added algebraically*, $6 + (-3) = 6 - 3 = 3$.

(c) $\dfrac{2.40 \times 10^5}{1.30 \times 10^3}$

SOLUTION Divide the decimals and then subtract the exponents algebraically, as

$$\frac{2.40 \times 10^5}{1.30 \times 10^3} = \frac{2.40}{1.30} \times \frac{10^5}{10^3} = 1.85 \times 10^{5-3} = 1.85 \times 10^2 \qquad Answer$$

(d) $\dfrac{2.40 \times 10^5}{1.30 \times 10^{-3}}$

SOLUTION

$$\frac{2.40 \times 10^5}{1.30 \times 10^{-3}} = \frac{2.40}{1.30} \times \frac{10^5}{10^{-3}} = 1.85 \times 10^{5-(-3)} = 1.85 \times 10^{5+3}$$

$$= 1.85 \times 10^8 \qquad Answer$$

Note that the exponents are *subtracted algebraically*, $5 - (-3) = 5 + 3 = 8$.

Exercise 1–7

Carry out the operation indicated on the following exponential numbers.

(a) $3.24 \times 10^4 \times 1.76 \times 10^{-2}$ 　　　 (b) $\dfrac{6.74 \times 10^{-8}}{2.12 \times 10^{-2}}$

(c) $\dfrac{2.97 \times 10^{-14}}{1.23 \times 10^2}$

Square Root of Exponential Numbers

To obtain the square root of a number, first express the number in exponential notation in which the power of 10 has an **even** exponent. To obtain the square root of the exponential number, obtain the square root of the decimal from a calculator (use the $\sqrt{\ }$ key) and obtain the square root of the power of 10 by dividing the exponent by 2. The reason the exponent was made even was to divide the exponent by 2 and obtain a whole number exponent.[3]

EXAMPLE 1–8

Determine the value of each of the following numbers.

(a) $\sqrt{4.00 \times 10^{-4}}$

---

[3] Calculators with the keys EXP, EE, or EEX automatically adjust the exponent, so you can read the square root directly from the display window without adjusting it. See Appendix 2 (Your Calculator).

SOLUTION   Take the square root of the decimal and then divide the exponents by 2.

$$\sqrt{4.00 \times 10^{-4}} = \sqrt{4.00} \times \sqrt{10^{-4}} = 2.00 \times 10^{-(4/2)}$$
$$= 2.00 \times 10^{-2} \quad \textit{Answer}$$

(b) $\sqrt{3.40 \times 10^{-3}}$

SOLUTION   Change the number 3.40×10^{-3} to a number with an *even* exponent, 34.0×10^{-4}, following guidelines previously mentioned. Then take the positive square root of 34.0×10^{-4} as

$$\sqrt{3.40 \times 10^{-3}} = \sqrt{34.0 \times 10^{-4}} = \sqrt{34.0} \times \sqrt{10^{-4}} = 5.83 \times 10^{-(4/2)}$$
$$= 5.83 \times 10^{-2} \quad \textit{Answer}$$

Exercise 1–8

Determine the value of each of the following numbers.

(a) $\sqrt{1.60 \times 10^3}$ (b) $\sqrt{1.60 \times 10^4}$ (c) $\sqrt{2.25 \times 10^{-7}}$

Positive Powers of Exponential Numbers

To raise an exponential number to a given positive power, first raise the decimal part to the power by using it as a factor the **number of times indicated by the power** and then multiply the exponent of 10 by the indicated power.

EXAMPLE 1–9

Perform the indicated operations on the following exponential numbers.

(a) Raise 2.45×10^4 to the second power.

SOLUTION   Multiply 2.45×2.45; then multiply the exponent (4) by 2 (the second power).

$$(2.45 \times 10^4)^2 = (2.45)^2 \times (10^4)^2 = 2.45 \times 2.45 \times 10^8$$
$$= 6.00 \times 10^8 \quad \textit{Answer}$$

(b) Raise 3.42×10^2 to the third power.

SOLUTION   Multiply $3.42 \times 3.42 \times 3.42$; then multiply the exponent (2) by 3 (the third power).

$$(3.42 \times 10^2)^3 = (3.42)^3 \times (10^2)^3 = 3.42 \times 3.42 \times 3.42 \times 10^6$$
$$= 40.0 \times 10^6 \text{ or}$$
$$4.00 \times 10^7 \quad \textit{Answer}$$

Exercise 1–9

Perform the indicated operations on the following exponential numbers.

(a) Raise 1.24×10^2 to the *second* power.
(b) Raise 1.24×10^2 to the *third* power.
(c) Raise 2.10×10^6 to the *third* power.

1–4 Scientific Notation

scientific notation A more systematic form of exponential notation in which the decimal part must have *exactly one* nonzero digit to the left of the decimal point.

Scientific notation is a more systematic form of exponential notation. In **scientific notation**, the decimal part must have *exactly one* nonzero digit to the left of the decimal point. Scientific notation is widely used by scientists.

EXAMPLE 1–10

Express the following in scientific notation to three significant digits.

(a) 6,780,000   (b) 2170
(c) 0.0756   (d) 10.7

SOLUTIONS Express the preceding in scientific notation by shifting the decimal place following the guidelines previously mentioned.

Number	Answer	
(a) 6,780,000	6.78×10^6	Move decimal point 6 places to the left.
(b) 2170	2.17×10^3	Move decimal point 3 places to the left.
(c) 0.0756	7.56×10^{-2}	Move decimal point 2 places to the right.
(d) 10.7	1.07×10^1	Move decimal point 1 place to the left.

Exercise 1–10

Express the following in scientific notation to three significant digits.

(a) 4,230,000   (b) 0.000653   (c) 7268

1–5 Linear Equations

A linear equation is an equation that usually contains one unknown (variable) whose highest power is equal to 1. A general example is $ax^1 = b$ where the "1" is the first power of the unknown (x), a is a coefficient ($a \neq 0$), and b is a number. We can calculate the oxidation numbers of elements in compounds and ions by solving linear equations (see Section 6–1).

To solve for the unknown quantity in a linear equation we need to carry out the following algebraic transformations or changes.

1. Clear any parentheses.

2. Collect similar terms. Place all unknowns on one side of the equation (usually the left) and all numbers on the other side (usually the right).

3. Rearrange. To place unknowns on one side of the equation and numbers on the other side, we must rearrange the equation as follows:

   (a) Adding or subtracting the same number on *both* sides of the equation does not change the equation.

   (b) Multiplying or dividing both sides of the equation by the same number does not change the equation.

4. Solve for *one* unit of the unknown.

EXAMPLE 1–11

Solve the following linear equations for the unknown (x).

(a) $2x = 4$

SOLUTION

Divide both sides of the equation by 2.

$$\frac{2x}{2} = \frac{4}{2}$$

$$x = 2 \qquad Answer$$

(b) $2x = 5 - 1$

SOLUTION   Collect similar terms.

$$2x = 5 - 1 = 4$$

Divide both sides of the equation by 2.

$$\frac{2x}{2} = \frac{4}{2}$$

$$x = 2 \qquad Answer$$

(c) $2x + 1 = 5$

SOLUTION   Subtract 1 from both sides of the equation.

$$2x + 1 - 1 = 5 - 1$$

$$2x = 4$$

Divide both sides of the equation by 2.

$$\frac{2x}{2} = \frac{4}{2}$$

$$x = 2 \qquad Answer$$

(d) $2(x + 1) = 8$

SOLUTION   Clear the parentheses.

$$2x + 2 = 8$$

Subtract 2 from both sides of the equation.

$$2x + \cancel{2} - \cancel{2} = 8 - 2$$

$$2x = 6$$

Divide both sides of the equation by 2.

$$\frac{\cancel{2}x}{\cancel{2}} = \frac{6}{2}$$

$$x = 3 \qquad Answer$$

(e) $3x = -18$

SOLUTION   Divide both sides of the equation by 3.

$$\frac{\cancel{3}x}{\cancel{3}} = \frac{-18}{3}$$

$$x = -6 \qquad Answer$$

A linear equation can be checked by substituting the value obtained for the unknown into the original equation. Checking the linear equations of **Example 1–11** gives the following solutions.

(a) $2(2) = 4$
$4 = 4$

(b) $2(2) = 5 - 1$
$4 = 4$

(c) $2(2) + 1 = 5$
$4 + 1 = 5$
$5 = 5$

(d) $2(3 + 1) = 8$
$2(4) = 8$
$8 = 8$

(e) $3(-6) = -18$
$-18 = -18$

Exercise 1–11

Solve the following linear equations for the unknown (x).

(a) $2x = 7 + 9$

(b) $5x + 4 = 4x + 1$

(c) $3(x + 4) = 3$

(d) $4(x + 7) = 2(x + 1)$

Water: a chemical that never goes out of style

Name: The chemical name for water is hydrogen oxide.

Appearance: Water is a colorless and odorless substance that can be found on the earth in three different forms (states) as shown in the picture. Without water, life on earth could not exist.

Water is a colorless and odorless substance that can be found on the earth in three different forms (states): solid ice, liquid water, and gaseous water (water vapor).

Occurrence: Water in its three states can be found all over the earth. Three-quarters of the surface of the earth is covered by water in the from of the oceans, the seas, and the polar ice caps. Solid water (ice) is found year round at the North and South Poles as well as at high altitudes and in cold climates. Water vapor is found in the atmosphere, and clouds are typically formed of water vapor.

Source: Water is generally obtained from rivers, streams, lakes, or underground wells. Such sources make up only about 2 percent of all water on the earth. Sea water can be made drinkable at moderate expense by the desalination (removal of salts) of sea water. As humans pollute these land-based sources, water for drinking and irrigation becomes even more scarce.

Its Role in Our World: Water is one of the most important chemical compounds for the survival of life forms. Humans need 2–3 quarts per day, and will perish in a matter of days if water is not available. Plants and animals need water, although there are species of each that have adapted to survive on very small amounts of water. The irrigation of crops is the single largest use of water, accounting for about two-thirds of society's total consumption. The need for clean water for drinking, agriculture, cleaning, and industrial uses has been a dominant concern of societies throughout history.

Water is recirculated on the earth among three great reservoirs by the processes of precipitation (rain, snow, sleet, etc.), runoff (in streams

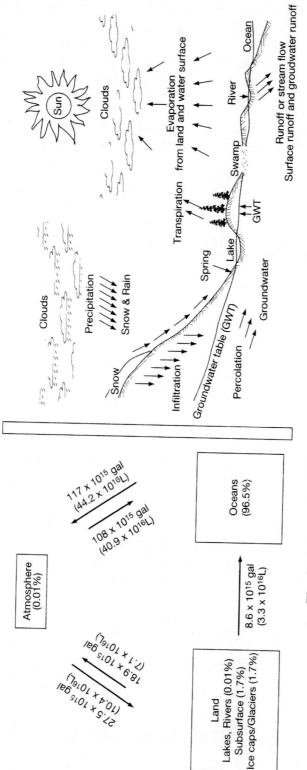

The Hydrologic Cycle. Water circulates among three great reservoirs, the oceans, the land, and the atmosphere. The percentages indicate the percentage of the total water on the earth in each reservoir. Note that the amount of water available for use (drinking, irrigation, etc.) is less than 2 percent of all water on the earth!

and rivers), and evaporation (see above). Most of the water on land (1.7 percent of all water) is located underground in the groundwater table. Wells are holes that go deep enough to penetrate the groundwater table, so that water gathers in the bottom of the well.

Water is extremely important in the body's chemical processes, too. All the chemical reactions that go on in our bodies rely on water as a medium. Water is also very important for its role in photosynthesis, the process by which green plants use carbon dioxide, water, and light to produce oxygen and sugar. The oxygen makes life on earth possible, and the sugar is the key source of energy for the plant kingdom. And since animals eat plants, animal life is completely dependent on the process of photosynthesis.

Unusual Facts: Most chemical compounds decrease in volume as they freeze; that is, they shrink. Water is very unusual in this respect, as it *expands* when it freezes. Because ice is larger in volume than the water from which it comes, ice floats in water. This isolated fact has enormous implications for the survival of life on the earth. For this reason, the ice at the North Pole floats and submarines can pass under the North Pole. (The South Pole actually has land underneath the ice.) More importantly, lakes freeze from the top down in winter, and liquid water exists below the surface of the lake. Thus, fish can live in lakes in the winter. If ice sank, lakes would freeze from the bottom up and might never thaw out during the summer. The earth's surface would be much colder and life would be very different, if in fact life existed.

Problems[4]

1. Determine the number of significant digits in the following numbers.

(a) 43$\underline{3}$ (b) 4,700,200

(c) 7$\overline{00}$ (d) 0.007070

2. Round off the following numbers to three significant digits.

(a) 12.89   (b) 1.3450

(c) 1.3550   (d) 1.3453

3. Perform the indicated mathematical operations and express your answer to the proper number of significant digits.

(a) $2.45 + 3.6452 + 0.04$ (b) $17.382 - 3.47$

(c) 8.45×6.0 (d) $\dfrac{13.7}{8.2}$

(e) $\dfrac{14.8}{6.7 \times 0.125}$

4. Express 625,000 as 6.25×10^n.

5. Express 0.000707 as 7.07×10^n.

---

[4] Answers and selected solutions to the problem sections at the ends of chapters will be found in Appendix 5.

6. Carry out the operations indicated on the following exponential numbers.

(a) $3.34 \times 10^3 + 2.40 \times 10^3$ (b) $4.76 \times 10^2 + 7.7 \times 10^1$

(c) $6.75 \times 10^3 - 1.74 \times 10^3$ (d) $3.74 \times 10^5 - 2 \times 10^3$

7. Carry out the operations indicated on the following exponential numbers.

(a) $6.42 \times 10^3 \times 1.41 \times 10^2$ (b) $3.82 \times 10^4 \times 1.24 \times 10^{-2}$

(c) $\dfrac{6.81 \times 10^6}{2.82 \times 10^2}$ (d) $\dfrac{7.62 \times 10^{-5}}{3.45 \times 10^{-2}}$

8. Determine the value of each of the following numbers.

(a) $\sqrt{9.00 \times 10^6}$ (b) $\sqrt{3.60 \times 10^5}$

(c) $\sqrt{4.90 \times 10^7}$ (d) $\sqrt{9.40 \times 10^{-5}}$

9. Perform the indicated operations on the following exponential numbers.

(a) Raise 1.74×10^3 to the second power.
(b) Raise 1.05×10^8 to the second power.
(c) Raise 2.04×10^2 to the third power.
(d) Raise 2.10×10^3 to the third power.

10. Express the following numbers in scientific notation to three significant digits.

(a) 6,720,000 (b) 0.00342

(c) 6272 (d) 0.06275

11. Solve the following linear equations for the unknown (x).

(a) $2x = 7 + 1$ (b) $3x + 7 = 16$

(c) $2(x + 2) = 14$ (d) $5(x + 2) = -20$

General Problems

12. Perform the indicated operations and express your answer in scientific notation to the proper number of significant digits.

(a) $\dfrac{700 - 32}{1.8}$ (b) $1.8 \times 35 + 32$

(c) $\dfrac{7.352 \times 14.1}{0.610}$ (d) $\dfrac{6.548}{12.1 \times 15.7}$

13. Carry out the operations indicated on the following exponential numbers and express your answer in scientific notation to the proper number of significant digits.

(a) $\dfrac{5.42 \times 10^3 + 6.3 \times 10^2}{18.7 \times 10^{10}}$ (b) $\dfrac{\sqrt{14.4 \times 10^3}}{16.2 \times 10^8}$

(c) $(2.24 \times 10^3)^2 + 6.5 \times 10^6$ (d) $(3.14 \times 10^3)^3 - 8.6 \times 10^9$

Answers to Exercises

1–1. (a) 3; (b) 4; (c) 3; (d) 3; (e) 3; (f) 5

1–2. (a) 262; (b) 435; (c) 61.2; (d) 61.3; (e) 61.4; (f) 63$\overline{0}$

1–3. (a) 8.82; (b) 5.71; (c) 33; (d) 6.85; (e) 1.0

1–4. 7.28×10^6

1–5. 5.43×10^{-4}

1–6. (a) 4.90×10^5; (b) 3.81×10^5; (c) 5.18×10^{-4}

1–7. (a) 5.70×10^2; (b) 3.18×10^{-6}; (c) 2.41×10^{-16}

1–8. (a) 4.00×10^1; (b) 1.26×10^2; (c) 4.74×10^{-4}

1–9. (a) 1.54×10^4; (b) 1.91×10^6; (c) 9.26×10^{18}

1–10. (a) 4.23×10^6; (b) 6.53×10^{-4}; (c) 7.27×10^3

1–11. (a) 8; (b) −3; (c) −3; (d) −13

QUIZ[5]

1. Determine the number of significant digits in the following numbers.

(a) 82.050   (b) 0.0607   (c) 700.02   (d) 67$\overline{00}$

2. Round off the following numbers to three significant digits.

(a) 27.68   (b) 42.65   (c) 7352   (d) 18.254

3. Express the following numbers in scientific notation to three significant digits.

(a) 2430   (b) 0.00752   (c) 76,550   (d) 0.035253

4. Perform the following operations and express your answer to three significant digits in scientific notation.

(a) $5.24 \times 10^4 + 72.2 \times 10^3$ (b) $\dfrac{6.75 \times 10^{-8}}{8.72 \times 10^6}$

(c) $\sqrt{9.24 \times 10^7}$ (d) Raise 1.23×10^3 to the third power

---

[5] Answers to Quizzes are found in Appendix 5.

MEASUREMENTS

2

COUNTDOWN

5 Determine the number of significant digits in the following numbers (Section 1–1).
(a) 0.0243   (3);   (b) 4.250   (4)
(c) 21$\bar{0}$ (3);   (d) 0.06080   (4)

4 Round off the following numbers to three significant digits (Section 1–2).
(a) 4.762   (4.76);   (b) 3.877   (3.88)
(c) 28.650   (28.6);   (d) 13,672   (13,700)

3 Perform the following operations and express your answer to three significant digits (Sections 1–1 and 1–2).
(a) 2.67
    0.128
    3.45

(b) $0.00246 \times 1247 =$

                    (6.25)   (3.07)

(c) $\dfrac{2.782}{0.645} =$

(d) $\dfrac{6.75 \times 4.287}{36.2 \times 0.785} =$

                    (4.31)   (1.02)

2 Perform the following operations and express your answer to three significant digits in scientific notation (Sections 1–1, 1–2, 1–3, and 1–4).
(a) $\dfrac{6.752 \times 10^6}{2.31 \times 10^4} =$

                                          (2.92×10^2)
(b) Raise 5.28×10^3 to the second power.

                                          (2.79×10^7)

1 Express the following numbers in scientific notation to three significant digits (Sections 1–1, 1–2, 1–3, and 1–4).
(a) 86,400 (8.64×10^4);   (b) 0.00623 (6.23×10^{-3})
(c) 0.05831 (5.83×10^{-2});   (d) 256,800 (2.57×10^5)

TASKS

1 Memorize the meaning of the metric unit prefixes given in Table 2–1 and the two equivalents for nanometers, microns, and angstroms given in Table 2–2.

2 Learn the *factor-unit* method of problem solving (Section 2–3).

3 Memorize the English-based system of measurements given in Table 2–3.

4 Memorize the metric-English equivalents given in Table 2–4.

OBJECTIVES

1 Give the distinguishing characteristics of each of the following terms:
   (a) matter (Section 2–1)
   (b) mass (Section 2–1)
   (c) weight (Section 2–1)
   (d) volume (Section 2–1)
   (e) temperature (Section 2–1 and 2–6)
   (f) density (Section 2–7)
   (g) specific gravity (Section 2–8)

2 Given a measurement in the metric system, convert it to any other related unit in the metric system (Examples 2–1, 2–2, and 2–3, Exercises 2–1, 2–2, and 2–3, Problems 1, 2, and 3).

3 Given a measurement in the metric system, convert it to the corresponding unit in the English-based system, or vice versa (Examples 2–4 to 2–9, Exercises 2–4 to 2–6, Problems 4, 5, and 6).

4 Given a temperature measurement in degrees Fahrenheit, convert it to degrees Celsius and kelvins, and vice versa (Examples 2–10 and 2–11, Exercises 2–7 and 2–8, Problems 7 to 12).

5 Given its mass and volume, calculate the density of a substance (Example 2–12, Exercise 2–9, Problem 13).

6 Given the density of a substance and its mass or volume, calculate the unknown volume or mass (Examples 2–13 and 2–14, Exercises 2–10 and 2–11, Problems 14 and 15).

7 Given the specific gravity of a substance and its mass or volume, calculate the unknown volume or mass (Examples 2–15 and 2–16, Exercises 2–12 and 2–13, Problems 16 and 17).

Almost daily we use some form of measurement. To enter college this semester you may have had to "measure" how much money you had in your bank account so that you could pay the tuition. If you are a married student, you had to "measure" the money your wife or husband would earn this year, so that you could afford to go to college and still support the family. Beside measuring money, you may also "measure" time, such as how long it will take you to read this chapter or do your homework.

Chemistry is concerned with the composition of substances and the changes those substances undergo. In the study of the composition of substances, chemists rely heavily on the measurement of quantities and characteristics of those substances. In this chapter we consider some measurements made in the study of chemistry.

2–1   Matter and Its Characteristics

matter Any substance that has mass and occupies space.

mass Quantity of matter in a particular object.

weight Measure of the gravitational force of attraction between the body's mass and the mass of the planet or satellite on which it is weighed.

Chemists refer to substances that make up our universe as **matter**—anything that has mass and occupies space. In addition, matter has measurable characteristics, such as weight, length, volume, temperature, and density.

We define **mass** as the quantity of matter in a particular body. *The mass of a body is constant and does not change*, whether it is measured in Colorado or New York or even the moon. In contrast, the **weight** of a body is the *gravitational force of attraction* between the body's mass and the mass of the planet or satellite on which it is weighed. Thus where you weigh matter affects its weight. An object that weighs 10 pounds (lb) on earth would weigh only 1 lb 11 oz on the moon. *The mass of the object has not changed*, but the weight (the attraction of the earth or moon for the object) *has changed*.

Figure 2–1 summarizes the relations among matter, mass, and weight. Other characteristics of matter can also fluctuate from object to object,

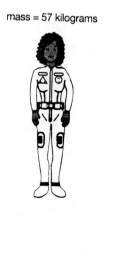

mass = 57 kilograms

(a)

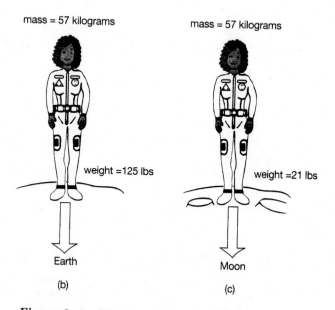

mass = 57 kilograms

weight =125 lbs

Earth

(b)

mass = 57 kilograms

weight =21 lbs

Moon

(c)

Figure 2–1 *Mass and weight: (a) an astronaut is composed of matter and has a mass that is a measure of that matter; (b) the same astronaut also has a weight that is a measure of the earth's gravitational attraction for her; (c) the same astronaut weighs less on the moon due to the moon's weaker gravity. Note that her mass has not changed!*

time to time, and place to place. Two otherwise identical steel girders might differ in length, for example. Matter takes up space—it occupies *volume*. The **volume** of a block of wood—the cubic space it takes up—is measured as its height times its width times its depth. The volumes of other shapes can also be measured. The **temperature** of an object—its degree of hotness—can also vary considerably, as you will see if you attempt to put hot water, tap water, or ice in your mouth.

volume Cubic space taken up by matter.

temperature Degree of hotness of matter.

2–2 Measuring Matter: The Metric System and the International System of Units

The International System of Units (SI units) is a system of units recommended for adoption throughout the world. It includes the metric system, but some of the basic units in the metric system are not used as the base in the SI. We shall first consider the metric system and then return to the SI, pointing out its basic units.

The metric system has as its basic units the gram (g, mass), the liter (or litre, L, volume), the meter (or metre, m, length), and the second (sec or s, time). The units for mass, volume, and length in the metric system are related in multiples of 10, 100, 1000, etc., like our monetary system. The prefixes used to define multiples or fractions of the basic units and their relation to our monetary system are shown in Table 2–1. Memorize the meaning of these metric unit prefixes (see Fig. 2–2).

The specific units of mass (gram), volume (liter), and length (meter) are shown in Table 2–2, which is an extension of Table 2–1 using the specific units. You must know these units and their equivalents in powers of 10 in order to work problems. For example, you must know that 100 cm = 1 m, and 1000 mL = 1 L, etc.

The SI has as its basic units the kilogram (kg, mass) and the meter (m, length). The derived unit, cubic meter (m^3) is the unit for volume, which is the volume of a cube that is 1 meter on each edge. One liter is defined as exactly equal to one cubic decimeter (1 L = 1 dm^3) or one cubic meter equals 1000 liters. The unit liter is considered a non-SI unit, while the cubic meter is an SI unit. Since non-SI units are still in use, we shall continue to use them in this book but also introduce the newer SI units.

TABLE 2–1   METRIC UNITS IN GENERAL VERSUS MONETARY SYSTEM

kilo	1000 units	vs. $1000	A grand
deci	1/10 unit	vs. 1/10 dollar	A dime
centi	1/100 unit	vs. 1/100 dollar	A cent
milli	1/1000 unit	vs. 1/1000 dollar	A mill
micro	1/1,000,000 unit	vs. No analogy	

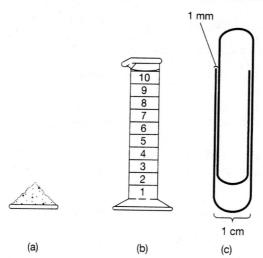

Figure 2-2   *Visual Concept of Mass, Volume, and Length (a) salt—1 g; (b) graduated cylinder—10 mL; (c) big paper clip—width 1 cm and thickness of wire 1 mm.*

TABLE 2-2   METRIC UNITS OF MASS, VOLUME, AND LENGTH

BASIC UNITS PER DERIVED UNIT			MASS		VOLUME		LENGTH[a]	
kilo	1000	(10^3)	kilogram	(kg)	kiloliter	(kL)	kilometer	(km)
basic unit	1	(10^0)	gram	(g)	liter	(L)	meter	(m)
deci-	0.1	(10^{-1})	decigram	(dg)	deciliter	(dL)	decimeter	(dm)
centi-	0.01	(10^{-2})	centigram	(cg)	centiliter	(cL)	centimeter	(cm)
milli-	0.001	(10^{-3})	milligram	(mg)	milliliter	(mL)[b]	millimeter	(mm)
micro-	0.000001	(10^{-6})	microgram	(μg)	microliter	(μL)	micrometer	(μm)
			or gamma	(γ)[c]			or micron	(μ)[c]

[a] Smaller units of length are

    (1) the nanometer [nm, 1/1000 micron (μ)]
    (2) the angstrom (Å, 1/10 nanometer).

We therefore have the following equivalents:

1000 nanometers (nm) = 1 micron (μ)

10 angstroms (Å) = 1 nanometer (nm)

(You must know these two equivalents in order to solve problems.)

[b] The milliliter (mL) and the cubic centimeter (cc, cm³) are exactly equivalent, since 1 L = 1 dm³ according to the SI definition.

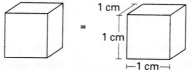

Volume of a cube = side × side × side
= cm × cm × cm
= cm³ or cc

1 milliliter (mL) = 1 cubic centimeter (cc)

[c] These are Greek letters. The gamma (γ) is written like an upside-down ℓ. The mu (μ) is written like a u with a tail at the beginning and is pronounced "mew."

2-3 Conversion within the Metric System. The Factor–Unit Method of Problem Solving

In considering conversion within the metric system, we shall describe a general method of problem solving called the *factor–unit* method.[1] This method is quite simple and is based on developing a relationship between different units expressing the same physical dimension. A simple problem will serve to illustrate our point. Consider the conversion of 707 mg to grams. We know the following relationship:

$$1000 \text{ mg} = 1 \text{ g} \quad (\text{see Table 2-2})$$

Dividing both sides of the equation by 1000 mg, we have

$$\frac{\cancel{1000 \text{ mg}}}{\cancel{1000 \text{ mg}}} = 1 = \frac{1 \text{ g}}{1000 \text{ mg}} \qquad \text{Factor } \mathbf{A}$$

which we shall call factor **A**. Now, dividing both sides of the equation (1000 mg = 1 g) by 1 g, we have

$$\frac{1000 \text{ mg}}{1 \text{ g}} = \frac{\cancel{1} \cancel{g}}{\cancel{1} \cancel{g}} = 1 \qquad \text{Factor } \mathbf{B}$$

which we shall call factor **B**. In the problem we have been given 707 mg, and we wish to express this number as the number of grams. We need to multiply the given quantity, 707 mg, by a factor, **A** or **B**, so that milligrams (mg) will cancel out and the number will be expressed as the number of grams (g).

$$707 \text{ mg} \times (\text{factor } \mathbf{A} \text{ or } \mathbf{B}) = \text{number of grams (g)}$$

Study hint

Choose the factor with the units you wish to remove in the *denominator*. Thus, factor **A** is correct because we want to remove the mg label.

We can multiply the 707 by 1 (both factors equal 1) and not change the value based on the unit multiplicative identity property from mathematics, but *only one* factor will give the correct units (labels), and hence the correct answer. Therefore, we choose factor **A**.

$$707 \cancel{\text{ mg}} \times \frac{1 \text{ g}}{1000 \cancel{\text{ mg}}} = 0.707 \text{ g} \qquad \textit{Answer}$$

If we had chosen factor **B** in error, then

$$707 \text{ mg} \times \frac{1000 \text{ mg}}{1 \text{ g}} = 707{,}000 \frac{\text{mg}^2}{1 \text{ g}}$$

which does not answer our original quesiton, and the units are also meaningless. The factor 1 g/1000 mg is not considered in significant digits, since it is

---

[1] This method has also been called "dimensional analysis," "unit conversion," and the "factor–label" method for problem solving.

an exact value and could be expressed as 1 g/$\overline{1000}$ mg; hence, the answer is expressed to three significant digits as found in the given, 707 mg.

Whenever possible in this book, we shall use the *factor–unit* method in problem solving.

Before we consider further conversions within the metric system, the following useful hints in problem solving are given.

1. First read the problem very carefully to determine what you are being asked to do.

2. Organize the data that are given, being sure to include *both* the *units of* the *given* and the *units* of the *unknown*.

3. **Write down the given and its units at the beginning of a line and the units of the unknown at the end of the line.**

4. Apply the principles you have learned to develop factors so that these factors, if used properly, will give the correct units in the unknown.

5. Check your answer to see if it is reasonable by checking both the mathematics and the units.

6. Finally, check the number of significant digits.

Now let us consider some more conversions within the metric system.

EXAMPLE 2–1

Convert 6.85 m to millimeters (mm).

SOLUTION

$$6.85 \text{ m} \times \text{(factor)} = \text{mm}$$

We know that 1000 mm = 1 m (see Table 2–2); hence, our factors are as follows:

$$\frac{1000 \text{ mm}}{1 \text{ m}} \; ; \quad \frac{1 \text{ m}}{1000 \text{ mm}}$$

$$\textbf{A} \qquad\qquad \textbf{B}$$

To obtain the correct units (mm), the factor we must use is factor **A**.

$$6.85 \; \text{m} \times \frac{1000 \text{ mm}}{1 \; \text{m}} = 6850 \text{ mm} \qquad \textit{Answer}$$

Multiplying 6.85 × 1000, we arrive at the answer 6850 expressed in mm. The number 6.85 is expressed to three significant digits, and hence our answer should be expressed accordingly. The factor 1000 mm/1 m is not considered in significant digits since it is an exact value and could be expressed as $\overline{1000}$ mm/1 m. The conversion factors are not used in determining significant digits.

EXAMPLE 2–2

The following masses were recorded in a laboratory experiment: 2.0000000 kg, 6.0000 g, 650.0 mg, 0.5 mg. What is the total mass in grams?

SOLUTION

$$2.0000000 \text{ kg} \times \frac{1000 \text{ g}}{1 \text{ kg}} = 2000.0000 \text{ g}$$

$$6.0000 \text{ g}$$

$$650.0 \text{ mg} \times \frac{1 \text{ g}}{1000 \text{ mg}} = 0.6500 \text{ g}$$

$$0.5 \text{ mg} \times \frac{1 \text{ g}}{1000 \text{ mg}} = \frac{0.0005 \text{ g}}{2006.6505 \text{ g}} \quad Answer$$

Note that all the masses in grams are recorded to the ten-thousandth of a gram.

EXAMPLE 2–3

Convert 78.1 mg to kilograms (kg). Express your answer in scientific notation.

SOLUTION

$$78.1 \text{ mg} \times (\text{factor}) = \text{kg}$$

We do not know a single factor converting mg to kg, but we do know factors that convert mg to g and g to kg:

$$1000 \text{ mg} = 1 \text{ g} \quad \text{and} \quad 1000 \text{ g} = 1 \text{ kg}$$

Using these factors, we have the following solution:

$$78.1 \text{ mg} \times \frac{1 \text{ g}}{1000 \text{ mg}} \times \frac{1 \text{ kg}}{1000 \text{ g}} = 0.0000781 \text{ kg}$$

$$= 7.81 \times 10^{-5} \text{ kg} \quad Answer$$

In carrying out this calculation divide 78.1 by 1000 and then by 1000 again to give 0.0000781.

EXAMPLE 2–4

Convert 6250 Å to microns (μ).

SOLUTION   From Table 2–2 there are 10 Å = 1 nm and 1000 nm = 1 μ. Using these factors with their units, we obtain the following solution:

$$6250 \text{ Å} \times \frac{1 \text{ nm}}{10 \text{ Å}} \times \frac{1 \text{ } \mu}{1000 \text{ nm}} = 0.625 \text{ } \mu$$

Exercise 2–1

Convert 175 mL to liters (L).

Exercise 2–2

The following lengths were recorded: 1.000000 km, 325.0 cm, 5.000 m. What is the total length in meters?

Exercise 2–3

Convert 625 cL to kiloliters (kL). Express your answer in scientific notation.

2–4 The English-Based Units

The English-based units are also used in the measurement of matter. The English-based units are used primarily in the United States, whereas the metric system is used throughout almost all of the rest of the world, including Great Britain.

The English-based units consist of the ounce, pound, and ton (mass); the fluid ounce, pint, quart, and gallon (volume); the inch, foot, yard, and mile (length); and the second (time). Table 2–3 summarizes the English-based units of measurement.

TABLE 2–3   SUMMARY OF THE ENGLISH-BASED UNITS OF MEASUREMENT

MASS		LENGTH		VOLUME	
16 ounces (oz)	= 1 pound (lb)	12 inches (in.)	= 1 foot (ft)	16 fluid ounces (fl oz)	= 1 pint (pt)
2000 pounds	= 1 ton	3 feet	= 1 yard (yd)	2 pints	= 1 quart (qt)
		5280 feet	= 1 mile (mi)	4 quarts	= 1 gallon (gal)

2–5 Conversion from Metric System to English-Based Units and Vice Versa

In order to convert from the metric system to English-based units and vice versa, we need to know certain conversion factors for mass, volume, and length. Table 2–4 gives these conversion factors. You must memorize these factors to solve problems that require conversion from one system to another.

EXAMPLE 2–5

Convert 275 g to pounds (lb).

SOLUTION   Using the conversion factor 454 g = 1 lb, we solve the

TABLE 2-4   METRIC-ENGLISH UNIT EQUIVALENTS

DIMENSION	METRIC UNIT	ENGLISH EQUIVALENT
Mass	454 grams[a] (g)	= 1 pound[a] (lb)
Volume	1 liter (L)	= 1.06 quarts (qt)
Length	2.54 centimeters (cm)	= 1 inch (in.)
Time	1 second (s)	= 1 second (sec)

[a] Although gram is a unit of mass and pound is a unit of weight, the two units are used interchangeably in most calculations in chemistry.

problem as follows:

$$275 \text{ g} \times \frac{1 \text{ lb}}{454 \text{ g}} = 0.606 \text{ lb} \quad Answer$$

EXAMPLE 2-6

Convert 1.25 gal to liters (L).

SOLUTION

$$1.25 \text{ gal} \times \frac{4 \text{ qt}}{1 \text{ gal}} \times \frac{1 \text{ L}}{1.06 \text{ qt}} = 4.72 \text{ L} \quad Answer$$

Notice that here two factors are needed since we do not know a direct conversion factor from gallons to liters: 4 qt = 1 gal and 1 L = 1.06 qt.

EXAMPLE 2-7

Convert 4.20 ft to meters (m).

SOLUTION To use the factor 2.54 cm = 1 in., the feet must be converted to inches by using 12 in. = 1 ft. The centimeters must then be converted to meters, by using the metric conversion factor 100 cm = 1 m. The solution is as follows:

$$4.20 \text{ ft} \times \frac{12 \text{ in.}}{1 \text{ ft}} \times \frac{2.54 \text{ cm}}{1 \text{ in.}} \times \frac{1 \text{ m}}{100 \text{ cm}} = 1.28 \text{ m} \quad Answer$$

EXAMPLE 2-8

A box has the following dimensions: 8.00 in., 20.0 cm, and 2.00 ft. Calculate its volume in cubic centimeters (cm³).

SOLUTION Convert the 8.00 in. and 2.00 ft to centimeters and then calculate the volume in cubic centimeters as side × side × side.

$$8.00 \text{ in.} \times \frac{2.54 \text{ cm}}{1 \text{ in.}} \times 20.0 \text{ cm} \times 2.00 \text{ ft} \times \frac{12 \text{ in.}}{1 \text{ ft}} \times \frac{2.54 \text{ cm}}{1 \text{ in.}}$$

$$= 24,800 \text{ cm}^3, \text{ to three significant digits.} \quad Answer$$

Note that $cm^1 \times cm^1 \times cm^1 = cm^3$, with the exponents being added algebraically as in multiplication of exponential numbers (see Section 1-3).

Study Hint

In cubing 2.54, remember to multiply it by itself *3 times* (2.54 × 2.54 × 2.54). The same applies to $(12)^3$ = 12 × 12 × 12 (see positive powers of exponential numbers, Section 1-3).

EXAMPLE 2-9

Convert a density of 7.50 g/cm³ to pounds per cubic feet (lb/ft³).

SOLUTION   The mass in grams must be converted to mass in pounds by using the factor 1 lb = 454 g, and the volume in cm³ must be converted to volume in ft³, which involves $(1 \text{ in.})^3 = (2.54 \text{ cm})^3$ and $(12 \text{ in.})^3 = (1 \text{ ft})^3$. The complete solution is as follows:

$$7.50 \ \frac{\text{g}}{\text{cm}^3} \times \frac{1 \text{ lb}}{454 \text{ g}} \times \frac{(2.54)^3 \ \text{cm}^3}{1 \text{ in.}^3} \times \frac{(12)^3 \ \text{in.}^3}{1 \text{ ft}^3} = 468 \text{ lb/ft}^3 \qquad Answer$$

Exercise 2-4

Convert 7.25 in. to meters (m).

Exercise 2-5

Convert 3.00 kg to pounds (lb).

Exercise 2-6

Convert 6.40 pt to milliliters (mL). Express your answer in scientific notation.

2-6 Temperature

There are three common temperature scales that we shall consider in this book. They are

1. The Fahrenheit scale—°F

2. The Celsius scale[2]—°C

3. The Kelvin scale—K (*Note:* no ° sign)

The Fahrenheit scale, named after the German physicist Gabriel Daniel Fahrenheit (1686–1736), is the scale most familiar to us. On this scale, the freezing point of pure water is 32° and the boiling point of water at 1 atmosphere (atm) pressure is 212°. On the Celsius scale, named after Swedish astronomer Anders Celsius (1701–1744), these points correspond to 0° and 100°, respectively.

In Fig. 2-3, these two scales are compared. On the Celsius scale there are 100° between the freezing point (fp) and the boiling point (bp) of water; however, on the Fahrenheit scale this difference corresponds to 180°. Thus, 180 divisions

---

[2] The Celsius scale is the same as the old centigrade scale with the same abbreviation: °C.

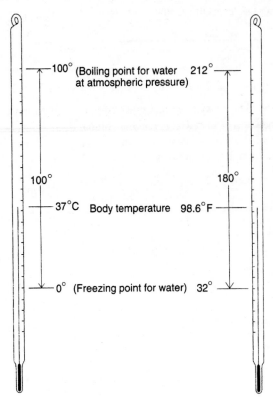

100° (Boiling point for water 212°
at atmospheric pressure)

100°   180°

37°C Body temperature 98.6°F

0° (Freezing point for water) 32°

Celsius (°C)   Fahrenheit (°F)

Figure 2–3 *Comparison of the Celsius and Fahrenheit scales.*

Fahrenheit equal 100 divisions Celsius—or, there are 1.8°F to 1°C. In addition, the freezing point of water is 0° on the Celsius scale and 32° on the Fahrenheit scale. To convert a given temperature form °C to °F, we need only to consider the preceding facts.

$$100 \text{ divisions } °C = 180 \text{ division } °F$$

or by dividing both sides of the equation by 20,

$$5 \text{ divisions } °C = 9 \text{ divisions } °F$$

To convert from °C to °F,

$$°C \times \frac{9 \text{ divisions } °F}{5 \text{ divisions } °C} = \frac{\text{divisions } °F \text{ above or below the}}{\text{freezing point of water}}$$

Since the freezing point of water is 32°F, we must add these divisions to 32 to get the temperature in °F:

$$(\tfrac{9}{5}°C) + 32 = °F \qquad\qquad (2\text{–}1)$$

$$\boxed{1.8°C + 32 = °F} \qquad\qquad (2\text{–}2)$$

(Note that the $\tfrac{9}{5}$ in Equation 2–1 is equal to 1.8 in Equation 2–2.)

Equation 2–2 may be rearranged as follows to convert °F to °C.[3]

$$1.8°C = °F - 32$$

$$\boxed{°C = \frac{(°F - 32)}{1.8}} \qquad (2\text{--}3)$$

In this case, we must remember to subtract 32 from the given temperature in °F to obtain the number of Fahrenheit divisions above or below the freezing point of water and then convert this to °C by dividing by 1.8. Using the number 1.8 in Equations 2–2 and 2–3 simplifies the calculation if you are using a calculator.

The Kelvin scale, named after the British physicist and mathematician William Thomson, Lord Kelvin (1824–1907), consists of a new scale with the zero point equal to −273°C (more accurately, −273.15°). To convert from °C to K we need only to add 273°. (In this book, we shall use 273 instead of 273.15 to simplify calculations.)

$$\boxed{K = °C + 273 \quad \text{or} \quad °C = K - 273} \qquad (2\text{--}4)$$

The lower limit of this scale is theoretically zero, with no upper limit. The temperature of some stars is estimated at many millions of kelvins.[4]

EXAMPLE 2–10

Convert 35°C to °F.

SOLUTION   Substituting into our derived Equation 2–2, we get

$$35°C = [(1.8 \times 35) + 32]°F = [63 + 32]°F = 95°F \qquad Answer$$

In regard to significant digits, express your answer to the smallest place (units, tenths, etc.) as given to you in the number in the problem. In this problem the degrees Celsius was given to the units place, so your answer in degrees Fahrenheit should be expressed to the units place. This is because the numbers 1.8 and 32 are considered to be exact.

Study Hint

There are *exactly* 5°C to each 9°F by definition. Thus, 1.8 is an exact number and does *not* affect the number of significant digits in a calculation.

Exercise 2–7

Convert −15°C to degrees Fahrenheit (°F).

---

[3] Other formulas are also useful, as $°F = \frac{9}{5}(°C + 40) - 40$ and $°C = \frac{5}{9}(°F + 40) - 40$. These formulas are based on the fact that Celsius and Fahrenheit are equal at −40°.

[4] One kelvin represents 1 unit or degree on the Kelvin temperature scale.

EXAMPLE 2–11

Convert −24°F to degrees Celsius (°C) and kelvins (K).

SOLUTION TO °C Substituting into our derived Equation 2–3, we have

$$-24°F = \frac{(-24 - 32)°C}{1.8} = \frac{(-56)°C}{1.8} = -31°C \quad Answer$$

Note that −24 was expressed to the units place; hence, the answer is expressed to the units place.

SOLUTION TO K Substituting into Equation 2–4, we get

$$-31°C = (-31 + 273)K = 242 \text{ K} \quad Answer$$

Again, the 273 is not considered in computing significant digits.

Exercise 2–8

Convert 68°F to degrees Celsius (°C) and kelvins (K).

2–7 Density

density Mass of a substance occupying a unit volume:

$$\text{Density} = \frac{\text{mass}}{\text{volume}}$$

Study Hint

The volumes must be the same in comparing densities. The density does not depend on just the amount.

One property of matter is that a given volume of different substances may have different masses: this property is measured by the density. **Density** is defined as the mass of a substance occupying a unit volume, or

$$\text{Density} = \frac{\text{mass}}{\text{volume}}$$

You know that certain substances are heavier than other substances, even though the volumes are the same. For example, balsa wood is much lighter than a lead brick of the same volume, as illustrated in Fig. 2–4.

The units of density used for solids and liquids are g/mL (g/cc), lb/ft³, and lb/gal, while the units generally used for gases are g/L and lb/ft³. In the SI the unit used for the density of solids, liquids, and gases is kg/m³. Density has the units mass/volume, and whenever the density of a substance is expressed, the particular unit of mass and volume *must* be given. For example, the density of water is 1.00 g/mL in the metric system and 1.00×10^3 kg/m³ in the SI. Thus, you should realize that it is insufficient to express the density of a substance as a pure number without units.

Density is often expressed as follows: $d^{20°} = 13.55$ g/mL for mercury. The 20° indicates the temperature in °C at which the measurement was taken; hence, mercury at 20°C has a density of 13.55 g/mL. The reason for recording the temperature is that almost all substances expand when heated and therefore the density of the material decreases as the temperature is increased; for example $d^{270°} = 12.95$ g/mL for mercury. Thus, the density of a material is dependent on its temperature.

Balsa wood

Lead brick

Figure 2–4   *Balsa wood is much lighter than a lead brick of the same volume.*

EXAMPLE 2–12

Calculate the density of a piece of metal that has a mass of 23.0 g and occupies a volume of 5.8 mL.

SOLUTION   Density is defined as the mass of a substance occupying a unit volume; hence, the solution is as follows:

$$\frac{23.0 \text{ g}}{5.8 \text{ mL}} = 4.0 \text{ g/mL, to two significant digits} \qquad Answer$$

5.8 has two significant digits)

Exercise 2–9

Calculate the density of a piece of metal that has a mass of 56.6 g and occupies a volume of 6.7 mL.

EXAMPLE 2–13

Calculate the volume in liters occupied by $88\overline{0}$ g of benzene at 20°C. $d^{20°} = 0.880$ g/mL.

SOLUTION   We have $88\overline{0}$ g of benzene and the density of benzene at 20°C is 0.880 g/mL. We are asked to calculate the volume in liters. In other words, we wish to convert a given amount of benzene from mass units to volume units. This is readily done by using the density of benzene as a conversion factor, since 1 mL = 0.880 g.

$$88\overline{0} \text{ g} \times \text{factor} = \text{volume units}$$

The choice of factors is as follows:

$$\frac{0.880\ g}{1\ mL}, \quad \frac{1\ mL}{0.880\ g}$$

$$\textbf{A} \qquad\qquad \textbf{B}$$

If we use factor **A**, our units would be g²/mL, which have no meaning and do not answer our question. But let us consider factor **B**:

$$880\ \cancel{g} \times \frac{1\ mL}{0.880\ \cancel{g}}$$

Conversion from mL to L yields the complete setup:

$$880\ \cancel{g} \times \frac{1\ \cancel{mL}}{0.880\ \cancel{g}} \times \frac{1\ L}{1000\ \cancel{mL}} = 1.00\ L \qquad \textit{Answer}$$

Exercise 2–10

Calculate the volume in milliliters occupied by 225 g of chloroform at 20°C. $d^{20°} = 1.49$ g/mL.

EXAMPLE 2–14

Calculate the mass in grams of a 25.0-mL volume of sulfuric acid at 18°C. $d^{18°} = 1.83$ g/mL.

SOLUTION In this problem the volume is given and we are asked to calculate the mass in grams. This is again done by using the density as a conversion factor, as follows:

$$25.0\ \cancel{mL} \times \frac{1.83\ g}{1\ \cancel{mL}} = 45.8\ g \qquad \textit{Answer}$$

Exercise 2–11

Calculate the mass in kilograms of a 155-mL volume of acetone at 20°C. $d^{20°} = 0.800$ g/mL.

2–8 Specific Gravity

specific gravity Density of a substance divided by the density of some substance taken as a standard, usually water at 4°C:

Specific gravity

$$= \frac{\text{density of substance}}{\text{density of water at 4°C}}$$

It is the ratio of the density of the substance to the density of the standard.

The **specific gravity** of a substance is the density of the substance divided by the density of some substance taken as a standard. For expressing the specific gravity of liquids and solids, water at 4°C is the standard with a density of 1.00 g/mL in the metric system.

$$\text{Specific gravity} = \frac{\text{density of substance}}{\text{density of water at 4°C}} \qquad (2\text{–}5)$$

$$\text{Density of substance} = \text{specific gravity} \times \text{density of water at 4°C} \qquad (2\text{–}6)$$

TABLE 2–5   SPECIFIC GRAVITY OF SOME SUBSTANCES

SUBSTANCE	SPECIFIC GRAVITY
Water	$1.00^{4°/4}$
Ether	$0.708^{25°/4}$
Benzene	$0.880^{20°/4}$
Acetic acid	$1.05^{20°/4}$
Chloroform	$1.49^{20°/4}$
Carbon tetrachloride	$1.60^{20°/4}$
Sulfuric acid (conc.)	$1.83^{18°/4}$

In calculating the specific gravity of a substance, we must express both densities in the *same units*. Specific gravity, therefore, has *no units*. To convert from specific gravity to density, we merely have to multiply specific gravity by the density of the reference substance (in most cases, water). We may thus find the density of any substance for which we have a reference density. Since the density of water in the metric system is 1.00 g/mL, the density of solids or liquids expressed in g/mL is numerically equal to their specific gravity.

Specific gravity is often expressed as follows:

$$\text{sp gr} = 0.708^{25°/4} \text{ of ether}$$

The 25° refers to the temperature in °C at which the density of ether was measured, and the 4 refers to the temperature in °C at which the density of water (the standard) was measured. Table 2–5 lists the specific gravity of a few substances.

EXAMPLE 2–15

Calculate the mass in grams of a $11\overline{0}$-mL volume of chloroform at 20°C.

SOLUTION   From Table 2–5, the specific gravity of chloroform is 1.49 at 20°. Hence, in the metric system the density is 1.00 g/mL × 1.49 = 1.49 g/mL at 20° (see Equation 2–6).

$$11\overline{0} \text{ mL} \times \frac{1.49 \text{ g}}{1 \text{ mL}} = 163.9 \text{ g, rounded off to three significant digits,}$$

is 164 g   *Answer*

Exercise 2–12

Calculate the mass in kilograms of a $25\overline{0}$-mL volume of acetic acid at 20°C. (Refer to Table 2–5.)

EXAMPLE 2–16

Calculate the volume in liters at 20°C of $84\overline{0}$ g of an organic liquid. The specific gravity of the organic liquid is $1.20^{20°/4}$.

SOLUTION The specific gravity is 1.20. Thus, in the metric system the density is 1.00 g/mL × 1.20 = 1.20 g/mL.

$$84\overline{0} \text{ g} \times \frac{1 \text{ mL}}{1.20 \text{ g}} \times \frac{1 \text{ L}}{1000 \text{ mL}} = 0.700 \text{ L} \qquad Answer$$

Exercise 2–13

Calculate the volume in milliliters of 325 g of carbon tetrachloride at 20°C. (Refer to Table 2–5.)

Sodium chloride: an old friend—salt

Salt can be obtained by the evaporation of seawater, or it can be mined directly from the ground in salt deposits.

Name: You may know it as salt or table salt, but it is also called rock salt, sea salt, or sodium chloride. As a mineral, sodium chloride is called halite.

Appearance: Sodium chloride (NaCl) is a crystalline solid, melting at a high temperature. Finely divided salt or small crystals of salt appear white, but large crystals are nearly colorless.

Occurrence: Sodium chloride is an essential part of the diet of all mammals (including humans). Salt can be found in the oceans, where it comprises about 2.8 grams out of every 100 grams of sea water (2.8 percent). Salt can also be found in huge deposits on land, which presumably resulted from the drying up of inland seas that became isolated from the oceans. Thus, all salt comes from the oceans in one way or another.

Source: Virtually all salt is obtained by one of two methods. It is either mined directly from salt deposits, or it is obtained by evaporation of sea water by the sun in large ponds. The salt in some deposits contains other compounds, such as calcium chloride ($CaCl_2$) or magnesium chloride ($MgCl_2$), but some deposits yield salt that is 99.8 percent pure sodium chloride.

Its Role in Our World: Sodium chloride is one of the principal raw materials of the chemical industry. Virtually all compounds containing sodium or chlorine are derived from salt in some way or another. Important chemicals produced by using sodium chloride include sodium metal, chlorine gas, hydrochloric acid (swimming pool acid), sodium carbonate (soda ash), sodium bicarbonate (baking soda), sodium hypochlorite (chlorine bleach), and sodium hydroxide (caustic soda or lye). In addition, salt is an important part of the manufacture of soaps and dyes; it is used in tin metallurgy, glazing pottery, and the tanning of leather.

Historically, salt has played an important role in the development of trade and the world economy. Because salt is an essential ingredient of human diets and proved to be a useful food preservative in ancient times, it was highly valued and was traded accordingly. The salt from central European mines was traded extensively throughout Europe, the Middle East, and the Far East. Early trade routes are sometimes referred to as the "salt routes." In fact, the word "salary" derives from the Latin word *salarium,* which referred to the money given to Roman soldiers for the purpose of buying salt. Although salt is an important part of the chemical industry, it is far less valuable today than in ancient times, since there are a variety of ways of obtaining salt now.

Unusual Facts: In more humid climates, salt absorbs moisture from the air and "cakes." Although these cakes are easy enough to break up, they can be a nuisance. This absorption of moisture is due to trace amounts of calcium chloride and magnesium chloride in the salt. Curiously, pure salt does not cake at all.

Problems

1. Carry out the following conversions. In all cases express your answer in scientific notation.

   (a) 2.75 km to cm
   (c) 225 mL to m^3

   (b) 625 cg to kg
   (d) 75.5 Å to μ

2. Add the following masses: 405 mg, 0.500 g, 0.002000 kg, 200.0 cg, 1.00 dg. What is the total mass in grams?

3. Add the following lengths: 2.0000000 km, 375.00 cm, 7.0000 m, 0.5 mm. What is the total length in meters?

4. Carry out the following conversions.

   (a) 8.00 in. to m
   (c) 1050 g to oz

   (b) 125 g to lb
   (d) 6.00 qt to mL

5. The 10,000-meter run is a track and field event. What is this distance in miles? Carry out your answer to two significant digits.

6. A certain distance runner runs 15 laps around a gym daily. One lap consists of $4\bar{0}0$ ft. What distance does this person run daily?

(a) in miles   (b) in meters

7. Convert each of the following temperatures to °F and K.

(a) 25°C   (b) $-6\bar{0}$°C

8. Convert each of the following temperatures to °C and K.

(a) 68°F   (b) -125°F

9. Liquid nitrogen has a boiling point of 77 K at 1 atm pressure. What is its boiling point on the Fahrenheit scale?

10. The official coldest temperature recorded in the United States was -79.8°F at Prospect Creek, Alaska, in January 1971. What is this temperature on the Celsius scale?

11. The highest recorded temperature in the world was 136.4°F at Azizia, Libya, in the Sahara Desert on September 13, 1922. What is this temperature in degrees Celsius (°C)?

12. For years, 98.6°F was considered normal body temperature. Recently, a study has shown that the average body temperature is 98.2°F. For men it is 98.1°F and for women it is 98.4°F. Convert these three *new* degrees Fahrenheit (°F) readings to degrees Celsius (°C).

13. Calculate the density in grams per milliliter for each of the following.

(a) a piece of metal having a volume of 45.0 mL and a mass of 425 g
(b) a substance occupying a volume of 84.5 mL and having a mass of 206 g

14. Calculate the volume in milliliters occupied by each of the following at 20°C.

(a) a sample of benzene having a mass of 335 g; $d^{20°} = 0.880$ g/mL
(b) a sample of carbon tetrachloride having a mass of 475 g; $d^{20°} = 1.60$ g/mL

15. Calculate the mass in grams of each of the following.

(a) a 755-mL volume of acetic acid at 20°C; $d^{20°C} = 1.05$ g/mL
(b) a 325-mL volume of benzene at 20°C; $d^{20°C} = 8.80 \times 10^2$ kg/m^3

16. Calculate the mass in grams of each of the following.

(a) a 35.0-mL volume of sulfuric acid (conc.) at 18°C
(b) a 1.75-L volume of carbon tetrachloride at 20°C
(Refer to Table 2–5.)

17. Calculate the volume in liters occupied by each of the following at 20°C.

(a) a sample of acetic acid having a mass of 285 g

(b) a sample of chloroform having a mass of 1.40 kg
(Refer to Table 2–5.)

General Problems

18. A cover has the dimensions 15.0 cm × 30.0 in. × 6.00 ft. Calculate its volume. In both cases express your answer in scientific notation.

(a) in cubic centimeters
(b) in cubic inches

19. Carry out the following conversions.

(a) a density of 4.70 g/mL to lb/gal
(b) a density of 5.00 g/cm³ to lb/ft³

20. Calculate the density in lb/ft³ of a piece of metal 15.0 cm × 30.0 in. × 5.00 ft having a mass of 1340 kg.

21. The air pressure in the tires on an automobile is 32.0 lb/in.² (psi). Express this pressure in kilograms per square meter (kg/m²). Express your answer in scientific notation.

22. Calculate the mass in kilograms of a rectangular piece of ice in a lake to two significant digits if the ice measures 15 feet long, 12 feet wide, and 11 inches thick. The density of ice at 0°C is 0.915 g/cm³.

Answers to Exercises

2–1. 0.175 L
2–3. 6.25 × 10⁻³ kL
2–5. 6.61 lb
2–7. 5°F
2–9. 8.4 g/mL
2–11. 0.124 kg
2–13. 203 mL

2–2. 1008.25 m
2–4. 0.184 m
2–6. 3.02 × 10³ mL
2–8. 20°C, 293 K
2–10. 151 mL
2–12. 0.262 kg

QUIZ

1. Convert 8.00 pt to milliliters (mL). Express your answer in scientific notation.

2. A child with the flu has a temperature of 39.8°C. What is his or her temperature in degrees Fahrenheit (°F)?

3. Calculate the mass in kilograms of 0.750 L of chloroform at 20°C ($d^{20°}$ = 1.49 g/mL).

4. Calculate the volume in liters at 20°C occupied by 155 g of concentrated nitric acid (sp gr = $1.42^{20°/4}$).

5. A piece of vanadium metal is machined into a cube that is 2.30 cm on an edge. The piece of metal has a mass of 72.5 g. Calculate the density of vanadium in grams per milliliter (g/mL).

3

MATTER

COUNTDOWN

5 Convert 2.50 L of water to quarts of water (Section 2–5).

(2.65 qt)

4 Convert a temperature of $-34°F$ to degrees Celsius (°C) and kelvins (K) (Section 2–6).

$(-37°C, 236 K)$

3 Calculate the density of a metal in grams per milliliter (g/mL) if the metal has a volume of 5.82 cm^3 and a mass of 65.8 g (Sections 2–2 and 2–7).

(11.3 g/mL)

2 Calculate the volume in liters occupied by 925 g of acetone at 20°C $[d^{20°} = 0.790$ g/mL] (Sections 2–2, 2–3, and 2–7).

(1.17 L)

1 Calculate the mass in kilograms occupied by $97\overline{0}$ mL of carbon tetrachloride. See Table 2–5 for additional data (Sections 2–2, 2–3, and 2–8).

(1.55 kg)

TASKS

1 Identify the three physical states of matter and give an example of each (Section 3–1).

2 Memorize the symbols for the 47 elements listed in Table 3–1.

OBJECTIVES

1 Give the distinguishing characteristics of each of the following terms:
   (a) matter (Section 2–1)
   (b) homogeneous matter (Section 3–2)
   (c) heterogeneous matter (Section 3–2)
   (d) pure substance (Section 3–2)
   (e) homogeneous mixture (Section 3–2)
   (f) solution (Section 3–2)

(g) mixture (Section 3–2)

(h) compound (Section 3–3)

(i) element (Section 3–3)

(j) physical properties (Section 3–4)

(k) chemical properties (Section 3–4)

(l) physical changes (Section 3–5)

(m) chemical changes (Section 3–5)

(n) atom (Section 3–6)

(o) formula unit (Section 3–7)

(p) molecule (Section 3–7)

(q) molecular formula (Section 3–7)

2 Given the name of a common substance, classify it as a compound, element, or mixture (Example 3–1, Exercise 3–1, Problem 1).

3 Given a property of a substance, classify it as a chemical or physical property (Example 3–2, Exercise 3–2, Problem 2).

4 Given a change of a substance, classify it as a chemical or physical change (Example 3–3, Exercise 3–3, Problem 3).

5 Given the formula of a compound, write the number of atoms, the name of each element, and the total number of atoms in one unit of the compound (Example 3–4, Exercise 3–4, Problem 4).

6 Given the number of atoms of each element in one molecule of a compound and the name of each of the elements, write the molecular formula of the compound (Example 3–5, Exercise 3–5, Problem 5).

In Chapter 2 we defined chemistry as concerned with the composition of substances and the changes it undergoes. Substances are matter, and matter is anything that has mass and occupies space. Your book is matter; your highlighter is matter; your paper is matter. Yes, you are matter. In Chapter 2 we were concerned with the composition of matter, which involves measurements. In this chapter we consider changes in matter. We consider whether the matter is shiny or dull, dense or light, and any changes or reactions the matter does or does not undergo. We also consider the various types of matter.

3–1 Physical States of Matter

The three physical states of matter are solid, liquid, and gas.[1] Matter exists either as a solid, liquid, or gas, depending on the temperature and pressure and the specific characteristics of the particular type of matter. Some matter exists in *all*

---

[1] Intermediate between solids and liquids are *liquid crystals.* Liquid crystals are substances whose properties are intermediate between those of a true crystalline solid and those of a true liquid as they go from the solid to the liquid physical state. One important property of liquid crystals is

three physical states, while other matter decomposes (breaks down) when an attempt is made to change its physical state. Water exists in all three physical states:

Solid: ice, snow
Liquid: water
Gas: water vapor and steam

The particular physical state of water is determined by the conditions (temperature and pressure) under which the observation is made. Common table sugar exists under normal conditions in only one physical state—solid. Attempts to change it to a liquid or a gas by heating at atmospheric pressure results in decomposition of the sugar, and the sugar turns caramel brown to black in color.

3–2 Homogeneous and Heterogeneous Matter. Pure Substances, Solutions, and Mixtures

homogeneous matter Matter uniform in composition and properties throughout.

heterogeneous matter Matter not uniform in composition and properties, and consisting of two or more physically distinct portions or phases unevenly distributed.

substance, pure Homogeneous matter characterized by definite and constant composition, and definite and constant properties under a given set of conditions.

homogeneous mixture Matter homogeneous throughout but composed of two or more pure substances whose properties may be varied in some cases *without limit.*

Matter is further divided into two major subdivisions: homogeneous and heterogeneous matter. **Homogeneous matter** is *uniform* in composition and properties throughout. It is the same throughout. **Heterogeneous matter** is *not uniform* in composition and properties and consists of two or more physically distinct *portions or phases* unevenly divided. A class including both men and women would be analogous to a heterogeneous mixture; the distinct phases would be women students and men students.

Homogeneous matter is divided into three categories: pure substances, homogeneous mixtures, and solutions. A **pure substance** is characterized by *definite* and *constant composition* and a pure substance has *definite* and *constant properties* under a given set of conditions. A pure substance obeys our definition of homogeneous matter not only in that it is uniform throughout in both composition and properties but also in that it has the additional requirement of definite and constant composition and properties. Some examples of pure substances are water, salt (sodium chloride), sugar (sucrose), mercuric or mercury(II) oxide, gold, iron, and aluminum.

A **homogeneous mixture** is homogeneous throughout, but it is composed of two or more pure substances whose proportions may be varied in some cases *without* limits. One example of a homogeneous mixture is air, which is a mixture of oxygen, nitrogen, and certain other gases. Mixtures of gases are generally called *homogeneous mixtures* but homogeneous mixtures composed of gases,

that some of them change color with a change in temperature. As a result of this property, liquid crystals are used to display numbers on wristwatches, pocket calculators, and electronic games, and to detect abnormal vascular diseases and cancer of the skin and breast. *Plasma,* sometimes considered to be a fourth state of matter, is similar to a gas, except that the particles are charged rather than neutral.

solution Homogeneous mixture composed of two or more pure substances; its composition can be varied usually *within certain limits.*

liquids, or solids dissolved in liquids are called *solutions.* A **solution** is homogeneous throughout and is composed of two or more pure substances; its composition can be varied usually *within certain limits.* Some common examples of solutions are sugar solution (sugar dissolved in water), a salt solution (salt dissolved in water), and carbonated water (carbon dioxide dissolved in water). Homogeneous mixtures (or solutions) differ from pure substances in that pure substances have *definite* and *constant compositions* throughout, whereas homogeneous mixtures consist of two or more pure substances in *variable* proportion. For example, a salt solution may consist of one gram of salt in a liter of water or 10 grams of salt in a liter of water.

mixture Matter composed of two or more substances, each of which retains its identity and specific properties.

Heterogeneous matter is also commonly called a **mixture**. This type of mixture is composed of two or more pure substances, each of which retains its identity and specific properties. In many mixtures, the substances can be readily identified by visual observations. For example, in a mixture of salt and sand, the white crystals of salt can be distinguished from the tan crystals of sand by the eye or a hand lens. Similarly, in a mixture of iron and sulfur, the yellow sulfur can be identified from the black iron by visual observation. Mixtures can usually be separated by a simple operation that does not change the composition of the several substances comprising the mixture. For example, a mixture of salt and sand can be separated by using water. The salt dissolves in water; the sand is insoluble. A mixture of iron and sulfur can be separated by dissolving the sulfur in liquid carbon disulfide (the iron is insoluble), or by attracting the iron to a magnet (the sulfur is not attracted).

3–3 Compounds and Elements

Pure substances are divided into two groups: compounds and elements. Figure 3–1 summarizes the classification of matter.

compound A pure substance that can be broken down by various chemical means into two or more different substances.

A **compound** is a pure substance that *can be broken down* by various chemical means into two or more different simpler substances. Of the pure substances mentioned in Section 3–2, water, table salt, mercuric or mercury(II) oxide, and sugar are compounds. The action of an electric current (electrolysis) breaks down both water (Equation 3–1) and molten table salt (Equation 3–2) to simpler substances. The action of heat decomposes both mercuric or mercury(II) oxide (Equation 3–3) and sugar (Equation 3–4).

$$\text{Water} \xrightarrow{\substack{\text{direct} \\ \text{electric} \\ \text{current}}} \text{hydrogen} + \text{oxygen} \qquad (3\text{–}1)$$

$$\text{Table salt (molten)} \xrightarrow{\substack{\text{direct} \\ \text{electric} \\ \text{current}}} \text{sodium} + \text{chlorine} \qquad (3\text{–}2)$$

$$\text{Mercuric or mercury(II) oxide} \xrightarrow{\text{heat}} \text{mercury} + \text{oxygen} \qquad (3\text{–}3)$$

$$\text{Sugar} \xrightarrow{\text{heat}} \text{carbon} + \text{water} \qquad (3\text{–}4)$$

(see Fig. 3–2)

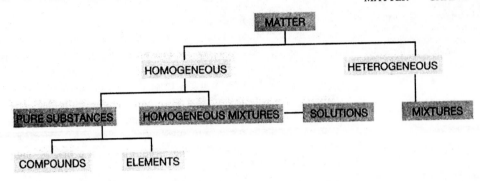

Figure 3-1   *Classification of matter.*

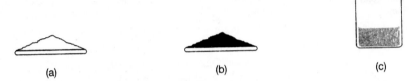

Figure 3-2   *Heat decomposes (a) the* compound *sugar to produce (b) the* element *carbon and (c) the* compound *water.*

element A pure substance that cannot be broken down into simpler substances by ordinary chemical means.

An **element** is a pure substance that *cannot be broken down* into simpler substances by ordinary chemical means. Of the pure substances mentioned in Section 3–2, gold, iron, and aluminum are elements. Among the pure substances produced in Equations 3–1 to 3–4, hydrogen, oxygen, sodium, chlorine, mercury, and carbon are all examples of elements. (In Equation 3–4, water is produced. Why is water not an element? See Equation 3–1.)

There are 109 known elements[2] at this writing of which 90 have been found to occur naturally so far. The remaining 19 have been produced solely through nuclear reactions in the laboratory. Minute amounts of some of these may exist in nature. The ranking and relative abundance (percent by mass) of the first ten elements in the earth's crust (upper 10 miles, including the oceans and atmosphere) are given in Table 3–1.

Each of the elements has a symbol that is an abbreviation for the name of

---

[2] Between 1981 and 1984 *three new elements* were discovered. They were elements 107, 108, and 109. The names and symbols for these elements are unnilseptium (Uns), element 107; unniloctium (Uno), element 108; and unnilennium (Une), element 109. The discoverers of these elements were a German nuclear research team. Recently (1992) the discoverers have proposed to change the names and symbols of these elements. The new proposed names and symbols are as follows: nielsbohrium (Ns), element 107; hassium (Hs), element 108; and meiterium (Mt), element 109. Nielsbohrium and meiterium are named after famous scientists, Niels Bohr and Lise Meitner. - Hassium is named after Hassia, the Latin name for the German state of Hesse, where the research was done. The international chemical community is now considering these name changes. Many chemists approve of them since they appear to put a place or human face on these elements. What do you think? For the present we will have to use the names unnilseptium (107), unniloctium (108), and unnilennium (109) until they are officially changed.

the element. All the elements and their symbols are listed inside the front cover of this book. Since many of the elements are rarely mentioned in most college chemistry courses, Table 3–1 condenses this list of 109 elements to 47 of the most common ones. You must know the names and symbols for all 47 elements given in Table 3–1. We suggest that you make flash cards with the name of the

TABLE 3–1   SOME COMMON ELEMENTS, THEIR SYMBOLS, AND RANKING OF RELATIVE ABUNDANCE (PERCENT BY MASS) FOR THE FIRST 10 ELEMENTS IN THE EARTH'S CRUST[a]

ELEMENT	SYMBOL[b]	RANKING (% MASS)	ELEMENT	SYMBOL	RANKING (% MASS)
Aluminum	Al	3 (8.3)	Lithium	Li	
Antimony	Sb		Magnesium	Mg	7 (2.3)
Argon	Ar		Manganese	Mn	
Arsenic	As		Mercury	Hg	
Barium	Ba		Neon	Ne	
Beryllium	Be		Nickel	Ni	
Bismuth	Bi		Nitrogen	N	
Boron	B		Oxygen	O	1 (46.4)
Bromine	Br		Phosphorus	P	
Cadmium	Cd		Platinum	Pt	
Calcium	Ca	5 (4.2)	Potassium	K	8 (2.1)
Carbon	C		Radium	Ra	
Chlorine	Cl		Selenium	Se	
Chromium	Cr		Silicon	Si	2 (28.2)
Cobalt	Co		Silver	Ag	
Copper	Cu		Sodium	Na	6 (2.4)
Fluorine	F		Strontium	Sr	
Gold	Au		Sulfur	S	
Helium	He		Tin	Sn	
Hydrogen	H	10 (0.1)	Titanium	Ti	9 (0.6)
Iodine	I		Uranium	U	
Iron	Fe	4 (5.6)	Xenon	Xe	
Krypton	Kr		Zinc	Zn	
Lead	Pb				

[a] Upper 10 miles, including the oceans and atmosphere.

[b] Some of these symbols do not appear to be related to the names of the elements. In these cases, the symbol used has been obtained from the Latin name by which the element was known for centuries.

NAME OF ELEMENT	LATIN NAME (SYMBOL)	
Antimony	Stibium	(Sb)
Copper	Cuprum	(Cu)
Gold	Aurum	(Au)
Iron	Ferrum	(Fe)
Lead	Plumbum	(Pb)
Mercury	Hydrargyrum	(Hg)
Potassium	Kalium	(K)
Silver	Argentum	(Ag)
Sodium	Natrium	(Na)
Tin	Stannum	(Sn)

element on one side and the symbol on the other side. Go over the cards until you know all of the elements and their symbols, and then study them some more periodically so that you do not forget any of them.

EXAMPLE 3–1

Classify each of the following as a compound, element, or mixture.

ANSWERS

(a)	sugar	compound	
(b)	aluminum	element	(see Table 3–1)
(c)	motor oil	mixture	(mixture of organic compounds plus additives)
(d)	hydrogen	element	(see Table 3–1)

Exercise 3–1

Classify each of the following as a compound, element, or mixture.

(a) iron   (b) uranium
(c) gasoline   (d) salt (sodium chloride)

physical properties Properties of a substance that can be observed without changing the composition of the substance.

chemical properties Properties of a substance that can be observed when the substance undergoes a change in composition.

3–4 Properties of Pure Substances

Just as each person has his or her own appearance and personality, each pure substance has its own properties, distinguishing it from other substances. The properties of pure substances are divided into physical and chemical properties.

Physical properties are those properties that can be observed without changing the composition of the substance. These properties include color, odor, taste, solubility, density, specific gravity, melting point, and boiling point. Physical properties of a pure substance are analogous to a person's appearance—the color of the person's hair and eyes or the person's height and weight (see Fig. 3–3).

Chemical properties are those properties that can be observed when a substance undergoes a change in composition. These properties include the fact that iron rusts, that coal or gasoline burns in air, that water undergoes electrolysis, and that chlorine reacts violently with sodium. Chemical properties of a pure substance are analogous to a person's personality, outlook on life, temperament, or reaction in various situations (see Fig. 3–3). Table 3–2 lists some physical and chemical properties of water (a pure substance) and iron (an element).

EXAMPLE 3–2

The following are properties of the element mercury; classify them as physical or chemical properties.

ANSWERS

(a) silvery-white liquid at room temperature   physical

	ANSWERS
(b) freezing point, $-38.9°C$	physical
(c) reacts with oxygen when heated	chemical
(d) reacts with nitric acid	chemical
(e) density at 20°C = 13.55 g/mL	physical
(f) reacts with hot concentrated sulfuric acid	chemical

Exercise 3-2

The following are properties of the element aluminum; classify them as physical or chemical properties.

(a) density at 20°C = 2.70 g/mL

(b) boiling point = 2327°C (1 atm pressure)

TABLE 3-2   SOME PHYSICAL AND CHEMICAL PROPERTIES OF WATER AND IRON

	PHYSICAL				
SUBSTANCE	COLOR	DENSITY (g/mL, 20°C)	MELTING POINT (°C)	BOILING POINT* (°C)	CHEMICAL
Water (liquid)	Colorless	0.998	0	100	Undergoes electrolysis; yields hydrogen and oxygen
Iron (solid)	Gray-white	7.874	1535	3000	Rusts; reacts with oxygen in air to form an iron oxide [ferric or iron(III) oxide]

ª At 1.00 atm presure.

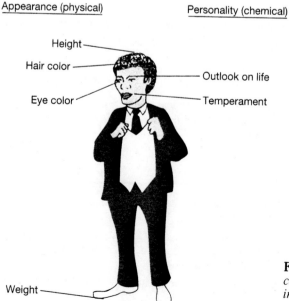

Appearance (physical)   Personality (chemical)

Height

Hair color

Eye color

Outlook on life

Temperament

Weight

Figure 3-3 *Physical and chemical properties are analogous to an individual's appearance and personality.*

(c) silvery-white shine

(d) reacts with concentrated base

3-5 Changes of Pure Substances

In determining the properties of pure substances, we shall observe certain changes or conversions from one form to another in these pure substances. These changes are divided into physical and chemical changes.

physical changes Changes in a substance that occur with no change in the composition of the substance taking place.

　　　Physical changes are those changes that can be observed without a change in the composition of the substance taking place. The changes in state of water from ice to liquid to water vapor (Equation 3–5) are examples of physical change, so melting, freezing, boiling, and condensing are physical changes.

$$\text{Ice} \rightleftharpoons \text{liquid water} \rightleftharpoons \text{water vapor} \qquad (3\text{--}5)$$

The difference between a property and a change should be noted here; a *property* distinguishes *one substance* from another substance, whereas a *change* is a *conversion* from one *form* or substance to *another*. Physical change in a substance is analogous to the change in a woman's appearance when she puts on eye shadow or lipstick.

chemical changes Changes in a substance that can be observed only when a change in the composition of the substance is occurring. *New substances are formed.*

　　　Chemical changes are those changes that can be observed only when a change in the composition of the substance is occurring. *New substances are formed.* The *new substances* have properties *different* from the old substances. In a chemical change a gas may be produced, heat energy may be given off (the flask gets hot), a color change may occur, or an insoluble substance may appear. We mentioned previously (Section 3–3) that the passage of electric current through liquid water yields the gases hydrogen and oxygen. Also, chlorine gas reacts violently with sodium metal to give sodium chloride, or common table salt. The change that occurs and that determines the chemical properties is a chemical change. These two changes are represented in Equations 3–1 and 3–6, respectively.

$$\text{Chlorine (Cl)} + \text{sodium (Na)} \longrightarrow \text{sodium chloride (NaCl)} \qquad (3\text{--}6)$$

Somewhat analogous to a chemical change would be a pugnacious person meeting another person—possibly of the opposite sex—and becoming more cordial—possibly becoming a "new" person!

　　　Table 3–3 lists various changes and classifies them as chemical or physical.

EXAMPLE 3–3

Classify the following changes as physical or chemical.

	ANSWER
(a) tearing up a piece of bread to put in hamburger	physical
(b) cooking a hamburger	chemical
(c) burning a piece of bread in a toaster	chemical
(d) cutting up an avocado	physical

TABLE 3–3   CLASSIFICATION OF CHANGES
AS PHYSICAL OR CHEMICAL

CHANGE	CLASSIFICATION
Boiling of water	Physical
Freezing of water	Physical
Electrolysis of water	Chemical
Reaction of chlorine with sodium	Chemical
Melting of iron	Physical
Rusting of iron	Chemical
Cutting of wood	Physical
Burning of wood	Chemical
Taking a bite of food	Physical
Digestion of food	Chemical

Exercise 3–3

Classify the following changes as physical or chemical.

(a) souring of milk
(b) tearing up your notes
(c) burning a piece of newspaper
(d) fermentation of grapes to produce wine

3–6   Elements and Atoms

atom The smallest particle of an element that can undergo chemical changes in a reaction.

In Section 3–3 we defined an element as a pure substance that cannot be decomposed into simpler substances by ordinary chemical means. *Elements are composed of atoms.* An **atom** is the smallest particle of an element that can undergo chemical changes in a reaction. The symbols for the elements represent not only the name of the element but also one atom of the element. For example, the symbol Na represents *one* atom of the element sodium, and the symbol H represents *one* atom of the element hydrogen.

3–7   Compounds, Formula Units, and Molecules

formula unit Generally, the smallest combination of *charged* particles (ions) in which the opposite charges present balance each other so that the overall compound has a net charge of zero, such as NaCl (Na^{1+}, Cl^{1-}).

In Section 3–3 we defined compounds as pure substances that can be broken down into two or more simpler substances by various chemical means. Compounds may be composed of *charged particles (ions)* or *molecules.*

In the case of an ionic compound (composed of charged particles or ions) it is convenient to represent the compound with a **formula unit** in which the opposite charges of the ions present balance each other so that the formula unit representing the compound has an overall charge of *zero.* Generally, the simplest formula unit possible is used. For example, sodium chloride is composed of sodium ions (Na^{1+}) and chloride ions (Cl^{1-}) in equal numbers, and the formula unit for this compound is NaCl.

molecule The smallest particle of a pure substance (element or compound) that can exist and still retain the physical and chemical properties of the substance, such as O_2 and H_2O.

For those compounds existing as molecules a molecular formula is used to represent the compound. The **molecule** is the smallest particle of the compound that can exist and still retain the physical and chemical properties of the compound. Molecules, like atoms, are the particles that undergo chemical changes in a reaction. These molecules are composed of atoms of elements, held together by chemical bonds; *hence these small particles called atoms are fundamental to all compounds.* Molecules may be composed of two or more *nonidentical* atoms—that is, atoms from different elements. Water molecules are composed of the nonidentical atoms of hydrogen and oxygen. Molecules may also be composed of one or more *identical* atoms, that is, atoms from the same element. Oxygen molecules are composed of two identical atoms of oxygen. For now, we consider only molecules of compounds. In Chapter 6, when we discuss bonding, we shall again refer to formula units and molecules from nonidentical and identical atoms.

molecular formula A formula composed of an appropriate number of symbols of elements representing *one* molecule of the given compound. Also defined as the formula containing the *actual* number of atoms of each element in *one* molecule of the compound.

An atom of an element is represented by a symbol, and a molecule of a compound is represented by a formula—more precisely, a **molecular formula**. Such a formula is composed of an appropriate number of symbols of elements representing one molecule of the given compound. For example, the molecular formula for water is H_2O (read "H-two, O"). The *subscripts* represent the *number of atoms* of the respective elements in *one* molecule of the compound. Where no subscript is given, the number of atoms is one. Hence, in one molecule of water there are 2 atoms of the element hydrogen and 1 atom of the element oxygen, resulting in a total of 3 atoms in one molecule of water.

EXAMPLE 3-4

Determine the number of atoms of each element and then write the name of the element and the total number of atoms in each of the following formulas.

ANSWERS

(a) H_2S (hydrogen sulfide)
    2 atoms hydrogen, 1 atom sulfur; total 3 atoms

(b) H_2O_2 (hydrogen peroxide)
    2 atoms hydrogen, 2 atoms oxygen; total 4 atoms

(c) CH_4O (methyl alcohol)
    1 atom carbon, 4 atoms hydrogen, 1 atom oxygen; total 6 atoms

(d) CO_2 (carbon dioxide)
    1 atom carbon, 2 atoms oxygen; total 3 atoms

(e) $Pb(C_2H_3O_2)_2$ [lead(II) acetate]
    1 atom lead, 4 atoms carbon, 6 atoms hydrogen, 4 atoms oxygen; total 15 atoms [*Note:* Clear the ( ) to get the number of atoms of carbon, hydrogen, and oxygen, the same as you did in solving linear equations (Section 1–5). The subscript 2 refers to everything in the ( ).]

Exercise 3–4

Determine the number of atoms of each element and then write the name of the element and the total number of atoms in each of the following molecular formulas.

(a) N_2O_3 (dinitrogen trioxide)
(b) $C_6H_{12}O_6$ (dextrose, glucose)

If the number of atoms of each element in one molecule of the compound is known, we can write the molecular formula of the compound.

EXAMPLE 3–5

From the number of atoms of each element in one molecule of the compound, write the molecular formula of the following compounds.

ANSWERS

(a) carbon monoxide (a deadly air pollutant); 1 atom carbon, 1 atom oxygen   CO

(b) ethyl alcohol; 2 atoms carbon, 6 atoms hydrogen, 1 atom oxygen C_2H_6O

(c) ethylene glycol (used as an antifreeze); 2 atoms carbon, 6 atoms hydrogen, 2 atoms oxygen $C_2H_6O_2$

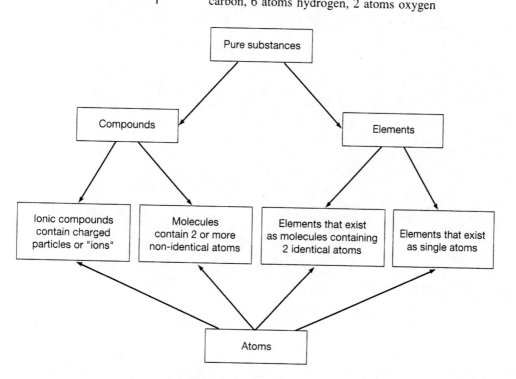

Figure 3–4 *Summary of atoms, molecules, charged particles (ions), elements, and compounds. Note that atoms are the fundamental components of both compounds and elements.*

ANSWERS

> (d) chlorophyll a; 55 atoms carbon, 72 atoms $C_{55}H_{72}MgN_4O_6$
>   hydrogen, 1 atom magnesium, 4 atoms nitro-
>   gen, 6 atoms oxygen
> (e) phosphoric acid; 3 atoms hydrogen, 1 atom H_3PO_4
>   phosphorus, 4 atoms oxygen

Exercise 3–5

From the number of atoms of each element in one molecule of the compound, write the molecular formula of the following compounds.

(a) sulfur trioxide; 1 atom sulfur, 3 atoms oxygen
(b) pyrophosphoric acid; 4 atoms hydrogen, 2 atoms phosphorus, 7 atoms oxygen

Figure 3–4 summarizes the relationships among atoms, molecules, formula units, elements, and compounds.

Gold: chemistry and the world economy

Gold is a bright and shiny metal that plays an important role in the world economy. Ingots of gold bullion in government vaults measure one aspect of a nation's wealth.

Name: The symbol for gold (Au) derives from the Latin *aurum,* meaning shining dawn. Gold was the first pure metal known to humans and has always been valued.

Appearance: Bright, lustrous, shiny yellow metal.

Occurrence: Gold is quite rare and makes up only 0.0000005 percent of the earth's crust.

Source: Gold occurs as a metal in nature due to its remarkable stability. Gold is also found to a limited extent as the ores *calaverite* ($AuTe_2$) and *sylvanite* [$(AuAg)Te_2$].

Its Role in Our World: The primary use of gold is as a currency standard for most nations of the world.

Gold and its alloys (mixtures, mostly with silver and copper) are used extensively in the making of fine jewelry.

Gold has proven to be quite versatile in the electronics industry due to its fine electrical conductivity, ductility, and resistance to corrosion.

Gold compounds are used in medicine in the treatment of arthritis, and gold alloys (mixture with other metals) are used in dentistry in gold inlays.

Unusual Facts: The alchemists tried in vain to search for ways to change lead into gold. In doing so, however, they developed techniques and apparatus that led to the development of chemistry as a science. Other elements can now be changed into gold through nuclear chemistry, but not in an economical manner.

Problems

1. Classify each of the following as a compound, element, or mixture.

   (a) distilled water
   (b) chlorine
   (c) dry ice (carbon dioxide)
   (d) salted buttered popcorn

2. The following are properties of the element thallium; classify them as physical or chemical properties.

   (a) oxidizes slowly at 25°C
   (b) bluish-white
   (c) malleable (can be shaped by beating with a hammer)
   (d) reacts with chlorine
   (e) reacts with nitric acid
   (f) melting point 303.5°C
   (g) easily cut with a knife

3. Classify the following changes as physical or chemical.

   (a) pumping of oil out of a well
   (b) separation of components of oil by distillation (heating)
   (c) burning of gasoline
   (d) grinding up of beef in a meat grinder
   (e) chewing the beef in the mouth
   (f) digestion of the beef
   (g) baking of bread
   (h) mixing of flour with yeast
   (i) fermentation to produce beer
   (j) smashing a car against a tree

4. Determine the number of atoms of each element, write the name of the element, and give the total number of atoms present in each of the following formulas.

(a) $C_{12}H_{22}O_{11}$ (sucrose, table sugar)
(b) NH_3 (ammonia)
(c) CCl_2F_2 (Freon-12)
(d) SnF_2 (stannous fluoride)
(e) $Ca_3(PO_4)_2$ (calcium phosphate)

5. From the number of atoms of each element in one molecule of the compound, write the molecular formula of the following compounds.

   (a) aspirin; 9 atoms carbon, 8 atoms hydrogen, 4 atoms oxygen
   (b) caffeine; 8 atoms carbon, 10 atoms hydrogen, 4 atoms nitrogen, 2 atoms oxygen
   (c) Mercurochrome; 20 atoms carbon, 8 atoms hydrogen, 2 atoms bromine, 1 atom mercury, 2 atoms sodium, 6 atoms oxygen
   (d) adenosine triphosphate (ATP); 10 atoms carbon, 16 atoms hydrogen, 5 atoms nitrogen, 13 atoms oxygen, 3 atoms phosphorus
   (e) 2,4,6-trinitrotoluene (TNT); 7 atoms carbon, 5 atoms hydrogen, 3 atoms nitrogen, 6 atoms oxygen

General Problems

6. The specific gravity $(20°/4)$ of a sample of concentrated nitric acid is 1.42. Calculate the volume in liters occupied by 165 g of this nitric acid at 20°C.

7. Calculate the volume in liters at 20°C occupied by 4.20 lb of iron (see Table 3–2).

8. Calculate the mass in pounds of 6.00 qt (20°C) of water (see Table 3–2).

9. A piece of metal, which is a cube 1.30 cm on each edge, has a mass of 5.94 g. Calculate the density of metal in grams per milliliter (g/mL).

10. Aluminum metal has a melting point of $66\overline{0}°C$. Convert this temperature to degrees Fahrenheit (°F) and kelvins (K).

Answers to Exercises

3–1. (a) and (b) element; (c) mixture; (d) compound

3–2. (a), (b), (c) physical; (d) chemical

3–3. (a) chemical; (b) physical; (c) and (d) chemical

3–4. (a) 2 atoms nitrogen, 3 atoms oxygen; total 5 atoms
     (b) 6 atoms carbon, 12 atoms hydrogen, 6 atoms oxygen; total 24 atoms

3–5. (a) SO_3; (b) $H_4P_2O_7$

QUIZ **1.** Give the symbol for each of the following elements:

   (a) gold   (b) radium

2. Three properties of the element sodium are listed below. Classify them as physical or chemical properties.

(a) reacts rapidly with oxygen from the air
(b) lustrous silver color
(c) soft, cuts easily with a knife

3. Isoamyl mercaptan ($C_5H_{12}S$) is the major constituent of skunk odor. Determine the number of atoms of each element, write the name of the element, and give the total number of atoms present in isoamyl mercaptan.

4. Benzocaine is used as a mild topical anesthetic in ointments for burns and wounds. From the number of atoms of each element in one molecule of benzocaine, write the molecular formula of the compound: 9 atoms carbon, 11 atoms hydrogen, 1 atom nitrogen, 2 atoms oxygen.

5. A procedure requires 10.5 g of bromoform ($CHBr_3$, $d^{20°} = 2.89$ g/mL). Calculate the number of milliliters of bromoform that must be measured out at 20°C to obtain the required mass of bromoform.

4

ATOMS

COUNTDOWN

5 Convert 6.75 Å to millimeters (mm). Express your answer in scientific notation [Sections 1–4 and 2–2 (Table 2–2)].

$$(6.75 \times 10^{-7} \text{ mm})$$

4 Convert the density of gold (Au), 19.3 g/cm³, to kilograms per cubic meter (kg/m³). Express your answer in scientific notation (Sections 1–4, 2–2, and 2–5).

$$(1.93 \times 10^4 \text{ kg/m}^3)$$

3 Convert 10.2 kg to pounds (lb) (Sections 2–2 and 2–5).

$$(22.5 \text{ lb})$$

2 Convert the melting point of gold (Au), 1064°C, to degrees Fahrenheit (°F) and kelvins (K) (Section 2–6).

$$(1947°F, 1337 \text{ K})$$

1 Calculate the volume in milliliters (mL) occupied by 0.650 kg of gold (Au).

$$d^{20°} = 19.3 \text{ g/cm}^3 \qquad (\text{Sections } 2\text{–}2 \text{ and } 2\text{–}7)$$

$$(33.7 \text{ mL})$$

TASKS

1 List and explain in your own words Dalton's basic proposals about atomic theory (Section 4–2).

2 Identify the three basic subatomic particles, their abbreviations, approximate masses in amu, and *relative* charges (Section 4–3).

OBJECTIVES

1 Give the distinguishing characteristics of each of the following terms:
(a) atomic mass scale (Section 4–1)
(b) electron (Section 4–3)
(c) proton (Section 4–3)

(d) neutron (Section 4–3)

(e) atomic number (Section 4–4)

(f) mass number (Section 4–4)

(g) isotope (Section 4–5)

(h) valence electrons (Section 4–7)

(i) rule of eight or octet rule (Section 4–7)

2 Given the symbol $_2^4E$, explain the meaning of each letter in the symbol (Section 4–4).

3 Given the atomic number and mass number of any element, calculate the number of protons and neutrons in the nucleus and the number of electrons outside the nucleus (Example 4–1, Exercise 4–1, Problem 1).

4 Given the exact atomic mass in amu for two or more isotopes and the percent abundance in nature for these isotopes, calculate the atomic mass of the element (Example 4–2, Exercise 4–2, Problems 2 and 3).

5 Given the principal energy levels, calculate the maximum number of electrons in each level (Example 4–3, Exercise 4–3, Problem 4).

6 Given the atomic number and mass number of any atom of atomic number 1 to 18, diagram the atomic structure of the atom by writing the number of protons and neutrons and by arranging the electrons in principal energy levels (Example 4–4, Exercise 4–4, Problem 5).

7 Given the symbol for any element of atomic number 1 to 18 and its atomic number, write the electron-dot formula for an atom of that element (Example 4–5, Exercise 4–5, Problem 6).

8 Given the atomic number of any element, (1) write the electronic configuration in sublevels, and (2) give the number of valence electrons for an atom of that element (Example 4–6, Exercise 4–6, Problem 7).

In Section 3–7 (Fig. 3–3) we showed that the small particles called atoms are fundamental to both compounds and elements. All matter on the earth and in the universe is composed of atoms. The explosion of the atomic bomb at the end of World War II vividly brought the term "atom" to the attention of the world. In this chapter we consider the structure of these minute atoms in more detail.

4–1 Atomic Mass

Atoms are very small. The diameter of an atom is in the range 1 to 5 angstroms (Å, see Table 2–2). If we were to place atoms of a diameter of 1 Å side by side, it would take 254,000,000 of them to occupy a 1-inch length, as illustrated in Fig. 4–1. That is a lot of atoms!

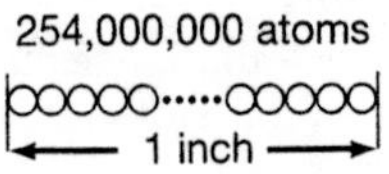

Figure 4-1 *Atoms are very small. If atoms of a diameter of 1 Å were placed side by side, it would take 254,000,000 atoms to occupy a 1-inch length.*

254,000,000 atoms

⟨︎oooo·····ooooo⟩

⟵ 1 inch ⟶

atomic mass scale Relative scale of atomic masses, based on an arbitrarily assigned value of exactly 12 atomic mass units (amu) for the *mass* of carbon-12.

The mass of an atom is also a very small quantity, too small to be determined on even the most sensitive balance. For example, by indirect methods the mass of a hydrogen atom is found to be 1.67×10^{-24} g; an oxygen atom, 2.66×10^{-23} g; and a carbon atom, 2.00×10^{-23} g. Since this mass is very small, chemists have devised a scale of relative masses of atoms called the **atomic mass** (*atomic weight*) **scale**. The scale is based on an arbitrarily assigned value of exactly 12 *atomic mass units*, abbreviated amu, for carbon-12 (Section 4-5 will discuss the nature of carbon-12). Hence, 1 atomic mass unit (amu) on the atomic mass scale is equal to 1/12 the mass of a carbon-12 atom. An atom that is twice as heavy as a carbon-12 atom would have a mass of 24 atomic mass units (amu).

All of the elements are listed on the inside front cover with their precise relative atomic masses in atomic mass units based on carbon-12. As you can see, some of these numbers are very exact and are carried out even to the ten-thousandths place, whereas others are expressed only to the units place. Therefore, for calculations that you will be doing in this book we have developed a Table of Approximate Atomic Masses. It is found on the inside back cover, and you should use it in all calculations in this text unless otherwise indicated.

4-2 Dalton's Atomic Theory

In the early part of the nineteenth century, the English scientist John Dalton (1767–1844) proposed an atomic theory based on experimentation and chemical laws known at that time. His proposals, after some modifications due to recent discoveries, still form the framework of our knowledge of the atom. His basic proposals were:

1. Elements are composed of tiny, discrete, indivisible, and indestructible particles called *atoms*. These atoms maintain their identity throughout physical and chemical changes.

2. Atoms of the same element are identical in mass and have the same chemical and physical properties. Atoms of different elements have different masses and different chemical and physical properties.

3. Chemical combinations of two or more elements consist in the uniting of the atoms of these elements in a simple numerical ratio such as *1 to 1*, or *1 to 2*, etc., to form a formula unit or molecule of a compound. In Section 3-7, for example, we mentioned that one molecule of water consists of *2* atoms of hydrogen and *1* atom of oxygen. Similarly, one formula unit of sodium chloride (NaCl, table salt) consists of *1* atom of sodium and *1* atom of chlorine.

4. Sometimes atoms of the different elements can unite in different ratios to form more than one compound. In the preceding case, 2 atoms of hydrogen united with 1 atom of oxygen to form a molecule of water, H_2O. Two atoms of hydrogen can also combine with 2 atoms of oxygen to form a molecule of hydrogen peroxide, H_2O_2. Other examples are carbon monoxide, CO, and carbon dioxide, CO_2.

Dalton's first proposal that atoms consist of tiny, discrete particles has been verified in that single, tiny atoms of both uranium and thorium have been photographed by using an instrument called the *electron microscope* (see Fig. 4–2). Dalton's first proposal has been modified in that atoms can be split and hence are not indestructible, as you may know from nuclear changes (radioactivity, atomic reactors, nuclear bombs). Also, in nuclear changes, the atoms change their identities. Dalton's second proposal states that all atoms of the same element have identical masses, but as you will learn in Section 4–5, isotopes (atoms of the same element with different masses) of elements exist. In general, with minor modifications, Dalton's proposals are considered to be valid today.

4–3   Subatomic Particles. Electrons, Protons, and Neutrons

Study hint

Just as atoms are the building blocks of compounds (see Fig. 3–4), protons, neutrons, and electrons are the building blocks of atoms.

electron A particle that has a relative charge of -1 (actual charge $= -1.602 \times 10^{-19}$ coulomb) and a mass of 9.109×10^{-28} g or 5.486×10^{-4} amu (relatively negligible).

proton A particle that has a relative charge of $+1$ (actual charge $= +1.602 \times 10^{-19}$ coulomb) and a mass of 1.6726×10^{-24} g or 1.0073 amu (approximately 1 amu).

Atoms are composed of subatomic particles. These subatomic particles are the **electron**, the **proton**, and the **neutron**. There are other subatomic particles, but these three form the basis for the model of the atom that we will consider in this text.

The **electron**, abbreviated e^-, is a particle having a relative unit negative[1] charge, and a mass of 9.109×10^{-28} g or 5.486×10^{-4} (0.0005486) amu. The mass of the electron is thus relatively small in terms of atomic mass units and is considered to be negligible for all practical purposes.

You cannot see electrons, but you have encountered electrons every day. When you comb your hair, electrons from your hair collect on the comb and can attract small pieces of paper. When you walk on a carpet and then approach certain objects, you get a shock. The electrons from the carpet accumulate in your body and you may be shocked when you touch certain objects. Both phenomena occur best when the humidity and temperature are low, and they are often described as the effects of static electricity.

The **proton**, abbreviated p or p^+, is a particle having a relative unit positive[2] charge, and a mass of 1.6726×10^{-24} g or 1.0073 amu. The mass of a proton

---

[1] The actual charge on an electron is -1.602×10^{-19} coulomb. The coulomb is a unit used for measuring electrical charge, but as you can see, the value of the charge of an electron in coulombs is quite awkward to handle. Because subatomic particles that are charged have charges that are the same or integral multiples of the charge of an electron, the *relative* charge of an electron may be chosen as -1.

[2] The actual charge on a proton is $+1.602 \times 10^{-19}$ coulomb. As you may note, this value is exactly the same as that on an electron, but opposite in sign; hence, the *relative* charge is considered to be $+1$.

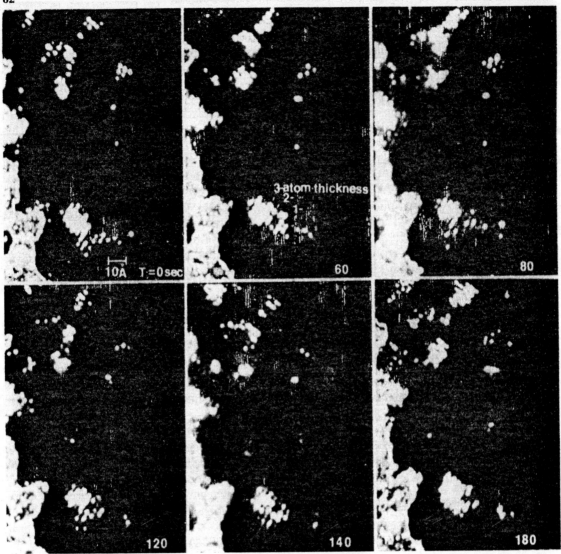

Figure 4–2 *Time-lapse micrographs of uranyl crystals in which uranium atoms can clearly be observed. These pictures were taken with a scanning transmission electron microscope. (Courtesy of Dr. Oscar H. Kapp, Department of Physics, University of Chicago.)*

is rounded off to 1 amu for most calculations because its mass is very close to 1 amu.

neutron A particle that has no charge and a mass of 1.6748×10^{-24} g or 1.0087 amu (approximately 1 amu).

The **neutron**, abbreviated n or n^0, is a particle having *no* charge and a mass of 1.6748×10^{-24} g or 1.0087 amu. The mass of a neutron is also rounded off to 1 amu for most calculations because its mass is very close to 1 amu too.

Table 4–1 summarizes the data for the subatomic particles. You must be able to identify these particles including their abbreviations, approximate masses in amu, and relative charges.

TABLE 4–1   SUMMARY OF SUBATOMIC
PARTICLES

PARTICLE (ABBREVIATION)	APPROXIMATE MASS (amu)	RELATIVE CHARGE
Electron (e^-)	Negligible	-1
Proton (p)	1	$+1$
Neutron (n)	1	0

4–4   General Arrangement of Electrons, Protons, and Neutrons. Atomic Number

Now, how are these three subatomic particles arranged in an atom? To answer this question, we must consider a few fundamental facts about the atom:

1. *All the protons and neutrons are found at the center of the atom in the nucleus.* Because most of the mass of the atom is concentrated in this very small region, the nucleus of the atom has a very high density (1.0×10^{14} g/mL). One milliliter of nuclear matter would have a mass of 1.1×10^8 tons! Also, since the protons are positively charged and the neutrons are neutral, the relative *charge* on the nucleus must be *positive* and *equal to the number of protons.*

2. *The number of protons* (mass of proton, 1 amu) *plus the number of neutrons* (mass of neutron, 1 amu) *equals the mass number of the atom,* which essentially equals the atomic mass in amu, since the mass of the electron is negligible. Hence, the number of neutrons present is equal to the mass number *minus* the number of protons (neutrons = mass number − protons).

3. *An atom is electrically neutral. The number of protons equals the number of electrons in a neutral atom.* If the number of electrons in an atom does not equal the number of protons, the atom is positively or negatively charged and is called an *ion.*

4. *Electrons are found outside the nucleus in certain energy levels.* In these energy levels, the electrons are dispersed at a relatively great distance from the nucleus. The nucleus has a diameter of approximately 1×10^{-5} Å, whereas the diameter of the entire atom is in the range 1 to 5 Å. Therefore, these electrons are dispersed at a distance that extends up to 100,000 times the diameter of the nucleus.

Before we look at some examples of the general arrangement of the subatomic particles in the atoms of some elements, we must consider the symbols used to describe the atom. The following is a general symbol for an element giving its mass number and atomic number:

$$A = \text{mass number in amu}$$

$$^A_Z E \qquad E = \text{symbol of the element}$$

$$Z = \text{atomic number}$$

atomic number Number of protons found in the nucleus of an atom of a given element.

The **atomic number** is *equal to the number of protons* found in the nucleus. The **mass number** is *equal to the sum of the number of protons and neutrons* in the nucleus.

mass number Sum of the number of protons and neutrons in the nucleus of an atom of an element.

EXAMPLE 4–1

For each of the following atoms, calculate the number of protons and neutrons in the nucleus and the number of electrons outside the nucleus.

(a) ^4_2He

SOLUTION

2 = atomic number = number of protons in nucleus
4 = mass number = sum of protons and neutrons
Hence, number of neutrons = 4 − 2 = 2 neutrons in the nucleus.
Number of electrons = number of protons = 2 electrons outside the nucleus.

$$\begin{array}{c} 2p \\ 2n \end{array} \qquad 2e^- \qquad Answer$$

Nucleus   Outside
                 nucleus

(b) $^{11}_5\text{B}$

SOLUTION

5 = atomic number = number of protons in nucleus
11 = mass number = sum of protons and neutrons
Neutrons = 11 − 5 = 6 neutrons in the nucleus.
Number of electrons = number of protons = 5 electrons outside the nucleus.

$$\begin{array}{c} 5p \\ 6n \end{array} \qquad 5e^- \qquad Answer$$

Nucleus   Outside
                 nucleus

(c) $^{24}_{12}\text{Mg}$

SOLUTION

12 = atomic number = number of protons in nucleus
24 = mass number = sum of protons and neutrons

Neutrons = 24 − 12 = 12 neutrons in the nucleus.

Number of electrons = number of protons = 12 electrons outside the nucleus.

$\begin{pmatrix} 12p \\ 12n \end{pmatrix}$ 12e^- *Answer*

Nucleus   Outside
                   nucleus

(d) $^{75}_{33}$As

SOLUTION

33 = atomic number = number of protons in nucleus

75 = mass number = sum of protons and neutrons

Neutrons = 75 − 33 = 42 neutrons in nucleus.

Number of electrons = number of protons = 33 electrons outside the nucleus.

$\begin{pmatrix} 33p \\ 42n \end{pmatrix}$ 33e^- *Answer*

Nucleus   Outside
                   nucleus

Exercise 4–1

For each of the following atoms, calculate the number of protons and neutrons in the nucleus and the number of electrons outside the nucleus.

(a) $^{7}_{3}$Li   (b) $^{35}_{17}$Cl

(c) $^{74}_{32}$Ge   (d) $^{107}_{47}$Ag

The law of electrostatics says that unlike charges attract each other. You might wonder then why the nucleus, which is positively charged because of the presence of protons, and the negatively charged electrons located outside the nucleus do not draw together and unite to neutralize the charges. To explain this, Danish physicist Niels Bohr[3] (1885–1962) proposed a theory that electrons in an atom have their energies restricted to certain energy values or *energy levels*, which we will consider in Section 4–6. For an electron to change its energy, it must shift from one energy level to another. To go to a higher energy level, a definite amount of energy is required equal to the energy difference between the

---

[3] We are also indebted to Niels Bohr for our description of electrons in sublevels (see Section 4–8). Many outstanding scientists have contributed to our knowledge of the atom. They are Sir William Crookes, Sir J. J. Thomson, Sir James Chadwick, Sir Ernest Rutherford—all British scientists; Eugen Goldstein, a German scientist; and Robert A. Millikan, an American scientist. (Although we listed Rutherford as a British scientist, he was born in New Zealand and educated primarily in that country.)

two levels. But to go to a lower energy level, a lower energy level must be available, and, if so, energy equal to the difference between the two levels is given off. If the electrons in an atom are already arranged in their lowest energy levels and no lower levels are available, then the electrons cannot drop in energy. The movement of electrons among energy levels is analogous to a person walking up a flight of stairs. To progress up the stairs you are restricted to certain levels of progress, each step. You cannot raise yourself up between the steps!

4–5 Isotopes

On close examination of the atomic masses of the element (inside front cover), you will note that the atomic masses of the elements are not whole numbers (carbon = 12.01115 amu and chlorine = 35.453 amu). Since the masses of the proton and neutron are nearly equal to 1, and since the mass of the electron is very slight, we would expect the atomic mass of an element to be very nearly a whole number—certainly not halfway between, as is the case with chlorine. The reason that many atomic masses are not even close to whole numbers is that all atoms of the same element do not necessarily have the same mass, a contradiction to Dalton's second proposal (Section 4–2). Atoms having different atomic masses or mass numbers, but the same atomic number, are called **isotopes**.

isotopes Atoms having different atomic masses or mass numbers but the same atomic number.

Carbon exists in nature as two isotopes: ^{12}C ($^{12}_6$C, exact atomic mass = 12.00000 amu, the atomic mass unit standard), and ^{13}C ($^{13}_6$C, exact atomic mass = 13.00335 amu). Structurally, the difference between these two isotopes is *one* neutron. ^{12}C has 6 neutrons, and ^{13}C has 7 neutrons, as follows:

Chlorine also exists in nature as two isotopes: ^{35}Cl ($^{35}_{17}$Cl, exact atomic mass = 34.96885 amu) and ^{37}Cl ($^{37}_{17}$Cl, exact atomic mass = 36.96590 amu). What is the difference in the number of neutrons in these two chlorine isotopes? *Isotopes of the same element have the same chemical properties but slightly different physical properties.* Hence both ^{35}Cl and ^{37}Cl have the same chemical properties but *slightly different* physical properties.

The atomic mass in amu for the elements C = 12.01115 and Cl = 35.453 is an *average mass* based on the *abundance of the isotopes in nature*. The atomic mass for the element may be obtained by multiplying the exact atomic mass of each isotope by the decimal of its percent abundance in nature and then taking the sum of the values obtained. This is similar to the calculation of your grade in a particular course. For example, if you get a score of 75 on an exam that counts 25 percent of your final grade and a score of 85 on an exam that counts 75 percent, your final average based on the "weight" of each exam is 82.5 (to three significant digits), *not* 80. The calculation is as follows:

$$75(0.25) + 85(0.75) = 18.75 + 63.75 = 82.5$$

Note that the percentage, meaning parts per 100, is converted to a decimal, meaning part per *one*, by dividing by 100. The following problem example illustrates the calculation of atomic masses for elements.

EXAMPLE 4–2

Calculate the atomic mass to four significant digits for carbon, given the following data.

ISOTOPE	EXACT ATOMIC MASS (amu)	ABUNDANCE IN NATURE (%)
^{12}C	12.00000	98.89
^{13}C	13.00335	1.110

SOLUTION   Convert the percentages (98.89 and 1.110) to decimal form by dividing by 100 to give 0.9889 and 0.01110, respectively. Therefore,

12.00000 amu (0.9889) + 13.00335 amu (0.01110)

$$= 12.01 \text{ amu} \quad \textit{Answer}$$

Based on the average mass, the atomic mass of carbon was found to be 12.01115 amu, but we would never find an atom of carbon that would have a relative mass of 12.01115 amu; it would have a relative mass of 12.00000 or 13.00335 amu, depending on the isotope with which we were working. But in general, for an ordinary-sized sample of carbon atoms containing the isotopes in the proportions given, we find it convenient to use the average mass, 12.01115 amu. The same reasoning applies to all the other elements and their atomic mass units, which are given inside the front cover of this text and are the average masses of the naturally occurring isotopes of the elements.

Exercise 4–2

Calculate the atomic mass to four significant digits for boron, given the following data.

ISOTOPE	EXACT ATOMIC MASS (amu)	ABUNDANCE IN NATURE (%)
^{10}B	10.013	19.60
^{11}B	11.009	80.40

4–6   Arrangement of Electrons in Principal Energy Levels

In Section 4–4 we did not specify how the electrons are arranged. We just said that they were outside the nucleus. In this section, we shall be more specific.

The electrons can exist at different specific energies in *principal energy levels*, which increase in energy as they increase in distance from the nucleus. That is, the nearer the electron is to the nucleus, the less energy the electron has; the farther away from the nucleus it is, the more energy it has. These principal energy levels are designated by whole numbers: as 1, 2, 3, 4, 5, 6, and 7. There is a maximum number of electrons that can exist in a given energy level. This number is found from the following equation:

$$\text{Maximum number of electrons in principal energy levels} = 2n^2$$
$$\text{where } n = \text{integers 1 to 7 of the principal energy levels.}$$

EXAMPLE 4–3

Calculate the maximum number of electrons that can exist in principal energy levels 1 and 2.

SOLUTION   For level 1—maximum number of electrons:

$$2 \times 1^2 = 2 \times 1 = 2 \quad \textit{Answer}$$

For level 2—maximum number of electrons:

$$2 \times 2^2 = 2 \times 4 = 8 \quad \textit{Answer}$$

Exercise 4–3

Calculate the maximum number of electrons that can exist in principal energy level 5.

Table 4–2 lists the principal energy levels and the maximum number of electrons at that level. These are the maximum numbers of electrons that can be accommodated at a given energy level, but an energy level may have *less than the maximum*.

Now let us consider the arrangement of the electrons in principal energy levels. Electrons go into the *lowest available energy level that has not been filled*. So begin by placing electrons into the lowest principal energy level and

TABLE 4–2   MAXIMUM NUMBER OF ELECTRONS IN PRINCIPAL ENERGY LEVELS

PRINCIPAL ENERGY LEVEL	MAXIMUM NUMBER OF ELECTRONS
1	2
2	8
3	18
4	32
5	50
6	72
7	98

(Increasing energy)

continue placing electrons into subsequent levels until the required number of electrons have been assigned [This method works well *only* to the element argon (Ar), atomic number 18; beyond that sublevels must be considered. See Section 4–8].

EXAMPLE 4–4

Diagram the atomic structure for each of the following atoms. Indicate the number of protons and neutrons, and arrange the electrons in principal energy levels.

(a) ^7_3Li

SOLUTION

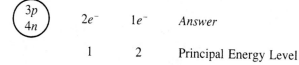

Note that here the maximum number of electrons in principal energy level 1 is 2, so to place 3 electrons outside the nucleus we go to the next higher energy level—principal energy level 2.

(b) $^{11}_5\text{B}$

SOLUTION

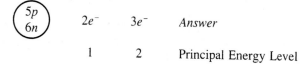

(c) $^{23}_{11}\text{Na}$

SOLUTION

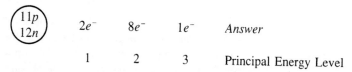

Pricipal energy level 2 can accommodate a maximum of 8 electrons, so to place 11 electrons outside the nucleus we must use not only principal energy levels 1 and 2 but also a higher energy level— principal energy level 3.

(d) $^{28}_{14}\text{Si}$

SOLUTION

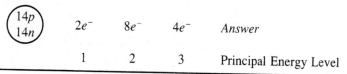

Exercise 4–4

Diagram the atomic structure for each of the followng atoms. Indicate the number of protons and neutrons, and arrange the electrons in principal energy levels.

(a) $^{12}_{6}C$ (b) $^{15}_{7}N$

(c) $^{27}_{13}Al$ (d) $^{31}_{15}P$

4–7 Electron-Dot Formulas of Elements

valence electrons Electrons occupying the highest principal level in an atom.

The electrons in the *highest principal energy level* in the preceding diagram of the atoms are usually called the valence energy level electrons or **valence electrons.** The remainder of the atom (nucleus and other electrons) is called the *core* (kernel). The electrons in the valence energy level are of higher energy than the inner electrons and are gained, lost, or shared when an atom of one element *reacts* with an atom of another element to form a molecule or ion. These valence electrons, due to their reactivity, are the ones depicted in electron-dot formulas. These electron-dot formulas do not represent a physical description of the electrons, but rather they serve as a convenient "bookkeeping" device.

To write electron-dot formulas of elements, we need to follow a few simple rules.

1. Write the symbol for the element to represent the *core*.

2. Assign a maximum of 2 electrons to each of the four sides of the symbol to give a total of 8 electrons around the symbol. A dot represents a single electron.

3. Arrange the valence electrons (highest principal energy level) around the four sides of the symbol, with 1 electron assigned to *each side up to a maximum of 4 electrons.*

4. If needed, pair up electrons on the four sides *up to a maximum of eight electrons.* Be sure not to exceed the actual number of valence electrons for the element (Helium is an exception, with both valence electrons shown on the same side of the symbol, since it has a completed principal energy level 1).

In writing the electron-dot formulas for atoms, be sure to determine the number of valence electrons *first.*

EXAMPLE 4–5

Write the electron-dot formulas for the following atoms.

(a) $^{7}_{3}Li$

    SOLUTION   1 valence electron, Li· or L̤i etc.   *Answer*

(b) $^{9}_{4}Be$

    SOLUTION   2 valence electrons, B̤e·   *Answer*

(c) $^{23}_{11}$Na

SOLUTION   1 valence electron, Na·   *Answer*

(d) $^{16}_{8}$O

SOLUTION   6 valence electrons, ·Ö: *Answer*

(e) $^{20}_{10}$Ne

SOLUTION   8 valence electrons, :N̈e: *Answer*

rule of eight or octet rule
In the formation of molecules from atoms, most molecules attempt to obtain this stable configuration of 8 electrons in the valence energy level of each atom.

In the preceding examples, you may have noted that 8 electrons filled all four sides, as in the case of neon (Ne). There is a specific rule governing this, the **rule of eight** or **octet rule**. In the formation of molecules from atoms, most molecules attempt to obtain this stable configuration of 8 electrons in the valence energy level of each atom. The elements helium (He), neon (Ne), argon (Ar), krypton (Kr), xenon (Xe), and radon (Rn) are called the *noble gases*. The noble gases are all relatively unreactive because they all have stable electronic configurations. All of the noble gases except helium have 8 valence electrons and satisfy the rule of eight or octet rule; helium has 2 valence electrons that *complete* its principal energy level 1, so it is also unreactive. In fact, the noble gases were once called the inert (unreactive) gases because of this lack of reactivity, but compounds containing the noble gases have now been prepared.

Exercise 4–5

Write the electron-dot formulas for the following atoms.

(a) $^{11}_{5}$B

(b) $^{15}_{7}$N

(c) $^{27}_{13}$Al

(d) $^{31}_{15}$P

4–8 Arrangement of Electrons in Sublevels

Experiments have shown that the arrangement of electrons is not quite as simple as we have described so far. In fact, the principal energy levels are divided further into sublevels. The sublevels, labeled s, p, d, and f, also have a limit to the number of electrons that they can contain. The s, p, d, and f sublevels may contain a maximum of 2, 6, 10, and 14 electrons, respectively, as shown below and in Table 4–3.

$$s = 2$$

$$p = 6$$

$$d = 10$$

$$f = 14$$

TABLE 4–3   MAXIMUM NUMBER OF ELECTRONS IN PRINCIPAL
ENERGY LEVELS 1 TO 7 AND THEIR RESPECTIVE SUBLEVELS

| | | MAXIMUM NUMBER OF ELECTRONS | |
PRINCIPAL ENERGY LEVEL	SUBLEVEL[a]	SUBLEVEL[a]	PRINCIPAL ENERGY LEVEL
1	s	2	2
2	s	2	8
	p	6	
3	s	2	18
	p	6	
	d	10	
4	s	2	32
	p	6	
	d	10	
	f	14	
5	s	2	50 (actually 32[b])
	p	6	
	d	10	
	f	14	
	(g)	(18)	
6	s	2	72 (actually 15[b])
	p	6	
	d	10	
	(f)	(14)	
	(g)	(18)	
	(h)	(22)	
7	s	2	98 (actually 2[b])
	(p)	(6)	
	(d)	(10)	
	(f)	(14)	
	(g)	(18)	
	(h)	(22)	
	(i)	(26)	

[a] Sublevel letters in parentheses, along with the maximum number of electrons in that sublevel, are not used in elements currently known.

[b] This is the actual maximum number of electrons found for the elements known at present; hence, these principal energy levels are incomplete.

(Note that 4 electrons are added each time.) As you see in Table 4–3, the number of sublevels equals the number of the principal energy level. For example, the first principal energy level has one sublevel (s), the second level has two sublevels (s and p), the third level has three sublevels (s, p, and d), and so on.

This arrangement of electrons in sublevels of principal energy levels is analogous to the arrangement of students in rooms in the various floors of the Electra Hostel. In the Electra Hostel, on the *first* (1) floor there is just one (1) room for guests, which can accommodate a maximum of two (2) students. On the *second* (2) floor there are two (2) rooms for guests. One room can accommodate a maximum of two (2) students and the other room can accommodate a maximum of six (6) students, with a total maximum of eight (8) students on the second floor. On the *third* (3) floor there are three (3) rooms for guests. One room can accommodate a maximum of two (2) students, the other room a maximum of six (6) students, and the third room a maximum of ten (10) students, with a total maximum of eighteen (18) students on the third floor. On the *fourth* (4) floor there are four (4) rooms for guests. One room can accommodate a maximum of two (2) students, the second room a maximum of six (6) students, the third room a maximum of ten (10) students, and the fourth room a maximum of fourteen (14) students, with a total maximum of thirty-two (32) students on the fourth floor.

Each of these sublevels, with its respective principal energy level, has a different energy. The order of the sublevels according to increasing energy, is as follows (the $<$ is read "less than"):

$$1s < 2s < 2p < 3s < 3p < 4s < 3d < 4p < 5s < 4d < 5p < 6s$$
$$< (4f < 5d) < 6p < 7s < (5f < 6d)$$

As you can see, there are places where a 4 sublevel is lower in energy than a 3 sublevel (4s versus 3d) or a 5 or 6 sublevel is lower in energy than a 4 sublevel (5s versus 4d, 6s versus 4f).

In filling each of these sublevels with electrons, the lowest unfilled energy sublevel is filled first. Figure 4–3 shows the relative energies of the various sublevels, and Fig. 4–4 gives a simplified way of remembering the order of filling. Follow the directions in Fig. 4–4 to remember the order for filling the sublevels until you know them well. You should note that the 4s sublevel fills before the 3d sublevel. You should also note that the 4f and 5d, and the 5f and 6d sublevels have been placed in parentheses in the above order. This is because the energies of the pairs of sublevels are quite close. From Fig. 4–4 we see that one electron is placed in the 5d sublevel before filling the 4f sublevel. The same is true for the 6d and 5f sublevels. After the 4f sublevel is filled, the rest of the 5d sublevel is filled to its maximum of 10 electrons. The same is done for the 6d sublevel after filling the 5f sublevel. Exceptions to the order of filling of sublevels do occur, but we shall not consider them in this text.

When writing the sublevel electron configuration of an atom, write the principal energy level *number* and the sublevel *letter*, followed by the number of electrons in the sublevel written as a superscript. The sublevels of a given principal energy level may be *grouped together*, or as *they are filled*. In generating these configurations, it is best to draw the diagram in Fig. 4–4 and then follow this diagram in filling the sublevels. Consider the ^{1}H atom:

$$^1_1H = 1s^1$$

number of electrons in *that* sublevel ⟵ sublevel ⟵ principal energy level

1 valence electron

EXAMPLE 4–6

(1) Write the electronic configuration in sublevels for the following atoms, and (2) give the number of valence electrons of each.

ANSWERS

(a) $^7_3$Li $1s^2, 2s^1$ (1)

(b) $^{14}_7$N $1s^2, 2s^2 2p^3$ (5)

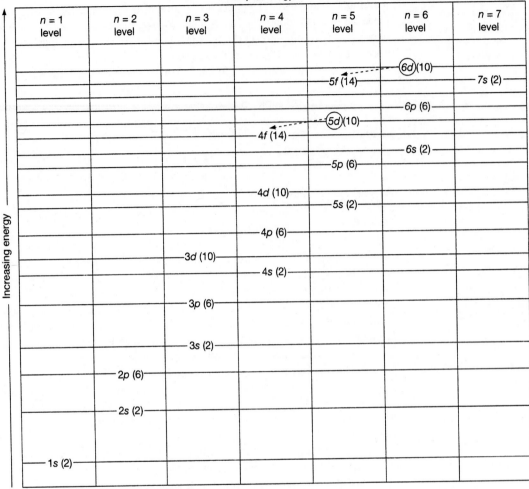

Figure 4–3 *Diagram showing the relative energies of the different electronic sublevels. The numbers in parentheses are the maximum number of electrons in the sublevel. Note the* circles *around the 5d and 6d sublevels. The reason for these circles is that* one *electron is placed in each of these sublevels before filling the 4f or 5f sublevels, respectively. Note the* dotted arrows *from the d sublevels to the f sublevels. After filling the* f *sublevels, the* d *sublevels fill to their maximum of 10 electrons.*

ANSWERS

(c) $^{37}_{17}\text{Cl}$ $1s^2,\ 2s^2 2p^6,\ 3s^2 3p^5$ (7)

(d) $^{69}_{31}\text{Ga}$ $1s^2,\ 2s^2 2p^6,\ 3s^2 3p^6 3d^{10},\ 4s^2 4p^1$ (3); the $3d$ sublevel fills after the $4s$ and before the $4p$. We can group the $3d$ sublevel with the other sublevels of principal energy level 3, regardless of the order of filling, or equally acceptable is the electronic configuration $1s^2,\ 2s^2,\ 2p^6,\ 3s^2 3p^6,\ 4s^2,\ 3d^{10},\ 4p^1$, following the order of filling.

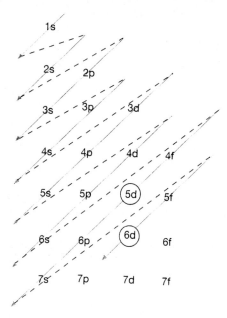

Figure 4–4 *Order of filling of sublevels. (a) Write down the principal energy levels with their sublevels to the f sublevel. (2) Draw diagonal lines which follow the order of filling. (The diagonal lines need not extend beyond the 6d sublevel since no elements at present have been discovered that have electronic configurations beyond the 6d sublevel.) (3) Circle the 5d and 6d sublevels to remind you that only one electron is placed in each of these sublevels before filling the 4f or 5f sublevels, respectively. After filling the f sublevels, the d sublevels fill to their maximum of 10 electrons.*

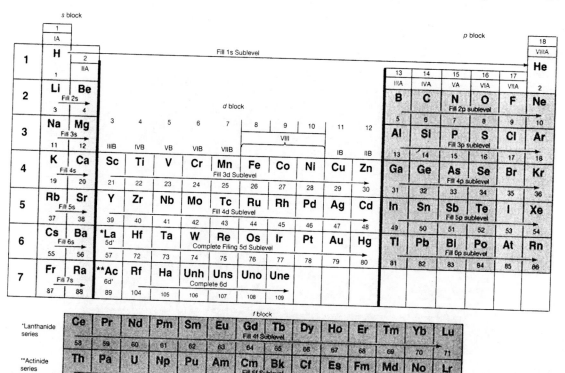

Figure 4–5 *The correlation of the filling of the sublevels with the Periodic Table. Note the groups of elements that fill the s, p, d, and f sublevels. The number at the bottom of each box refers to the atomic number of the element.*

Exercise 4-6

(1) Write the electronic configuration in sublevels for the following atoms, and (2) give the number of valence electrons of each.

(a) ^9_4Be (b) $^{28}_{14}Si$

(c) $^{40}_{20}Ca$ (d) $^{75}_{33}As$

The filling of the sublevels correlates with the Periodic Table (a table of the elements) as shown in Fig. 4–5. Chapter 5 discusses the Periodic Table in more detail. Note that there are blocks of elements that fill just the s sublevels, those that fill just the p sublevels, those that fill just the d sublevels, and finally those that fill just the f sublevels. When an element is in a particular block, this means that the last electron placed in that atom occupies the sublevel corresponding to that block. For example, titanium ($^{48}_{22}Ti$) has the electronic configuration $1s^2$, $2s^22p^6$, $3s^23p^6$, $4s^2$, $3d^2$. Titanium is the second element in the d block, so its last electron is the $3d^2$ electron. Hence, you can check that you filled the sublevels correctly by comparing your answer to the Periodic Table. Just remember that there are exceptions to the order of filling sublevels, so that not all elements work out in this manner.

Silver: pretty as a picture

Silver compounds play a major role in our everyday life because they are key ingredients in photographic film.

Name: Silver is one of the oldest known metals (the others are gold and copper). The origins of the name for silver are not clear, but they probably derive from the German *silber* and the Old English *siolfor*. The symbol (Ag) comes from the Latin *argentum*.

Appearance: A lustrous bright silver metal that is an excellent conductor of heat and electricity.

Occurrence: Sometimes silver occurs in the pure (uncombined) state in nature, although most of these deposits have been exhausted. Argentite (Ag_2S) is a valuable silver-rich ore.

Source: Currently, most silver is obtained as a by-product during the processing of copper, lead, and zinc containing ores [*argentites* (Cu, Fe, Zn, Ag)$_{12}Sb_4S_{13}$].

Its Role in Our World: The most important use of silver containing salts is in the photographic industry. Silver bromide (AgBr) and silver iodide (AgI) are used extensively in preparing photographic films.

The second biggest silver consumer is the electronics industry, where silver and silver alloys (silver mixed with cadmium, copper, palladium, or gold) are used as electrical contacts, solders, and brazing alloys.

Silver salts are also used in the manufacture of silver-based batteries. Such diverse devices as torpedoes, watches, aircraft, and rockets use batteries containing silver oxide (Ag_2O) and zinc.

Silver is used to make jewelry, sterling silverware, mirrors, and in dental fillings. While silver was once used in coins and as an exchange medium, coins now account for less than 5 percent of world silver consumption.

Unusual Facts: Very dilute solutions containing silver salts are excellent disinfectants. Historically, humans have taken advantage of this property by storing food in silver containers to retard the rate of spoilage.

Problems

1. For each of the following atoms, calculate the number of protons and neutrons in the nucleus and the number of electrons outside the nucleus.

   (a) $^{39}_{19}K$ (b) $^{46}_{22}Ti$

   (c) $^{59}_{27}Co$ (d) $^{69}_{31}Ga$

   (All the atoms listed above exist, although they may not be the most abundant isotope in nature.)

2. Calculate the atomic mass to four significant digits for gallium, given the following data.

ISOTOPE	EXACT ATOMIC MASS (atm)	ABUNDANCE IN NATURE (%)
^{69}Ga	68.9257	60.40
^{71}Ga	70.9249	39.60

3. Calculate the atomic mass to four significant digits for antimony, given the following data.

ISOTOPE	EXACT ATOMIC MASS (atm)	ABUNDANCE IN NATURE (%)
^{121}Sb	120.9038	57.25
^{123}Sb	122.9041	42.75

4. Calculate the maximum number of electrons that can exist in the following principal energy levels.

   (a) 1   (b) 2

   (c) 6   (d) 7

5. Diagram the atomic structure for each of the following atoms. Indicate the number of protons and neutrons, and arrange the electrons in principal energy levels.

(a) ^9_4Be (b) $^{19}_9F$

(c) $^{24}_{12}Mg$ (d) $^{32}_{16}S$

6. Write the electron-dot formulas for the following atoms.

(a) ^4_2He (b) ^7_3Li

(c) $^{14}_7N$ (d) $^{38}_{18}Ar$

7. (1) Write the electronic configuration of sublevels for the following atoms, and (2) give the number of valence electrons of each.

(a) ^6_3Li (b) $^{15}_7N$ (c) $^{27}_{13}Al$

(d) $^{84}_{36}Kr$ (e) $^{64}_{30}Zn$ (f) $^{138}_{56}Ba$

General Problems

8. The element osmium (Os) has the following physical properties:

$$mp = 3045°C, \quad bp = 5027°C, \quad density = 22.57 \text{ g/cm}^3$$

(a) Calculate its melting point in °F.
(b) Calculate its density in kg/m³.
(c) Calculate its density in lb/ft³.

9. Write the electronic configuration in sublevels for an isotope of osmium (Os, atomic number = 76) having a mass number of 192.

10. Write the electronic configurations in sublevels for $^{12}_6C$ and $^{28}_{14}Si$ and determine the number of valence electrons for each. Note that $^{12}_6C$ is directly above $^{28}_{14}Si$ in the Periodic Table. What can you conclude from these observations?

11. A single proton has a mass of 1.67×10^{-24} g. Calculate the number of protons in 1.00 lb of protons.

Answers to Exercises

4-1. (a) ⊙ 3p 4n $3e^-$ (b) ⊙ 17p 18n $17e^-$

(c) ⊙ 32p 42n $32e^-$ (d) ⊙ 47p 60n $47e^-$

4-2. 10.81 amu   4-3. 50 electrons

4-4. (a) ⊙ 6p 6n $2e^-$ $4e^-$
                  1   2

(b) ⊙ 7p 8n $2e^-$ $5e^-$
                1   2

(c) $\left(\begin{array}{c}13p \\ 14n\end{array}\right)$ $2e^-$ $8e^-$ $3e^-$
                         1   2   3

(d) $\left(\begin{array}{c}15p \\ 16n\end{array}\right)$ $2e^-$ $8e^-$ $5e^-$
                         1   2   3

4–5. (a) $\dot{B}\cdot$; (b) $\cdot\dot{N}\cdot$; (c) $\dot{A}l\cdot$; (d) $\cdot\dot{P}\cdot$

4–6. (a) $1s^2, 2s^2$ (2)
(b) $1s^2, 2s^22p^6, 3s^23p^2$ (4)
(c) $1s^2, 2s^22p^6, 3s^23p^6, 4s^2$ (2)
(d) $1s^2, 2s^22p^63d^{10}, 4s^24p^3$ (5), or $1s^2, 2s^22p^6, 3s^23p^6, 4s^2, 3d^{10}$ $4p^3$ (5)

QUIZ

1. Diagram the atomic structure for the following atoms. Indicate the numbe of protons and neutrons, and arrange the electrons in principal energy levels

(a) $^{19}_9F$ (b) $^{37}_{17}Cl$

2. Calculate the atomic mass to four significant digits for copper from th following data:

ISOTOPE	EXACT ATOMIC MASS (amu)	ABUNDANCE IN NATURE (%)
^{63}Cu	62.9298	69.09
^{65}Cu	64.9278	30.91

3. Calculate the number of electrons that can exist in the following principa energy levels:

(a) 3   (b) 6

4. Write the electron-dot formulas for the following atoms:

(a) $^{12}_6C$ (b) $^{40}_{18}Ar$

5. For the following atoms, (1) write the electronic configuration in sublevels and (2) give the number of valence electrons for each.

(a) $^{39}_{19}K$ (b) $^{69}_{31}Ga$

THE PERIODIC CLASSIFICATION OF THE ELEMENTS

COUNTDOWN

5 Convert 125 mL to cubic meters (m^3). Express your answer in scientific notation (Sections 1–4, 2–2, and 2–5).

$$(1.25 \times 10^{-4} \ m^3)$$

4 The following are properties of the element tungsten (W); classify them as physical or chemical properties (Section 3–4).

(a) melting point, 3410°C (physical)
(b) oxidizes in air (chemical)
(c) steel gray in color (physical)
(d) sp gr = $19.3^{20°/4}$ (physical)

3 For each of the following atoms, calculate the number of protons and neutrons in the nucleus and the number of electrons outside the nucleus (Section 4–4).

(a) $^{56}_{26}Fe$ $\left(\overset{26p}{\underset{30n}{\bigcirc}} \quad 26e^- \right)$

(b) $^{107}_{47}Ag$ $\left(\overset{47p}{\underset{60n}{\bigcirc}} \quad 47e^- \right)$

2 Diagram the atomic structure for each of the following atoms. Indicate the number of protons and neutrons, and arrange the electrons in principal energy levels (Section 4–6).

$$\left(\qquad\qquad 1 \qquad 2 \qquad 3 \qquad \begin{array}{l} \text{Principal} \\ \text{Energy Levels} \end{array} \right)$$

(a) $^{27}_{13}Al$ $\left(\overset{13p}{\underset{14n}{\bigcirc}} \quad 2e^- \quad 8e^- \quad 3e^- \right)$

(b) $^{31}_{15}P$ $\left(\overset{15p}{\underset{16n}{\bigcirc}} \quad 2e^- \quad 8e^- \quad 5e^- \right)$

1 (1) Write the electronic configuration in sublevels for the following atoms and (2) give the number of valence electrons of each (Section 4–8).

(a) $^{19}_{9}F$ $\qquad\qquad\qquad\qquad\qquad\qquad$ $[1s^2, 2s^22p^5\ (7)]$

(b) $^{48}_{22}Ti$ $\qquad\qquad\qquad\qquad\qquad$ $[1s^2, 2s^22p^6, 3s^23p^63d^2, 4s^2$

$\qquad\qquad\qquad$ or $1s^2, 2s^22p^6, 3s^23p^6, 4s^2, 3d^2\ (2)]$

OBJECTIVES

1 Give the distinguishing characteristics of each of the following terms:

(a) metals (Section 5–1)

(b) nonmetals (Section 5–1)

(c) Periodic Law (Section 5–2)

(d) periods (Section 5–3)

(e) groups (Section 5–3)

(f) representative elements (Section 5–3)

(g) transition elements (Section 5–3)

(h) metalloid (Section 5–4)

2 Given the Periodic Table, point to the elements in a period and in a group (Section 5–3).

3 Given the Periodic Table and any element in the table, determine:

(a) whether that element is a metal, a nonmetal, or a metalloid (Sections 5–1 and 5–4, Problems 1 and 2).

(b) the number of valence electrons if that element belongs to an A group (Example 5–1, Exercise 5–1, Problems 3 and 4).

4 Given the electronic configuration of a number of elements, group the elements together according to those you would expect to show similar chemical properties (Example 5–2, Exercise 5–2, Problems 5 and 6).

5 Given the Periodic Table, determine the trends in metallic and nonmetallic properties in the A group elements as you move down a given group (Example 5–3, Exercise 5–3, Problems 7 and 8).

6 Given the Periodic Table, determine the trend in atomic radii as you move down a given group (Example 5–4, Exercise 5–4, Problems 9 and 10).

There appears to be a human tendency to want to group things together in some order. By placing things in groups, we organize them so that we may find them or remember them more easily. We group books in a library by subject or author; we group food in the market by type; we group classes in college catalogs by subject matter; and we group students by giving them grades! In biology we group animals together according to whether they are single or multicell organisms, or whether they have a backbone.

In chemistry in the early nineteenth century, there were approximately 55 known elements. Chemists began attempting to group these known elements, and the eventual result was the Periodic Table. You have already been introduced to the Periodic Table in Section 4–8 when we discussed the arrangement of electrons in sublevels. In this chapter we consider the classification of the elements in the Periodic Table and some general characteristics of the different groups of elements.

5–1 Grouping the Elements: Metals and Nonmetals

One of the simplest ways of categorizing the elements involves grouping them as metals or nonmetals, depending on whether they have metallic qualities.

In general, **metals** have the following physical and chemical properties:

metals Elements that have a high luster, conduct electricity and heat well, are malleable and ductile, have high densities and melting points, are hard, and do not readily combine with one another.

■ They have a high luster (shine), as in the case of silver.

■ They conduct electricity and heat well, as in the case of copper.

■ They are malleable (can be shaped by beating with a hammer), as in the case of tin.

■ They are ductile (can be drawn out into a thin wire), as in the case of gold.

■ Most have high densities, as in the case of lead ($d^{20°} = 11.34$ g/mL).

■ Many have high melting points, as in the case of iron (mp = 1535°C). Hence, metals are generally solids at room temperature. The exceptions to this are mercury (Hg), gallium (Ga), and cesium (Cs), which are liquids at ordinary temperatures.

■ Most are hard, as in the cases of iron, tungsten, and chromium. However, a few, such as sodium and lead, are soft.

In general, metals have the following chemical properties:

■ They do not readily combine chemically with each other.

■ They do combine chemically with nonmetals and hence are normally found in nature in the combined form. Iron is found combined with oxygen or sulfur, and aluminum is found combined with oxygen and silicon or just oxygen. A few relatively unreactive metals are found in nature in the *free state*, that is, not combined with any other element. Gold, silver, copper, and platinum are often found in the free state.

nonmetals Elements that generally are dull, do not conduct electricity and heat well, are brittle, have low densities and melting points, are soft, and combine readily with metals and with other nonmetals.

The physical and chemical properties of the metals just listed are *general* properties that vary from metal to metal. Metals usually exhibit many, but *not necessarily all*, of these properties.

In contrast, the physical properties of the **nonmetals** generally differ from those of the metals. The general physical properties of the nonmetals are as follows:

- They generally are not lustrous but rather dull, as in the cases of sulfur and carbon (graphite).

- They are usually poor conductors of heat and electricity, as in the case of sulfur.

- They are not ductile nor malleable, but rather brittle, as in the case of carbon.

- They have low densities, as in the case of the gases nitrogen and oxygen.

- They have low melting points, so that at least one nonmetal exists in each of the three physical states of matter. Sulfur, phosphorus, carbon, and iodine are solids, bromine is a liquid, and fluorine, chlorine, nitrogen, and oxygen are gases at room temperature and atmospheric pressure.

- They are soft, as in the cases of sulfur and phosphorus. One exception, however, is diamond, which is a form of carbon. Diamond is one of the hardest known materials.

Nonmetals also generally have the following chemical properties:

- They combine with metals. A few exist in nature in the free state (uncombined); these are oxygen and nitrogen (both in air), sulfur, and carbon (coal, graphite, and diamond).

- They may also combine with each other. Carbon dioxide, carbon monoxide, silicon dioxide (sand), sulfur dioxide, and carbon tetrachloride are examples of compounds formed from two nonmetals.

5–2 The Periodic Law

As more elements were being discovered, chemists in the early nineteenth century attempted to *classify* the *elements* that had similar properties into groups or families, the way the various mammals may be classified. For example, one classification of mammals is the cat family, whose characteristics are a round head, 28 to 30 teeth, eyes with vertically slit pupils, retractable claws, etc. The cat family includes not only domestic house cats but also lions, tigers, leopards, jaguars, and bobcats, to name just a few. All have the *same* general characteristics mentioned. Similarly, many of the elements also have general characteristics that can be used to classify them as belonging to a particular group or family.

   In the middle to late nineteenth century, two chemists independently classified the elements known at that time and their classifications are the basis of the present one. Lothar Meyer (1830–1895), a German chemist, devised an incomplete Periodic Table and published it in a book in 1864; he extended it to include a total of 56 elements in 1869. Also in 1869, a Russian chemist, Dmitri Mendeleev (1834–1907), presented a paper describing a Periodic Table (see Fig. 5–1). Mendeleev went further than Meyer in that he left gaps in his table and predicted that new elements would be discovered to fill them. He also predicted

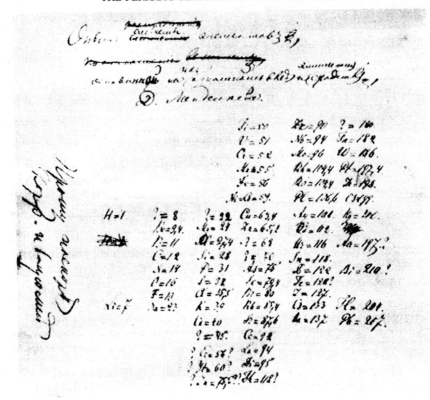

Figure 5–1 *Section of a manuscript of "Essay on the System of Elements," by Dmitri I. Mendeleev, dated February 17, 1869 (Julian Calendar). In this preliminary version the periods were vertical and the groups were horizontal. Note the absence of the noble gases, which were unknown at the time. The elements were placed in order of the atomic masses, but in the case of discrepancies, similarities in chemical properties were used to place the elements. (Note that Te and I lines begin with O = 16 and F = 19, respectively.) Note also the various question marks entered.*

the properties of these yet undiscovered new elements—truly a bold undertaking in science. Mendeleev lived to see the discovery of some of the elements he predicted, with properties similar to those he had forecast.

Since both Meyer's and Mendeleev's Periodic Tables ordered the elements by *increasing atomic* **masses**, several discrepancies occurred in their tables. After the discovery of the proton, Henry G. J. Moseley (1888–1915), a British physicist, studied and determined the nuclear charge on the atoms of the elements and concluded that the elements should be arranged by *increasing atomic* **number**. With this new arrangement of the Periodic Table, the discrepancies disappeared.

The elements are arranged in order of **increasing atomic number**, and elements with similar chemical properties recur at definite intervals (see Fig. 5–2). In Fig. 5–2, you will note that all the elements with the same number and kind of valence electrons are located in the same vertical column. For example,

H 1							He 2
Li 3	Be 4	B 5	C 6	N 7	O 8	F 9	Ne 10
Na 11	Mg 12	Al 13	Si 14	P 15	S 16	Cl 17	Ar 18

Figure 5–2 *An abbreviated periodic classification of the elements, based on atomic number. Similar chemical properties recur at definite intervals. (The numbers 1 to 18 represent the atomic numbers of the elements.)*

Be and Mg both have **2** valence electrons in an *s* sublevel (see Section 4–8, Be: $1s^2, 2s^2$; Mg: $1s^2, 2s^2 2p^6, 3s^2$). The noble gases (He, Ne, Ar: see Section 4–7) all appear in the same vertical column and all have 8 electrons in their highest energy level (rule of eight or octet rule), except helium, with 2 (a completed first-energy level). The basis of the Periodic Law is the classification of elements by increasing atomic number. Therefore, the **Periodic Law** states that the elements with similar chemical properties appear at regular intervals in the Periodic Table when they are listed in order of increasing atomic number.

Periodic Law Elements with similar chemical properties appear at regular intervals in the Periodic Table when they are listed in order of increasing atomic number.

5–3   The Periodic Table. Periods and Groups

Following the Periodic Law and completing our abbreviated classification of the elements begun in Fig. 5–2, we obtain a complete Periodic Table, as shown in Fig. 5–3 and inside the front cover. This Periodic Table, the one we now use, was first proposed in 1895 by Julius Thomsen (1826–1909), a Danish chemist.

The Periodic Table is arranged in 7 horizontal rows called **periods** or *series* and 18 vertical columns called **groups** or *families*. Recently this table has been modified by labeling the 18 columns with numbers in addition to the Roman numerals previously used. These numbers are shown above the Roman numerals in the Periodic Table (see Fig. 5–3 and inside the front cover). When we refer to a group (vertical column) in this text, we will use the Roman numeral followed by the column number in parentheses.

periods The seven horizontal rows in the Periodic Table.

groups The 18 vertical columns in the Periodic Table.

The elements from left to right in a given period vary gradually from very metallic properties, such as sodium (Na), to nonmetallic properties, such as chlorine (Cl). At the end of each period is a member of group VIIIA (18), the noble gases, which are relatively inert (unreactive). The elements in a given group resemble each other in that they have similar chemical properties.

Now, let us consider in detail each of the seven periods (horizontal rows). Follow this discussion by studying the Periodic Table (Fig. 5–3). Follow the electronic structures of the elements by studying Fig. 5–4.

Period 1 contains only two elements—hydrogen (H) and helium (He). In this period, the first principal energy level is being filled (1*s* sublevel). The first energy level is filled with two electrons and helium is placed in group VIIIA (18), the noble gases. The number of the period gives the principal energy level number that the electrons *begin* to fill.

Period 2 contains eight elements from lithium (Li) to neon (Ne). In this period, the second principal energy level is being filled (2*s* and 2*p* sublevels), resulting in a completely filled second energy level in neon.

GROUPS

PERIODS	1 IA	2 IIA	3 IIIB	4 IVB	5 VB	6 VIB	7 VIIB	8 VIII	9 VIII	10 VIII	11 IB	12 IIB	13 IIIA	14 IVA	15 VA	16 VIA	17 VIIA	18 VIIIA
1	H 1																	He 2
2	Li 3	Be 4											B 5	C 6	N 7	O 8	F 9	Ne 10
3	Na 11	Mg 12					TRANSITION ELEMENTS						Al 13	Si 14	P 15	S 16	Cl 17	Ar 18
4	K 19	Ca 20	Sc 21	Ti 22	V 23	Cr 24	Mn 25	Fe 26	Co 27	Ni 28	Cu 29	Zn 30	Ga 31	Ge 32	As 33	Se 34	Br 35	Kr 36
5	Rb 37	Sr 38	Y 39	Zr 40	Nb 41	Mo 42	Tc 43	Ru 44	Rh 45	Pd 46	Ag 47	Cd 48	In 49	Sn 50	Sb 51	Te 52	I 53	Xe 54
6	Cs 55	Ba 56	*La 57	Hf 72	Ta 73	W 74	Re 75	Os 76	Ir 77	Pt 78	Au 79	Hg 80	Tl 81	Pb 82	Bi 83	Po 84	At 85	Rn 86
7	Fr 87	Ra 88	**Ac 89	Rf 104	Ha 105	Unh 106	Uns 107	Uno 108	Une 109									

*Lanthanide series	Ce 58	Pr 59	Nd 60	Pm 61	Sm 62	Eu 63	Gd 64	Tb 65	Dy 66	Ho 67	Er 68	Tm 69	Yb 70	Lu 71
**Actinide series	Th 90	Pa 91	U 92	Np 93	Pu 94	Am 95	Cm 96	Bk 97	Cf 98	Es 99	Fm 100	Md 101	No 102	Lr 103

Figure 5–3 *The Periodic Table of the elements. (The numbers below the symbol of the elements represent the atomic numbers of the elements.)*

Period 3 also contains eight elements—from sodium (Na) to argon (Ar), with the third principal energy level being filled (3s and 3p sublevels *only*). Argon, the last element in the period, has eight electrons in its third energy level. Periods 2 and 3 are called the *short periods*, because they contain only eight elements each.

Period 4 contains 18 elements—from potassium (K) to krypton (Kr). In this period, the 4s and 4p energy levels are filling and the **3d** sublevel is being filled from scandium (Sc) to zinc (Zn). Consider the element Sc (atomic number 21). From the diagram in Fig. 4–4, the 3d fills after the 4s and before the 4p: $1s^2$, $2s^2 2p^6$, $3s^2 3p^6$, $4s^2$, $3d^1$. The 3d continues to fill to the element Zn (atomic number 30).

Period 5 contains 18 elements—from rubidium (Rb) to xenon (Xe). In this period, the 5s and 5p energy levels are filling and the **4d** sublevel is being filled from yttrium (Y) to cadmium (Cd).

Period 6 consists of 32 elements—from cesium (Cs) to radon (Rn). In this period, the 6s and 6p energy levels are filling. In the process, the **5d** and **4f** sublevels are also being filled. Elements 58 to 71, cerium (Ce) to lutetium (Lu), are called the *lanthanide* series and correspond to the filling of the **4f** sublevel. These elements are placed at the bottom of the table for convenience,

Study hint

The Periodic Table organizes a large body of information. Even if you know nothing about an element, you may learn something about its properties by examining its position in the Table. The Periodic Table is your friend. It is the best "cheat sheet" you can get.

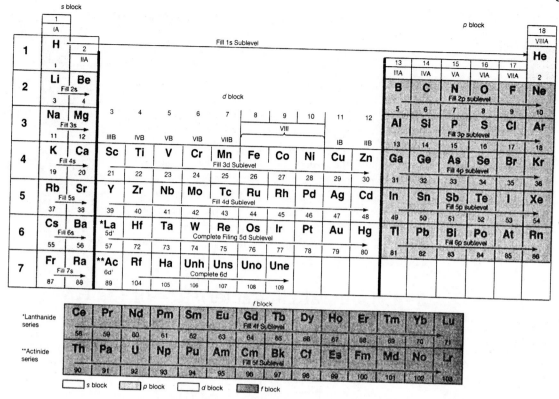

Figure 5–4 *The correlation of the filling of the sublevels with the Periodic Table. Note the groups of elements that fill the s, p, d, and f sublevels. The number at the bottom of each box refers to the atomic number of the element.*

since if they were placed in the main body, the table would be extremely wide and cumbersome.

Period 7 consists at present of 23 elements—from francium (Fr) to unnilennium (Une). In this period, the 7s energy level is filled and the *6d* and *5f* sublevels are being filled. Elements **90** to **103**, thorium (Th) to lawrencium (Lr), are called the *actinide* series and correspond to the filling of the *5f* sublevel. Again, for convenience these elements are placed at the bottom of the table. This period is incomplete and could end with element 118, which would be one of the noble gases and which should have properties like those of radon (Rn). Periods 4, 5, 6, and 7 are called the *long periods* because they contain more elements than the other periods.

Most of the 18 groups or families (vertical columns) are classed as A or B based on their Roman numeral designation. The **representative elements** consist of the A group elements. The **transition elements** include all the B group elements and the group VIII elements (three vertical columns in this group, 8, 9, and 10). Lanthanum (La) plus the lanthanide series and actinium (Ac) plus the actinide series are classed as transition elements in group IIIB (3). The

representative elements
The elements in the A group elements (1, 2, 13, 14, 15, 16, 17, 18).

transition elements The elements in the B group elements and the group VIII elements (3, 4, 5, 6, 7, 8, 9, 10, 11, 12).

gradual change from metallic to nonmetallic properties from left to right within a given period is more evident in the representative elements than in the transition elements. The transition elements are all metals and have one or two electrons in their outermost level. In addition, they also have valence electrons in the next lower *d* sublevel or the *f* sublevel, which lies below that. In this respect they differ markedly from the representative elements, which have all of their valence electrons in their outermost level. There are three complete transition series corresponding each to the filling of the 3*d*, 4*d*, and 5*d* sublevels and two series involving the filling of the 4*f* and 5*f* sublevels, respectively.

Because the groups or families have similar properties, they also have special names (remember the cat family in our classification of mammals). Group IA (1) elements (except hydrogen) are called the *alkali metals*. Hydrogen, although present in group IA (1), is not considered with the alkali metals because not all of its properties resemble those of the alkali metals. The elements in group IIA (2) are called the *alkaline earth metals*; those in group VIIA (17) are called the *halogens*; and those in group VIIIA (18) are called the *noble gases*.

5-4 General Characteristics of the Groups

The use of the Periodic Table to correlate general characteristics of the elements is one of the fundamental principles of chemistry. There are five general characteristics of groups that we shall consider here:

1. The Periodic Table separates the *metals* from the *nonmetals*, as shown in Fig. 5-3 by a heavy **black** stair-step line. To the right of this line are the nonmetals and to the left are the metals, with the more metallic metals on the *extreme left*. As you can see, most of the elements are considered to be metals, and even some of the so-called nonmetals, such as silicon (Si), phosphorus (P), arsenic (As), and selenium (Se), have considerable metallic properties. Elements that lie on the heavy **black** stair-step line are called *metalloids* (not aluminum). **Metalloids** have both metallic and nonmetallic properties. Examples are boron, silicon, germanium (Ge), arsenic, antimony, tellurium (Te), polonium (Po), and astatine (At), but not aluminum. Aluminum is considered to be a metal rather than a metalloid because it has mostly metallic properties. The elements in group VIIIA (18) consist of a special group of nonmetals called the *noble gases*.

metalloid Elements (except aluminum) that lie on the heavy **black** stair-step line in the Periodic Table; they have both metallic and non-metallic properties.

2. In the **A group** elements (representative elements) the number of valence electrons (see Section 4-8) is given by the *group Roman numeral* or the *units digit* in the vertical column number. For example, sodium is in group IA (1) and hence it has 1 valence electron ($1s^2$, $2s^2$, $2p^6$, $3s^1$). Sulfur is in group VIA (16) and hence it has 6 valence electrons ($1s^2$, $2s^2$, $2p^6$, $3s^23p^4$). The number of valence electrons is 8 for all of the elements in group VIIIA (18), except helium, which has only 2. This general characteristic does not hold for the transition elements (B group elements and group VIII elements, 8, 9, and 10) since they *usually* have 1, 2 or 3 valence electrons, but the number may vary considerably.

EXAMPLE 5–1

Using the Periodic Table, indicate the number of valence electrons for the following elements.

ANSWERS

(a) magnesium

2. Magnesium is in group IIA (2); hence, it has 2 valence electrons ($1s^2$, $2s^22p^6$, $3s^2$).

(b) oxygen

6. Oxygen is in group VIA (16); therefore, it has 6 valence electrons ($1s^2$, $2s^22p^4$).

(c) krypton

8. Krypton is in group VIIIA (18); thus, it has 8 valence electrons ($1s^2$, $2s^22p^6$, $3s^23p^6$, $4s^2$, $3d^{10}$, $4p^6$).

(d) silicon

4. Silicon is in group IVA (14); hence, it has 4 valence electrons ($1s^2$, $2s^22p^6$, $3s^23p^2$).

Exercise 5–1

Using the Periodic Table, indicate the number of valence electrons for the following elements.

(a) calcium   (b) phosphorus
(c) selenium   (d) xenon

3. Elements in the same group have *similar chemical properties and similar electronic configurations.* For example, all the alkali metals (group IA, 1) react rapidly with chlorine to form the metal chloride (see Section 3–5). All members of the alkali metals have the same electronic configuration in the valence energy level, the difference being the addition of principal energy levels.

Li $1s^2$, $2s^1$
Na $1s^2$, $2s^22p^6$, $3s^1$
K $1s^2$, $2s^22p^6$, $3s^23p^6$, $4s^1$
Rb $1s^2$, $2s^22p^6$, $3s^23p^63d^{10}$, $4s^24p^6$, $5s^1$
Cs $1s^2$, $2s^22p^6$, $3s^23p^63d^{10}$, $4s^24p^64d^{10}$, $5s^25p^6$, $6s^1$
Fr $1s^2$, $2s^22p^6$, $3s^23p^63d^{10}$, $4s^24p^64d^{10}4f^{14}$, $5s^25p^65d^{10}$,
       $6s^26p^6$, $7s^1$

EXAMPLE 5–2

Group the following electronic configurations of elements together according to those you would expect to show similar chemical properties.

(a) $1s^2$, $2s^22p^2$
(b) $1s^2$, $2s^22p^6$, $3s^23p^1$
(c) $1s^2$, $2s^22p^1$
(d) $1s^2$, $2s^2p^6$, $3s^23p^63d^{10}$, $4s^24p^1$

Answers (b), (c), and (d) have the s^2p^1 electronic configuration and, hence, would be expected to show similar chemical properties.

Exercise 5–2

Group the following electronic configurations of elements together according to those you would expect to show similar chemical properties.

(a) $1s^2$, $2s^2 2p^6$, $3s^2 3p^2$
(b) $1s^2$, $2s^2 2p^6$, $3s^2$
(c) $1s^2$, $2s^2 2p^2$
(d) $1s^2$, $2s^2 2p^6$, $3s^2 3p^6 3d^{10}$, $4s^2$

Because the electronic configurations of the elements in a group are similar, the formulas of compounds of elements in that group are also similar. Sodium hydroxide has the formula NaOH; hence, the formula for cesium (Cs) hydroxide is CsOH, because cesium is in the same group as is sodium. If there is any exception to this similarity of chemical properties in a given group, it is usually in the first element of the group. For example in group IA (1, not hydrogen), lithium is not as similar to sodium in chemical properties as sodium is to potassium. In group IIIA (13), boron is not as similar to aluminum as aluminum is to gallium (Ga). In other words, if one of the elements in a group is "out of step," it is usually the first element in the group.

4. In the *A group* elements, the *metallic properties increase* within a given group with *increasing atomic numbers,* and the *nonmetallic properties decrease.* In group VA (15), the first member of the group is nitrogen, considered to be a nonmetal; the last member of the group is bismuth, with very definite metallic properties. Since the more metallic metals are on the extreme left of the table, and the metallic properties increase with increasing atomic number in a given A group, the most metallic stable (nonradioactive) element would be found in the lower left-hand corner and would be cesium (Cs).[1] The most nonmetallic element [excluding the relatively unreactive group VIIIA (18), the noble gases] would be found in the upper right-hand corner and would be fluorine.

EXAMPLE 5–3 ───────────

Using the Periodic Table, indicate which element in the following pairs of elements is more metallic.

	ANSWERS
(a) barium or strontium	barium
(b) rubidium or cesium	cesium
(c) aluminum or boron	aluminum
(d) chlorine or iodine	iodine

---

[1] Francium (Fr) is radioactive and is unstable, decomposing to other elements. It is not considered here because of its instability.

TABLE 5–1   SOME PHYSICAL PROPERTIES OF THE HALOGENS[a]

ELEMENT	mp (°C)	bp[b] (°C)	DENSITY[c] (g/mL)	ATOMIC RADIUS[d] (Å)
F	−219.6	−188.1	1.11 at bp	0.72
Cl	−101.0	−34.6	1.56 at bp	0.99
Br	−7.2	58.8	2.93 at bp	1.14
I	113.5	184.4	4.93 at 20°C	1.33

[a] Although astatine (At) is a halogen, it is not considered in this table because it is radioactive and so unstable that it is not found in nature. Hence, an insufficient amount of it is present at any one time to allow studying its properties in detail.

[b] At 1.00 atm pressure.

[c] All densities are for the liquid state, except iodine, which is given for the solid state.

[d] Determined by dividing the observed distance between centers of identical adjacent atoms in the elemental state by 2.

Exercise 5–3

Using the Periodic Table, indicate which element in the following pairs of elements is more metallic.

(a) phosphorus or antimony
(b) antimony or bismuth
(c) tellurium or selenium
(d) iodine or bromine

5. There is a somewhat *uniform gradation of many physical and chemical properties* within a given group with increasing atomic number. In group VIIA (17), the halogens (see Table 5–1), the melting and boiling points, the densities, and the atomic radii of the elements increase as the atomic number increases. The increase in atomic radii with an increase in atomic number within a given group is true for all the elements, since a new principal energy level is being added as you go down the group to the next period. Thus, the atomic radius of the atom is increased, as shown in Fig. 5–4. With regard to chemical reactivity of the halogens, fluorine is the most reactive, then chlorine, followed by bromine and iodine in that order.

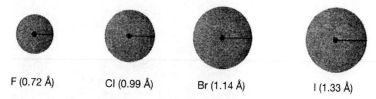

F (0.72 Å)   Cl (0.99 Å)   Br (1.14 Å)   I (1.33 Å)

Figure 5–5 *Atomic radii of group VIIA (17) elements (except astatine). As the atomic number increases in a given group, the atomic radii of the atoms increase.*

EXAMPLE 5–4

Using the Periodic Table, indicate which element in the following pairs of elements has the greater atomic radius.

(a) fluorine or chlorine

SOLUTION   Both fluorine (F) and chlorine (Cl) are in the same group—group VIIA (17). The atomic number of chlorine (17) is greater than that of fluorine (9); therefore, *chlorine* would have the greater radius because of the addition of a new principal energy level.

(b) sulfur or oxygen

SOLUTION   Both sulfur (S) and oxygen (O) are in the same group—group VIA (16). The atomic number of sulfur (16) is greater than that of oxygen (8); therefore, *sulfur* would have the greater radius because of the addition of a new principal energy level.

Exercise 5–4

Using the Periodic Table, indicate which element in the following pairs of elements has the greater atomic radius.

(a) barium or strontium   (b) zinc or mercury

Mercury: Quicksilver

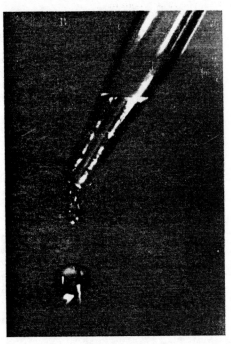

Mercury is a mobile silver liquid at ordinary temperatures. It is one of only four elements that are liquid at room temperature.

Name: *Symbol derives from hydrargyrum* (Latin), meaning liquid silver. Named after the Roman god Mercury, the messenger of the gods, who was clever and fast. Also: quicksilver.

Appearance: Bright, silver, metallic liquid.

Occurrence: Mercury occurs in small amounts in the earth's crust combined with other elements, such as oxygen and sulfur.

Source: Obtained commercially by heating cinnabar (a combination of mercury and sulfur) in oxygen to 600°C. The products of this process are metallic mercury and sulfur dioxide gas.

Its Role in Our World: Mercury is used extensively in the electronics industry, because it is one of the few liquid materials that is an excellent conductor of electricity.

Mercury is used in batteries for hearing aids, calculators, watches, and pacemakers.

Mercury is used in fluorescent lamps for home, industrial, and street lighting. These lamps produce a characteristic yellow light.

Mercury amalgams (alloy with silver and other metals) are used in dental fillings.

Mercury is used in a variety of measuring instruments including thermometers (see Section 2–6) and barometers (see Section 12–2), because it has a high density and remains liquid over a large range of temperatures.

Unusual Facts: Mercury and its compounds are quite toxic (poisonous). For years, mercury was used to extract gold and silver, and mercury compounds were used as pesticides, antiseptics, and explosives. Today these functions are largely carried out in less hazardous ways. Felt used to be made by softening leather with mercury salts, and hat-makers often exhibited the effects of the constant exposure to mercury-containing substances. The Mad Hatter in *Alice in Wonderland* is based on this historical premise.

Problems

(If in some of the following problems you are not familiar with the symbols for the elements, look them up inside the front cover.)

1. Using the Periodic Table, classify the following elements as metals, nonmetals, or metalloids.

   (a) rubidium
   (c) tellurium

   (b) iridium
   (d) selenium

2. Using the Periodic Table, classify the following elements as metals, nonmetals, or metalloids.

   (a) bromine
   (c) germanium

   (b) thallium
   (d) tin

3. Using the Periodic Table, indicate the number of valence electrons for the following elements.

   (a) cesium
   (c) tellurium

   (b) germanium
   (d) neon

4. Using the Periodic Table, indicate the number of valence electrons for the following elements.

   (a) krypton
   (c) gallium

   (b) astatine
   (d) arsenic

5. Group the following electronic configurations of elements in pairs according to those you would expect to show similar chemical properties.

(a) $1s^2$, $2s^22p^6$, $3s^23p^63d^{10}$, $4s^24p^64d^{10}$, $5s^25p^4$
(b) $1s^2$, $2s^22p^6$, $3s^23p^6$, $4s^2$
(c) $1s^2$, $2s^22p^6$, $3s^23p^4$
(d) $1s^2$, $2s^22p^6$, $3s^2$

6. Group the following electronic configurations of elements in pairs according to those you would expect to show similar chemical properties.

(a) $1s^2$, $2s^22p^6$, $3s^23p^3$
(b) $1s^2$, $2s^22p^6$, $3s^23p^63d^{10}$, $4s^24p^64d^{10}$, $5s^25p^6$, $6s^1$
(c) $1s^2$, $2s^22p^6$, $3s^23p^6$, $4s^1$
(d) $1s^2$, $2s^22p^6$, $3s^23p^63d^{10}$, $4s^24p^64d^{10}4f^{14}$, $5s^25p^65d^{10}$, $6s^26p^3$

7. Using the Periodic Table, indicate which element in the following pairs of elements is more metallic.

(a) phosphorus or arsenic   (b) cesium or sodium
(c) silicon or aluminum   (d) lead or germanium

8. Using the Periodic Table, indicate which element in the following pairs of elements is more metallic.

(a) barium or calcium   (b) magnesium or phosphorus
(c) silicon or lead   (d) oxygen or polonium

9. Using the Periodic Table, indicate which element in the following pairs of elements has the greater atomic radius.

(a) barium or magnesium   (b) copper or silver
(c) nitrogen or phosphorus   (d) lead or tin

10. Using the Periodic Table, indicate which element in the following pairs of elements has the greater atomic radius.

(a) selenium or oxygen   (b) arsenic or nitrogen
(c) gallium or indium   (d) rubidium or potassium

General Problems

11. Prior to the discovery of germanium in 1886, Mendeleev predicted in 1869 the properties of this element. Using the Periodic Table, determine the following for germanium (atomic number 32).

(a) Would this element be classified as a metal, a nonmetal, or a metalloid?
(b) How many valence electrons would it have?
(c) Write the electronic configuration in sublevels for both germanium and its precursor silicon.
(d) Would it be more metallic or more nonmetallic than its precursor silicon?
(e) Mendeleev predicted a density for what he called "eka-silicon," now called germanium, of 5.5 g/mL. The actual density for germanium was

found to be 5.3 g/mL. Convert the predicted and actual values into the SI unit, kg/m³. Express your answer in scientific notation. (*Hint*: See Sections 2–5 and 2–7.)

12. One form of elemental selenium melts at 144°C. What is the melting point of this form of selenium in degrees Fahrenheit (°F) and kelvins (K)?

13. The element bromine (Br) has an atomic radius of 1.14 Å. Calculate its atomic radius in centimeters (cm). Express your answer in scientific notation.

Answers to Exercises

5–1. (a) 2; (b) 5; (c) 6; (d) 8

5–2. (a) and (c); (b) and (d)

5–3. (a) antimony; (b) bismuth; (c) tellurium; (d) iodine

5–4. (a) barium; (b) mercury

QUIZ   You may use the Periodic Table and the List of Elements with their symbols.

1. Classify the following elements as metals, nonmetals, or metalloids.

(a) selenium   (b) indium

(c) arsenic   (d) osmium

2. Indicate the number of valence electrons for each of the following elements:

(a) rubidium   (b) gallium

(c) arsenic   (d) astatine

3. Group the following electronic configurations of elements in pairs according to those you would expect to show similar chemical properties.

(a) $1s^2, 2s^2 2p^6, 3s^2$
(b) $1s^2, 2s^2 2p^6, 3s^2 3p^6 3d^{10}, 4s^2 4p^4$
(c) $1s^2, 2s^2 2p^4$
(d) $1s^2, 2s^2 2p^6, 3s^2 3p^6, 4s^2$

4. Indicate which element in each of the following pairs is more metallic:

(a) sodium or rubidium   (b) sulfur or tellurium

5. Indicate which element in each of the following pairs has the greater atomic radius:

(a) iodine or astatine   (b) chromium or tungsten

6

COMPOUNDS

5 Solve the following linear equations for the unknown (x) (Section 1–5).
(a) $1 + x - 6 = 0$ (5);   (b) $2x + 3(-2) = -2$ (2)

4 Convert 3.50 μ to angstroms (Å). Express your answer in scientific notation [Sections 1–4, 2–2 (Table 2–2), and 2–3].

$(3.50 \times 10^4 \text{ Å})$

3 Diagram the atomic structure for each of the following atoms. Indicate the number of protons and neutrons, and arrange the electrons in principal energy levels (Section 4–6).

$$\left(\begin{array}{cccc} & 1 & 2 & 3 & \text{Principal} \\ & & & & \text{Energy Level} \end{array} \right)$$

(a) $^{16}_{8}\text{O}$ $\left(\boxed{\begin{array}{c} 8p \\ 8n \end{array}} \quad 2e^- \quad 6e^- \right)$

(b) $^{40}_{18}\text{Ar}$ $\left(\boxed{\begin{array}{c} 18p \\ 22n \end{array}} \quad 2e^- \quad 8e^- \quad 8e^- \right)$

2 Write the electron-dot formulas for the following atoms (Section 4–7).
(a) $^{16}_{8}\text{O}$ $(:\dot{\underset{..}{O}}\cdot)$;   (b) $^{40}_{18}\text{Ar}$ $(:\dot{\underset{..}{Ar}}:)$

1 Using the Periodic Table, indicate the number of valence electrons for the following elements (Section 5–4).
(a) silicon   (4);   (b) iodine   (7)

TASKS

1 Learn the rules for calculating oxidation numbers (Section 6–1).

2 Memorize the seven elements that exist as diatomic molecules (Section 6–4).

3 Begin memorizing the names and formulas of ions in Table 6–1 (cations), Table 6–2 (anions), and Table 6–4 (polyatomic ions), unless directed otherwise by your instructor. You must know these for Chapter 7, Nomenclature of Inorganic Compounds.

4 Memorize the partial order of electronegativities (Section 6–5), unless directed otherwise by your instructor. You must know these for Chapter 7, Nomenclature of Inorganic Compounds.

OBJECTIVES

1 Give the distinguishing characteristics of each of the following terms:
- (a) chemical bonds (Section 6–1)
- (b) rule of eight or octet rule (Section 6–1)
- (c) rule of two (Section 6–1)
- (d) ion (Section 6–2)
- (e) cation (Section 6–2)
- (f) anion (Section 6–2)
- (g) oxidation number (Section 6–3)
- (h) ionic charge (Section 6–3)
- (i) ionic bond (Section 6–4)
- (j) calorie (Section 6–4)
- (k) covalent bond (Section 6–5)
- (l) orbital (Section 6–5)
- (m) bond length (Section 6–5)
- (n) electronegativity (Section 6–5)
- (o) coordinate covalent bond (Section 6–6)
- (p) Lewis structures (Section 6–7)
- (q) structural formula (Section 6–7)
- (r) bond angle (Section 6–7)
- (s) double bond (Section 6–7)
- (t) triple bond (Section 6–7)
- (u) polyatomic ion (Section 6–7)

2 Using the rules for calculating oxidation numbers, calculate the oxidation numbers for any element in any compound or any ion (Example 6–1, Exercise 6–1, Problem 1).

3 Given the formula of a molecule or ion and the atomic number and symbol of each atom in the compound or ion, write the electron-dot and structural formula for the molecule or ion (Examples 6–2 and 6–3, Exercises 6–2 and 6–3, Problems 2 and 3).

4 Given the formulas for two ions, write the correct formula unit for a compound containing those ions (Example 6–4, Exercise 6–4, Problem 4).

5 Given the Periodic Table, determine the maximum positive oxidation number of any element and the maximum negative oxidation number of any nonmetal (Example 6–5, Exercise 6–5, Problem 5).

6 Given the Periodic Table and any A group elements, determine the correct formula of a binary compound (Example 6–6, Exercise 6–6, Problem 6).

7 Given the Periodic Table and numerical values of various properties of some elements in a group, predict the value of the comparable property of another element in the same group (Example 6–7, Exercise 6–7, Problem 7).

8 Given the Periodic Table and the formula of a compound, predict the formula of a second compound containing an element that differs from, but is in the same group as, an element in the first compound (Example 6–8, Exercise 6–8, Problem 8).

9 Given the Periodic Table and the formula of a compound, predict the type of bonding in binary and ternary compounds (Example 6–9, Exercise 6–9, Problem 9).

The Periodic Table is a table of *elements*, but we usually do not see elements in the pure state. In our everyday life we encounter compounds, like water (H_2O), salt (NaCl), sugar ($C_{12}H_{22}O_{11}$), and the like. These compounds are formed from atoms of the elements. In this chapter we consider how atoms are bonded together to form compounds. We also consider the Periodic Table and use it to make predictions about elements.

6–1 Chemical Bonds

chemical bonds The attractive forces that hold atoms together as compounds.

rule of eight or **octet rule** In the formation of molecules from atoms, most molecules attempt to obtain this stable configuration of 8 electrons in the valence energy level of each atom.

rule of two An exception to the rule of eight (octet rule). A completed first principal energy level is also a stable configuration. Helium atoms and hydrogen atoms in the combined state obey this rule.

The attractive forces that hold atoms together are called **chemical bonds**. There are two general types of bonds between atoms in a compound: (1) ionic (or electrovalent) bonds and (2) covalent bonds. *These bonds are formed through interactions among the valence electrons of the atoms in the compound.* To understand these interactions, recall the **rule of eight** or **octet rule** (Section 4–7). According to this rule, a stable configuration is achieved in many cases if 8 electrons are present in the valence energy level surrounding each atom. The atoms achieve these complete energy levels by *gaining, losing,* or *sharing valence electrons.* One exception to the rule of eight is helium, whose first principal energy level is completed with just 2 electrons. This exception gives us the **rule of two**: *a completed first principal energy level is also a stable configuration.*

In general, atoms having 1, 2, or 3 valence electrons tend to *lose* these electrons to become positively charged ions (cations). Metals show this type of behavior. Alternatively, atoms with 5, 6, or 7 valence electrons tend to *gain* electrons and become negatively charged ions (anions). Many nonmetals fall into this category. Many nonmetals may also *share* electrons to obtain 8 electrons in their valence energy level. Those elements with 4 valence electrons, for example, carbon, are most apt to share their valence electrons.

6–2 Charged Species: Cations and Anions

As stated in Section 6–1, the formation of compounds from elements often involves the gaining and losing of electrons. When an atom gains or loses an

Ion Atom or group of atoms in which the number of electrons does not equal the number of protons; it carries either a positive or negative charge.

cation Ion with a positive charge.

electron, the number of electrons no longer matches the number of protons in the nucleus, and the resulting species has either a positive or a negative charge (Section 4-3). Such species are called **ions** (Section 4-4).

When sufficient energy is added to an atom, an electron may be knocked out of the atom. When an atom loses an electron in this way, the remaining portion of the atom is called a **cation**, a positively charged ion, because the atom now has *more protons than electrons*.

$$atom + energy \longrightarrow cation + electron$$

For example, consider the formation of a sodium ion:

$$Na + energy \longrightarrow Na^{1+} + electron$$

Similarly, if enough energy is added, two electrons may be removed as in the formation of a magnesium ion:

$$Mg + energy \longrightarrow Mg^{2+} + 2\ electrons$$

Note that the positive charge on the cation matches the number of electrons knocked out of the atom.

Many cations of the elements exist, and the most common ones are summarized in Table 6-1. You should learn these ions because they will be used to describe many compounds.

Atoms may also gain electrons. When an atom gains an electron, energy is often released. When an atom gains an electron in this way, the remaining portion of the atom is called an **anion**, a negatively charged ion, because the atom now has *more electrons than protons*. As an example, consider the formation of a chloride ion:

anion Ion with a negative charge.

$$Cl + electron \longrightarrow Cl^{1-} + energy$$

Anions with charges of -2 or -3 may also exist in compounds. Table 6-2 contains the most common anions. Learn these ions now; you will need this information in the chapters to come.

6-3 Oxidation Numbers. Calculating Oxidation Numbers

Before we consider how atoms bond together to form compounds and the structures of these compounds, we must understand the meaning of a new term, "oxidation number." The term *oxidation number* is used to describe the relative numbers of atoms that combine to form a compound. For example, in H_2O, there are 2 hydrogens for every oxygen.

oxidation number A positive or negative whole number used to describe the combining capacity of an element in a compound.

Oxidation number (ox no) is usually a positive or negative whole number used to describe the combining capacity of an element in a compound.[1] A

---

[1] Fractional oxidation numbers for atoms in compounds do exist, but they are not common for inorganic compounds. An example is $Na_2S_4O_6$ (sodium tetrathionate), where the sulfur atom has an average oxidation number of $2\frac{1}{2}^+$.

TABLE 6-1 SOME COMMON METALS WITH THE FORMULAS OF THE CATIONS AND THEIR NAMES

METAL (SYMBOL)	CATION[a]	NAME OF CATION[b]
1+ Ionic Charge		
Hydrogen[c] (H)	$*H^{1+}$	Hydrogen
Lithium (Li)	$*Li^{1+}$	Lithium
Potassium (K)	$*K^{1+}$	Potassium
Silver (Ag)	Ag^{1+}	Silver
Sodium (Na)	$*Na^{1+}$	Sodium
2+ Ionic Charge		
Barium (Ba)	$*Ba^{2+}$	Barium
Cadmium (Cd)	Cd^{2+}	Cadmium
Calcium (Ca)	$*Ca^{2+}$	Calcium
Magnesium (Mg)	$*Mg^{2+}$	Magnesium
Nickel (Ni)	Ni^{2+}	Nickel(II)
Strontium (Sr)	$*Sr^{2+}$	Strontium
Zinc (Zn)	Zn^{2+}	Zinc
3+ Ionic Charge		
Aluminum (Al)	$*Al^{3+}$	Aluminum
1+ and 2+ Ionic Charges		
Copper (Cu)	Cu^{1+}	Copper (I) or cuprous
	Cu^{2+}	Copper(II) or cupric
Mercury (Hg)	Hg_2^{2+}	Mercury(I) or mercurous[d]
	Hg^{2+}	Mercury(II) or mercuric
2+ and 3+ Ionic Charges		
Iron (Fe)	Fe^{2+}	Iron(II) or ferrous
	Fe^{3+}	Iron(III) or ferric
2+ and 4+ Ionic Charges		
Lead (Pb)	Pb^{2+}	Lead(II) or plumbous
	Pb^{4+}	Lead(IV) or plumbic
Tin (Sn)	Sn^{2+}	Tin(II) or stannous
	Sn^{4+}	Tin(IV) or stannic

[a] In the cations marked with an asterisk (*), you can determine the ionic charge using the Periodic Table. You must memorize the ionic charge on all other cations.

[b] The Roman numeral in parenthesis indicates the ionic charge on each atom in the ion.

[c] Not a metal, but often reacts as a metal.

[d] Experimental evidence indicates that mercury(I) or mercurous ion exists as a dimer (two units) with an ionic charge of $1+$ on *each* atom $[Hg^{1+}]_2 = Hg_2^{2+}$.

Study hint

For the cations marked with an asterisk (*), the Roman group numeral in the Periodic Table represents the positive ionic charge on the element.

TABLE 6–2   SOME COMMON NONMETALS
WITH THE FORMULAS OF THE ANIONS AND
THEIR NAMES

NONMETALS (SYMBOL)	ANION	NAME OF ANION
	1^- Ionic Charge	
Bromine (Br)	Br^{1-}	Brom*ide* ion
Chlorine (Cl)	Cl^{1-}	Chlor*ide* ion
Fluorine (F)	F^{1-}	Fluor*ide* ion
Hydrogen (H)	H^{1-}	Hydr*ide* ion
Iodine (I)	I^{1-}	Iod*ide* ion
	2^- Ionic Charge	
Oxygen (O)	O^{2-}	Ox*ide* ion
Sulfur (S)	S^{2-}	Sulf*ide* ion
	3^- Ionic Charge	
Nitrogen (N)	N^{3-}	Nitr*ide* ion
Phosphorus (P)	P^{3-}	Phosph*ide* ion

Study hint

The ionic charge on the anions except hydride (H^{1-}) ion can be determined by subtracting 8 from the Roman group number in the Periodic Table. The hydride ion can be determined by subtracting 2 (completed energy level 1) from the Roman group number (IA, 1): $1 - 2 = -1$.

compound will usually contain elements with both positive and negative oxidation numbers, and the *sum of the oxidation numbers of all of the atoms in a compound is zero.* This principle applies to both ionic compounds (Section 6–3) and covalent compounds (Section 6–4).

The oxidation number is an arbitrary assignment based on certain rules (see below) that provides chemists with a method of electronic "bookkeeping." For example, in HCl, we might assign H an oxidation number of $+1$, which would mean that Cl has an oxidation number of -1. Now, we could look at the compounds of several metals with chlorine, NaCl, $MgCl_2$, and $AlCl_3$. If Cl has an oxidation number of -1 (from HCl example), we can immediately assign oxidation numbers of $+1$, $+2$, and $+3$ to Na, Mg, and Al in the series of compounds above.

Some of the elements have only one oxidation number or oxidation state. Sodium (Na), magnesium (Mg), and aluminum (Al) are examples of such elements. Other elements may have more than one oxidation state. An example is oxygen: in water (H_2O), oxygen has an oxidation number of -2 (remember, hydrogen is $+1$), while in hydrogen peroxide (H_2O_2) oxygen has an oxidation number of -1.

For ions containing a single atom, such as Na^{1+} or Cl^{1-}, the oxidation number of the element is the same as the charge on the ion, or the **ionic charge**. Cations have positive oxidation numbers and anions have negative oxidation numbers. *In general,* metals will have positive oxidation numbers and nonmetals will have negative oxidation numbers, if they are combined.

The following rules are used for assigning or determining oxidation numbers:

ionic charge The charge on an ion. The ion may consist of a single atom or a group of atoms bonded together.

1. The algebraic sum of the oxidation numbers of all the atoms in the formula for a compound is *zero.*

2. The oxidation number of an element in the *elemental* or *free* (uncombined) state is *zero.*

3. The oxidation number of a *monatomic ion* (an ion containing a single atom) is the same as its ionic charge. The algebraic sum of the oxidation numbers of all the atoms in a *polyatomic ion* (an ion containing many atoms) is equal to the ionic charge on the polyatomic ion.

4. In compounds containing two unlike atoms, a negative oxidation number is assigned to the more electronegative atom (see Section 6–5 and Fig. 6–5), whereas a positive oxidation number is assigned to the less electronegative atom. For example, in hydrogen chloride (HCl) the oxidation numbers of hydrogen and chlorine are 1^+ and 1^-, respectively, because chlorine is more electronegative than hydrogen (3.0 vs. 2.1, Fig. 6–5). Similarly, in water (H_2O) the oxidation numbers of hydrogen and oxygen are 1^+ and 2^-, respectively, because oxygen (3.5) is more electronegative than hydrogen (2.1).

5. In most compounds containing hydrogen, the oxidation number of hydrogen is 1^+. The exceptions to this rule are the hydrides of metals, where hydrogen has an oxidation number of 1^- (NaH, LiH, CaH_2, AlH_3, etc.). Note that here the H atom is written second in these formulas. In forming hydrides, hydrogen has acted as a nonmetal.

6. In most oxygen compounds, the oxidation number of oxygen is 2^-. The exceptions to this rule are the peroxides, in which oxygen has an oxidation number of 1^- (Na_2O_2, H_2O_2, BaO_2, etc.).[2]

Note that in Table 6–1, the charge on each cation (except Hg_2^{2+}) is equal to the oxidation number of the atom. Similarly, Table 6–2 lists the charges of the common anions, which is the same as the oxidation number of those atoms.

EXAMPLE 6–1

Calculate the oxidation number for the element indicated in each of the following compounds or ions.

(a) N in HNO_3

SOLUTION   The oxidation numbers (ox nos) of H and O in the compound are 1^+ and 2^- (see rules 5 and 6, respectively). The sum

---

[2] Other exceptions do exist. Some of these exceptions are as follows: OF_2, in which oxygen has an oxidation number of 2^+, since fluorine is more electronegative than oxygen (see Fig. 6–5); O_2F_2, in which oxygen has an oxidation number of 1^+; and the superoxides, such as KO_2, in which oxygen has an average oxidation number of $\frac{1}{2}^-$. In this book these exceptions will not be considered.

of the oxidation numbers of all the elements in the compound must equal zero. Therefore,[3]

$$+1 + \text{ox no of N} + 3(-2) = 0$$

$$+1 + \text{ox no of N} - 6 = 0$$

$$\text{ox no of N} - 5 = 0$$

$$\text{ox no of N} = +5 \text{ or } 5^+ \qquad \textit{Answer}$$

(b) N in $NO_2{}^{1-}$

SOLUTION The oxidation number of oxygen is 2^- and the sum of the oxidation numbers of all of the elements in the ion *must equal the charge on the ion or 1^-* (see rule 3). Therefore,

$$\text{ox on N} + 2(-2) = -1$$

$$\text{ox no N} - 4 = -1$$

$$\text{ox no N} = +4 - 1$$

$$\text{ox no N} = +3 \text{ or } 3^+ \qquad \textit{Answer}$$

(c) N in N_2O_5

SOLUTION The oxidation number of oxygen is 2^-. The sum of the oxidation numbers of all the elements in the compound must equal zero. There are two atoms of nitrogen, but we must solve for *one* atom of nitrogen; therefore,

$$2(\text{ox no N}) + 5(-2) = 0$$

$$2(\text{ox no N}) - 10 = 0$$

$$2(\text{ox no N}) = 10$$

$$\text{ox no N} = \frac{10}{2}$$

$$\text{ox no N} = +5 \text{ or } 5^+ \qquad \textit{Answer}$$

---

[3] In solving for the oxidation numbers of elements, x may be substituted for "ox no of the element." The equation is then solved as a linear equation (see Section 1–5). Hence, Example 6–1(a) could be solved as follows:

$$+1 + x + 3(-2) = 0$$

$$+1 + x - 6 = 0$$

$$x - 5 = 0$$

$$x = +5 \text{ or } 5^+ \qquad \textit{Answer}$$

Exercise 6–1

Calculate the oxidation number for the element indicated in each of the following compounds or ions.

(a) S in H_2SO_4

(b) S in SO_3^{2-}

(c) S in $S_2O_3^{2-}$

(d) C in CO_3^{2-}

We now consider general types of bonding—the ionic, the covalent, and the coordinate covalent bonds—and the compounds in which they are involved.

6–4 The Ionic Bond

ionic bond A bond formed by the transfer of one or more electrons from one atom to another. The resulting positively and negatively charged species attract each other and form a bond.

The **ionic bond** is formed by the transfer of one or more electrons from one atom to another. The bond formed between the two oppositely charged species is based on the attraction of a positively charged particle for a negatively charged particle (the law of electrostatics, Section 4–4). *Unlike particles attract each other and like particles repel each other.* This results in an attraction that is weak by human standards, but very strong by atomic standards. Further, in the crystal there are many such forces that sum up to a very strong bonding force overall. Compounds formed by the transfer of electrons from one atom to another atom are called *ionic compounds.*

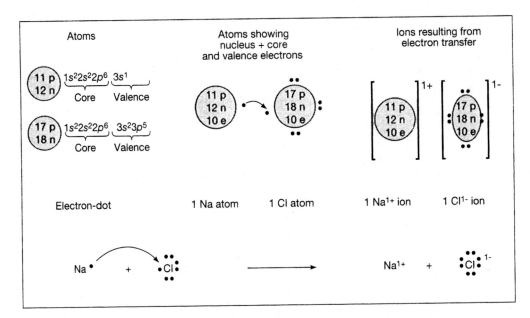

Figure 6–1 *Formation of sodium chloride NaCl, from a sodium atom and a chlorine atom, is an example of a compound formed by ionic bonding. The negative chloride ion attracts the positive sodium ion.*

Let us consider an example of an ionic compound. Sodium chloride, NaCl, is formed when a sodium atom combines with a chlorine atom, as Fig. 6–1 shows. The sodium atom has 1 valence electron and the chlorine atom has 7 valence electrons. The 1 valence electron from sodium is lost to the chlorine atom, giving 8 electrons in the highest energy level of the sodium ion (neon noble-gas configuration) and 8 electrons in the highest energy level of the chloride ion (argon noble-gas configuration). The rule of eight or octet rule is satisfied for both the sodium and chloride ions. The bond now formed between the positive sodium ion and the negative chloride ion is an *ionic bond*.

There are five important points to consider regarding the formation of all ionic compounds. *First*, the transfer of electrons can result in great changes in properties. For example, the sodium atoms and the chlorine atoms differ considerably from sodium chloride (sodium ions and chloride ions). Sodium, composed of sodium atoms, is a soft metallic solid and can be cut with a knife, whereas chlorine, composed of chlorine molecules (Cl_2), is a greenish gas with a strong, irritating odor. Sodium chloride is edible, but both sodium metal and chlorine gas are poisonous. Sodium reacts explosively with water while sodium chloride dissolves easily in water. The transfer of an electron from one atom to another always produces a drastic change in the properties of the newly formed compound. (Table 6–3 lists some physical properties of sodium, chlorine, and sodium chloride.)

Second, the charge of the ion is related to the numbers of *protons* and *electrons* in the ion. In the sodium atom, there are 11 protons in the nucleus and 11 electrons about the nucleus; the atom is neutral. There are still 11 nuclear protons in the ion but only 10 electrons, since 1 electron was lost to the chlorine atom. The result is a *net of one proton* or *one positive charge* in excess, giving a charge or oxidation number on the sodium ion of 1^+. In the chlorine atom there are 17 nuclear protons and 17 orbital electrons; thus, the atom is neutral. After an electron is received from the sodium atom, there are 18 electrons and 17 nuclear protons, resulting in a *net of 1 electron* or *one negative charge* in excess, and giving a charge or oxidation number on the chloride ion of 1^-. Therefore, the charges on the ions are directly related to their atomic structures.

Third, the radii of the ions differ from those of the atoms, as shown in Fig. 6–2. The radius of the sodium atom is 1.86 Å, whereas the radius of the

TABLE 6–3   PROPERTIES OF SODIUM, CHLORINE, AND SODIUM CHLORIDE

ELEMENT OR COMPOUND	APPEARANCE AT ROOM TEMPERATURE	MELTING POINT (°C)	BOILING POINT[a] (°C)
Sodium	Soft, silvery, solid cut with a knife	98	892
Chlorine	Greenish gas; strong irritating odor	−101	−35
Sodium chloride	Colorless crystalline solid	801	1413

[a] At 1.00 atm pressure.

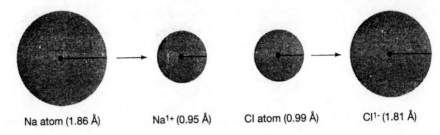

Na atom (1.86 Å)   Na^{1+} (0.95 Å)   Cl atom (0.99 Å)   Cl^{1-} (1.81 Å)

Figure 6–2   *The radii of ions differ from those of the atoms as shown by a sodium atom and ion, and by a chorine atom and ion.*

sodium ion is only 0.95 Å. This decrease in radius results from (1) the loss of an energy level, for principal energy level 3 in the sodium atom is no longer occupied after the transfer of that electron to the chlorine atom, and (2) a further decrease in size because of a greater nuclear attraction of the 11 positively charged protons on the remaining 10 electrons. The radius of the chlorine atom is 0.99 Å, whereas the radius of the chloride ion has increased to 1.81 Å. This increase in radius of the chloride ion over that of the chlorine atom is partly due to a smaller nuclear attraction (17 protons) on the 18 orbital electrons, causing an expansion of the radius of the energy level.

> **calorie** (cal) A unit of measurement of heat energy. It is defined as the amount of heat required to raise the temperature of 1.00 g of water from 14.5°C to 15.5°C. A kilocalorie is equal to 1000 calories.

Fourth, energy is given *off* in *bond formation*. In the formation of 1.00 g of sodium chloride, 2.63 kilocalories of heat energy is evolved. Therefore, to "break" the ionic bonds in 1.00 g of sodium chloride and to form the sodium and chlorine atoms, 2.63 kilocalories of energy would be required.

Fifth, the smallest unit of an *ionic compound* is called a *formula unit* (empirical formula unit), since it is a combination of *ions* and *not* discrete molecules (see Section 3–7). Hence, one formula unit of NaCl consists of 1 sodium ion and 1 chloride ion.

Some properties of ionic compounds are as follows:

1. They have relatively high melting points (above 300° C).

2. They conduct an electric current in the liquid state or in aqueous solution.

6–5   The Covalent Bond

> **covalent bond** A bond formed by the sharing of electrons between atoms.

The **covalent bond** is formed by the *sharing* of electrons between atoms. Compounds formed by the sharing of electrons are called *covalent compounds*. The smallest unit of a *covalent compound* is called a *molecule* (see Section 3–7); in an *ionic compound*, the smallest unit is a *formula unit* (see Section 6–4). The term *molecule* is used for compounds consisting primarily of *covalent bonds*, whereas *formula unit* is used for compounds consisting primarily of *ionic bonds*. A formula unit is *not* a molecule, since a formula unit does not really exist as a discrete entity but as ions. In Section 6–9 we show how to predict the predominant type of bonding in compounds by using the Periodic Table. Let us first consider some examples of covalent compounds.

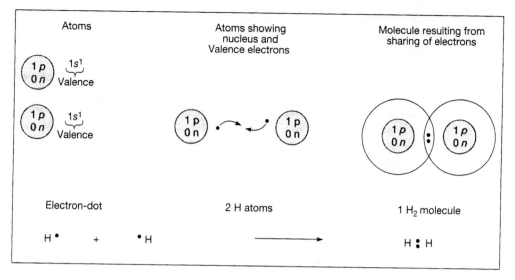

Figure 6–3 *The formation of hydrogen, H₂, from two hydrogen atoms is an example of covalent bonding.*

The hydrogen molecule, H_2, is a simple example of a *covalent* substance as shown in Fig. 6–3. The hydrogen atom as such is relatively unstable, since it has only one valence electron. By sharing its valence electron with another hydrogen atom, however, it completes the first principal energy level and gives a stable electronic configuration to the molecule.

An alternative way of viewing the H_2 molecule involves the concept of *orbitals.* An **orbital** is a region in space about the nucleus where a particular electron might be expected to be found. For example, a $1s$ orbital is that region about the nucleus where as $1s$ electron is likely to be found. A hydrogen $1s$ orbital has a spherical shape (Fig. 6–4). When two hydrogen atoms share electrons to form a covalent bond, the two $1s$ orbitals are pushed together to form a peanut shaped *molecular orbital* where the two bonding electrons are expected to be found (Fig. 6–4).

In the hydrogen molecule, as in all covalent substances, there are four important facts to remember. *First*, as with ionic compounds, the individual

orbital A region in space about the nucleus where a particular electron might be expected to be found. A $1s$ orbital is that region about the nucleus where a $1s$ electron is most likely to be found.

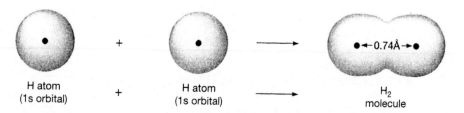

Figure 6–4 *Molecular orbital representation of a hydrogen, H₂, molecule. (A dot represents the nucleus of an atom.)*

uncombined atoms differ markedly from the molecules. In fact, individual hydrogen atoms are so unstable that they exist for only a very short time. Thus, when we write the formula for *hydrogen* we must write it as H_2 (2 atoms of hydrogen—a diatomic molecule) and not as H.

Second, the two positive nuclei attract each of the two electrons to produce a molecule more stable than the separate atoms. This attraction by the nuclei for the two electrons counterbalances the repulsion of the two positive nuclei for each other; the greatest probability of finding the electrons is somewhere *between* the two nuclei. A simple analogy may help to illustrate this point. Suppose we consider the nuclei of the two hydrogen atoms as "old potbelly stoves" and the two electrons as children running around each of these "stoves" trying to keep warm (see Fig. 6–4). When two atoms come together, the children (electrons) now have two sources of heat (nuclei), and these children can now run *between* the "stoves" and keep *all* parts of their body, front and back, warm. The children (electrons) are now warmer and happier than they were when they had just one "stove" (nucleus), and a stable molecule results.

Third, the distance between the nuclei is such that the $1s$ orbitals of the hydrogen atoms have the maximum overlap, without having the nuclei so close that they repel each other (causing the molecule to fly apart). In the hydrogen molecule, the distance between the nuclei is 0.74 Å, as shown in Fig. 6–4. The distance between the nuclei of covalently bonded atoms is called the **bond length**.

Fourth, during the process of covalent bond formation, energy is evolved. In this case, 52.0 kilocalories of heat energy are evolved in the formation of 1.0 g of gaseous hydrogen, H_2. Therefore, to "break" all the covalent bonds in 1.0 g of gaseous hydrogen and to form the hydrogen atoms, 52.0 kilocalories of heat energy would be required.

Besides H_2, other elements are not stable as single atoms and also exist as diatomic molecules. These molecules are F_2, Cl_2, Br_2, I_2, O_2, and N_2. Hence, when we write the *formulas of these elements, we do not write them as single atoms but as diatomic molecules*.

In the preceding example, the electrons have been shared *equally* by both atoms. This principle of equal sharing is not generally found in molecules that contain different atoms, because some atoms have a greater attraction for electrons than others. The tendency for an atom to attract a pair of electrons in a covalent bond is measured by its **electronegativity**. Linus C. Pauling developed a series of electronegativities for the elements. A partial series of decreasing electronegativities is F > O > Cl = N > Br > I = C = S > P = H > B > Si. The assigned values of the Pauling electronegativities of a number of elements are given in the Periodic Table in Fig. 6–5. Note that the metals generally have low electronegativities while the nonmetals have relatively high electronegativities. See Fig. 6–5.

Let us consider an example of a molecule in which there is an *unequal* sharing of electrons in the covalent bond due to the difference in electronegativity of the atoms in the molecule. A typical example is hydrogen chloride gas, shown in Fig. 6–6. The electronegativity of hydrogen is 2.1, whereas that of chlorine is 3.0 (see Fig. 6–5). Hence, the more electronegative chlorine atom would have a greater attraction for the pair of electrons in the covalent bond than would the

---

Study hint

Look at the Periodic Table. Notice the *six* elements, N, O, F, Cl, Br, and I trace the numeral 7, with the top of the 7 pointing toward the *seventh* element H!

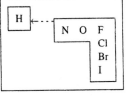

bond length The distance between the nuclei of covalently bonded atoms.

H_2
F_2
Cl_2
Br_2
I_2
O_2
N_2

F
O
Cl, N
Br
I, C, S
P, H
B
Si

electronegativity A measure of the tendency of an atom to attract a pair of electrons in a covalent bond.

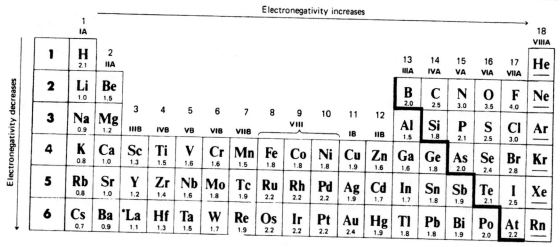

Figure 6-5   *The electronegativities of a number of the elements.*

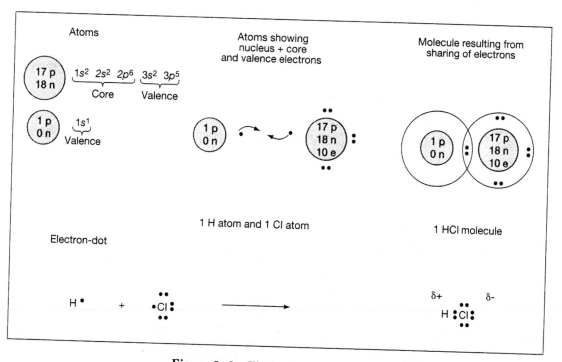

Figure 6-6   *The formation of hydrogen chloride, HCl, from one hydrogen atom and one chlorine atom is an example of an unequal sharing of electrons in a covalent bond.*

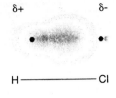

δ+   δ-

H ————————— Cl

Figure 6–7 *The hydrogen chloride molecule, showing the greater attraction for the electron pair in the covalent bond by the more electronegative chlorine atom. Compare this unequal sharing of electrons with the equal sharing of electrons shown in Fig. 6–4 with hydrogen. (The dots represent the nuclei of the atoms.)*

hydrogen atom in a molecule of hydrogen chloride gas. Thus, the bonding electrons would be found closer to the chlorine atom on the average. The molecule would appear as shown in Fig. 6–7. This unequal sharing of electrons in a covalent bond is often shown by placing a δ^- (lowercase Greek letter delta, δ, meaning partially charged) above the relatively negative atom, and a δ^+ above the partially positive atom. Hydrogen chloride gas would be depicted as

$$\overset{\delta^+}{H} \overset{\delta^-}{:\ddot{C}l:}$$

Unequal sharing of electrons in a covalent bond occurs whenever the atoms differ in electronegativity. The greater the differences in electronegativities, the more unequal the sharing of electrons in the covalent bond. This type of covalent bond is often referred to as a *polar covalent bond* or *polar bond.*

Unequal sharing of electrons in a covalent bond acts as a transition from equal sharing of electrons in covalent bonding to purely ionic bonding when the difference in electronegativities is great enough, as shown in Fig. 6–8.

Covalently bonded compounds have different properties than do ionic bonded compounds. Covalent compounds have relatively lower melting points (less than 300°C) and do not conduct an electric current when liquid or in aqueous solution, as ionic compounds do.

Equal sharing in
covalent bonding

Unequal sharing in
covalent bonding

Ionic or
electrovalent bonding

Figure 6–8 *The transition from equal sharing in covalent bonding to ionic bonding is bridged by unequal sharing of electrons in a covalent bond. When the difference in electronegativities is sufficiently large, the more electronegative atom gains essentially full possession of the shared pair, and ions result.*

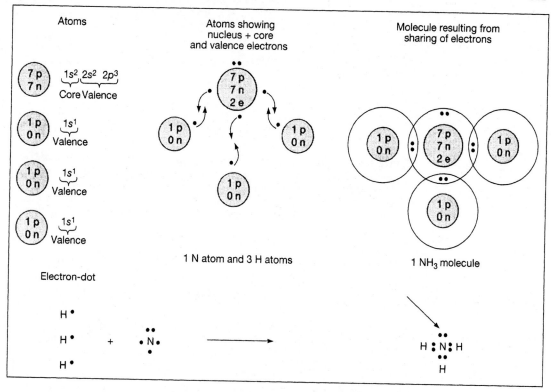

Figure 6–9   *The formation of ammonia, NH_3, from three hydrogen atoms and one nitrogen atom. (Note in the electron-dot formula the unshared pair of electrons is shown by the arrow, ↓ .)*

6–6   The Coordinate Covalent Bond

coordinate covalent bond A bond formed when *both* the electrons in an electron-pair bond are supplied by *one* atom.

In a covalent bond, each atom contributes one electron to form an electron pair between the *two* atoms. In **coordinate covalent bonding**, also called *coordinate bonding*, one atom supplies *both* of the bonding electrons.[4]

An example of a species containing a coordinate covalent bond is the ammonium ion ($NH_4{}^{1+}$). The ammonium ion is formed from a proton or hydrogen ion $\left(\text{H atom without an electron, } \begin{pmatrix} 1p \\ 0n \end{pmatrix}^{1+} \right)$ and ammonia.

First let us consider the formation of ammonia. The ammonia molecule (NH_3) is formed from 3 hydrogen atoms and 1 nitrogen atom, as shown by Fig. 6–9. The nitrogen atom, with 5 electrons in its second principal energy level, shares 1 electron from each of the 3 hydrogen atoms to give a total of 8 electrons around

---

[4] As you may have noted, the covalent bond and the coordinate covalent bond are formed in a different manner, but once they are formed there is no difference between them.

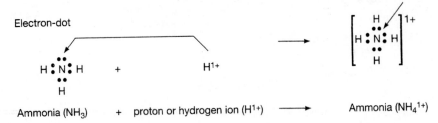

Figure 6–10 *The formation of an ammonium ion, NH_4^{1+}, from the molecule of ammonia, NH_3, and a proton or hydrogen ion, H^{1+}, is an example of coordinate covalent bonding. The coordinate covalent bond is shown by the large arrow (↙) and is indistinguishable from the other three covalent bonds. The positive one ionic charge is dispersed over the entire ion.*

the nitrogen atom. Each hydrogen atom with its 1 electron shares 1 electron with the nitrogen atom to give 2 electrons around the hydrogen atom, completing its first principal energy level. In the electron-dot diagram for NH_3 (Fig. 6–9), a pair of electrons not involved in the bonding to the hydrogen atoms (no atom attached to this pair of electrons) is called an *unshared pair of electrons*.

When a proton (hydrogen ion) is added to ammonia, the proton becomes attached to this unshared pair of electrons to form a coordinate covalent bond, as shown in Fig. 6–10. The hydrogen ion now shares 2 electrons with the nitrogen and is stabilized. The unshared pair of electrons acts to form the coordinate covalent bond with the proton. The new *ion* that is formed, the ammonium ion— NH_4^{1+}—is charged, because the ammonia was uncharged and the proton had a charge. The sum of these charges, 1^+, is now dispersed over the entire ion.

The descriptions of ionic, covalent, and coordinate covalent bonds are only models that we use to visualize the phenomenon of chemical bonding. No real difference exists between covalent and coordinate covalent bonds, but rather a difference in the way we perceive their formation. Thus, the four N—H bonds of the NH_4^{1+} are chemically the same. A coordinate covalent bond is *identical* to a covalent bond; only its "history" is different.

> **Study hint**
>
> The unshared pair of electrons on the nitrogen act like an electric plug or glue and form a coordinate covalent bond by "plugging" or "gluing" to the hydrogen. The 1^+ charge is over the entire ion, just as jelly becomes smeared over a young child's face when he or she eats jelly on his or her bread!

6–7   Lewis Structures and Structural Formulas of Molecules. Polyatomic Ions

In the discussion of bonding, we have used electron-dot formulas to depict the various types of bonding. The electron-dot formulas of molecules are also called *Lewis Structures* after Gilbert N. Lewis (1875–1946), an American chemist, who proposed the theory of covalent bonding in 1916. All the elements in these formulas had a total of 8 electrons (rule of eight or octet rule) in their last principal energy level, except the element hydrogen (rule of two). Both hydrogen and elemental helium are satisfied with just two electrons because 2 electrons

Lewis structures A method of expressing the covalent bond among atoms in a molecule using the rule of eight or octet rule and dots (:) to represent bonds.

completes their outermost principal energy level. Hence, in writing electron-dot formulas of compounds, we shall follow the rule of eight or octet rule for all atoms except hydrogen. Therefore, we can define **Lewis structures** (electron-dot formulas of molecules) as a method of expressing the covalent bonds among atoms in a molecule using the rule of eight or octet rule and dots (:) to represent bonds. Compounds do exist in which the elements in these compounds do not obey the rules of eight, but we shall not consider electron-dot formulas of these compounds in this book.

Lewis structures are of extreme importance in depicting the reactions of molecules to form new compounds. Therefore, we should consider them in detail.

1. Write the electron-dot formulas for the elements that occur in the molecule. (Review Section 4–7.)

2. Arrange the atoms so that each atom obeys the rule of eight or octet rule, with hydrogen obeying the rule of two.

3. In molecules containing three different atoms, the "central atom" acts as the starting point, with the other atoms arranged around this central atom. The central atom is generally the least electronegative atom (excluding hydrogen) because the least electronegative atom is the atom most willing to share electrons with several other atoms.

EXAMPLE 6–2

Write the Lewis structures and structural formulas for the following molecules.

(a) H_2O (^1_1H and $^{16}_8O$)

SOLUTION

H·

$^x_xO^{xx}_x$

H·

Electron-dot formulas of elements. There are 1 and 6 valence electrons for H and O, respectively (see Sections 4–7 and 5–3). Hydrogen must gain 1 electron and oxygen must gain 2 electrons to complete their valence energy levels.

$H^x_xO^{xx}_x$
$\cdot^x$
H

Answer

Lewis structure. Arranging the hydrogen atoms around the oxygen atom gives 8 electrons about the oxygen and 2 about each of the hydrogens. (All the bonding electrons about oxygen and hydrogen are equivalent, and although we use different symbols to identify the electrons from the various atoms, all the bonding electrons in this molecule are equivalent.)

To simplify an electron-dot formula, draw a dash (—) to *denote a single covalent bond for each pair of electrons shared between atoms.* These formulas are called structural formulas. A **structural formula** is a formula showing the arrangement of atoms within the molecule, using a dash (—) for each pair of electrons shared between

structural formula Formula showing the arrangement of atoms within a molecule, using a dash (—) for each pair of electrons shared between atoms.

bond angle The angle defined by two bonds that connect three atoms in a molecule.

atoms. Hence, the structural formula for water is as follows:

H—O
105°
    H

Answer

The angle between the O—H bonds is called the *bond angle*. In water, this angle has been found to be 105°. A **bond angle** is defined by two covalent bonds that connect three atoms in a molecule. The unshared pairs of electrons are usually not shown in a structural formula. Fig. 6–11 shows models of the water molecule.

(b) CO_2 carbon dioxide ($^{12}_6C$ and $^{16}_8O$)

SOLUTION

$^{xx}_x O ^x_x$

·Ċ·

$^{xx}_x O ^x_x$

Electron-dot formulas of elements. There are 4 and 6 valence electrons for C and O, respectively. Carbon must gain 4 electrons and oxygen must gain 2 electrons to complete their valence energy levels.

$^{xx}_x O ^x_x$:C: $^{xx}_x O ^x_x$

Answer

Lewis structure. The carbon atom is placed in the center because it has the lowest electronegativity value (2.5 vs. 3.5 for oxygen). If carbon accepts 2 electrons from each oxygen (by a sharing process), and at the same time shares 2 of its valence electrons with each of the two oxygen atoms, all the atoms in the molecule will have achieved the completed valence energy level of "eight." In each bond between carbon and oxygen, 4 electrons are being shared, 2 having been donated by the carbon and 2 by the oxygen. Such a bond of 4 electrons being shared is called a **double bond**.

double bond Chemical bond in which two atoms share two pairs of electrons (four electrons).

O=C=O

Answer

Structural formula. Draw a dash for each pair of shared electrons; hence, there are two dashes connecting the carbon atom to each of the oxygen atoms—a *double bond*.

(c) HCN, hydrogen cyanide[5] (^1_1H, $^{12}_6C$, $^{14}_7N$)

SOLUTION

H⊗ ·Ċ· $^{xx}_x N ^x$

Electron-dot formulas of elements. There are 1, 4, and 5 valence electrons for H, C, and N, respectively. Hydrogen must gain 1 electron, carbon must gain 4 electrons, and nitrogen must

---

[5] Other electron-dot formulas can be drawn for hydrogen cyanide and the sulfuric acid molecule, but based on the observed properties of hydrogen cyanide and sulfuric acid the Lewis structures here account for most of these properties.

gain 3 electrons to complete their valence energy levels.

H ⊗ Ċ ·   ×ᴺˣ

H ⊗ C ×××N×

Answer

triple bond Chemical bond in which two atoms share three pairs of electrons (six electrons).

Lewis structure. The carbon atom is placed in the center because it has the lowest electronegativity value (2.5 vs. 3.0 for nitrogen; exclude hydrogen, Section 6–6, guideline 3). Bonding the hydrogen with the carbon by a covalent bond gives 2 electrons about the hydrogen. To get a total of 8 electrons around the nitrogen, we need to find 3 more electrons. These must come from the carbon atom by covalent sharing. Thus, moving the remaining 3 electrons from carbon and 3 electrons from nitrogen between the carbon and nitrogen atoms gives 8 electrons each about carbon and nitrogen. There are now 6 electrons between the carbon and nitrogen atoms. Such an arrangement is called a **triple bond**.

H—C≡N

Answer

Structural formula. For each pair of electrons, draw a dash to the other atom; hence, three dashes connect the carbon and nitrogen atoms—a *triple bond*.

(d) H_2SO_4, sulfuric acid[5] (^1_1H, $^{32}_{16}S$, $^{16}_8O$)

SOLUTION

H⊗

·S̈·

H⊗

×ᴼˣ
×ᴼˣ
×ᴼˣ
×ᴼˣ

Electron-dot formulas of elements. There are 1, 6, and 6 valence electrons for H, O, and S, respectively. Hydrogen must gain 1 electron, and oxygen and sulfur must gain 2 electrons each to complete their valence energy levels.

H⊗Ö×S̈×Ö⊗H

×Ö×

H⊗Ö×S̈×Ö⊗H

×Ö×

or

×Ö×

H⊗Ö×S̈×Ö×

×Ö×
⊗
H

Answer

Lewis structure. The sulfur atom is placed in the center because it has the lowest electronegativity value (2.5 vs 3.5 for oxygen; exclude hydrogen, Section 6–6, guideline 3). Bond 2 oxygen atoms to the sulfur atom with covalent bonds, and then bond the 2 hydrogen atoms to those oxygen atoms using covalent bonds. This gives 8 electrons about the sulfur and oxygen atoms, and 2 electrons around the hydrogen atoms. We must also account for the remaining 2 oxygen atoms. We can place them on the sulfur atom using coordinate covalent bonds and still obey the rule of eight or octet rule for both oxygen and sulfur. The oxygens

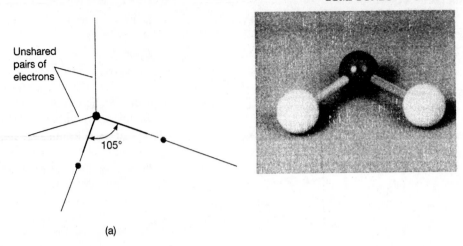

Unshared
pairs of
electrons

105°

(a)

Figure 6–11   *Molecular models of water, H_2O. (a) Prentice-Hall model, (b) Ball-and-stick model.*

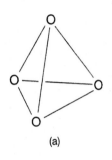

(a)

Figure 6–12   *The structure of sulfuric acid, H_2SO_4. (a) The tetrahedral arrangement of oxygen atoms in sulfuric acid. (b) A ball-and-stick model of sulfuric acid.*

are arranged in a *tetrahedral* configuration (a triangular-based pyramid; see Fig. 6–12a,b).

Structural formula. Draw a dash (—) for each pair of electrons. Both structural formulas at the left are equivalent, as you may note from the three-dimensional structure in Fig. 6–12b.

```
         O
         |
H—O—S—O—H
         |
         O
```

or

```
         O
         |
H—O—S—O
         |
         O
         |
         H
```

Answer

Exercise 6–2

Write the Lewis structures and structural formulas for the following molecules.

(a) HCl ($^1_1$H and $^{35}_{17}$Cl)

(b) H$_2$S ($^1_1$H and $^{32}_{16}$S)

(c) C$_2$H$_6$ ($^{12}_6$C and $^1_1$H)

(d) CS$_2$ ($^{12}_6$C and $^{32}_{16}$S)

The Lewis structures previously considered have all been for molecules. Following the general guidelines previously established for molecules, we can draw Lewis structures and structural formulas for ions containing more than one atom. For negative ions, depending on the net negative charge on the ion, an excess of an electron or electrons must be present. Ions consisting of two or more covalently bonded atoms with a net negative or positive charge on the ion are called **polyatomic ions** (many atom ions). The charge on the polyatomic ion is called the *ionic charge*. The term *oxidation number* is used to describe the oxidation state of *each* of the *atoms* comprising the polyatomic ion.

polyatomic ions Ions consisting of two or more covalently bonded atoms that possess a net negative or positive charge on the ion.

EXAMPLE 6–3

Write the Lewis structures and structural formulas for the following polyatomic ions.

(a) OH^{-1}, hydroxide ion ($^{16}_8$O and $^1_1$H)

SOLUTION

H· $^{XX}_{XX}$ $\overset{XX}{\underset{XX}{\text{x O x}}}$ *Electron-dot formulas of elements.*

$-\otimes\overset{XX}{\underset{XX}{\text{O}}}{}^X_{\,}\text{H}$ **Lewis structure of the polyatomic ion.** Arrange the hydrogen atom with the oxygen atom and add an electron (⊗) for the 1⁻ ionic charge to give 8 electrons about oxygen and 2 about hydrogen.

Answer

[O—H]$^{1-}$ **Structural formula** with the ionic charge (1⁻) dispersed over the *entire* ion.

Answer

(b) NO$_3$$^{1-}$, nitrate ion ($^{14}_7$N and $^{16}_8$O)

SOLUTION

·N̈· $\overset{XX}{\underset{XX}{\text{x O x}}}$
$\overset{XX}{\underset{XX}{\text{x O x}}}$
$\overset{XX}{\underset{XX}{\text{x O x}}}$ *Electron-dot formulas for the elements.*

$\begin{array}{l}\text{x O x}\\ \text{N: x O}\\ \text{x O x}\\ -\otimes\text{x}\end{array}$ **Lewis structure of the polyatomic ion.** Arrange the oxygen atoms about the central nitrogen atom because nitrogen is less electronegative than oxygen and add 1 electron (⊗) for the ionic charge. The formation of a double bond, a covalent bond, and a coordinate cova-

Answer

lent bond gives 8 electrons around each of the oxygen atoms and the nitrogen atom.

$$\left[\begin{array}{c} O \\ \diagdown \\ O \end{array} N = O \right]^{1-}$$

Structural formula with the ionic charge (1^-) dispersed over the entire ion. The $O—N—O$ bond angle in the nitrate ion is $120°$.

Answer

Exercise 6–3

Write the Lewis structures and structural formulas for the following poly-atomic ions.

(a) ClO^{1-} ($^{35}_{17}Cl$ and $^{16}_{8}O$)   (b) ClO_4^{1-} ($^{35}_{17}Cl$ and $^{16}_{8}O$)

(c) SO_4^{2-} ($^{32}_{16}S$ and $^{16}_{8}O$)   (d) CN^{1-} ($^{12}_{6}C$ and $^{14}_{7}N$)

TABLE 6–4   SOME COMMON POLYATOMIC IONS AND THEIR FORMULAS

FORMULA OF POLYATOMIC ION	NAME OF POLYATOMIC ION
	1^+ Ionic Charge
NH_4^{1+}	Ammonium
	1^- Ionic Charge
$C_2H_3O_2^{1-}$	Acetate
ClO_3^{1-}	Chlorate
ClO_2^{1-}	Chlorite
CN^{1-}	Cyanide
HCO_3^{1-}	Hydrogen carbonate or bicarbonate
HSO_4^{1-}	Hydrogen sulfate or bisulfate
HSO_3^{1-}	Hydrogen sulfite or bisulfite
OH^{1-}	Hydroxide
ClO^{1-}	Hypochlorite
NO_3^{1-}	Nitrate
NO_2^{1-}	Nitrite
ClO_4^{1-}	Perchlorate
MnO_4^{1-}	Permanganate
	2^- Ionic Charge
CO_3^{2-}	Carbonate
CrO_4^{2-}	Chromate
$Cr_2O_7^{2-}$	Dichromate
$C_2O_4^{2-}$	Oxalate
SO_4^{2-}	Sulfate
SO_3^{2-}	Sulfite
	3^- Ionic Charge
PO_4^{3-}	Phosphate

The ammonium ion NH_4^{1+}, considered in Section 6–6, is another polyatomic ion. Table 6–4 lists the various polyatomic ions. The first negative polyatomic ion in this table is acetate. The formula for the acetate ion is $C_2H_3O_2^{1-}$ and it is read "C-two, H-three, O-two one minus." You should know the names and formulas of these polyatomic ions so that you can use them to write formulas of compounds. Again we suggest that you make flash cards.

6–8 Writing Formulas

We shall now use the names and formulas of the cations (Table 6–1), anions (Table 6–2), and polyatomic ions (Table 6–4) to write formulas of compounds. To write the correct formula for a compound, you *must know* or have given to you the ionic charges of the cations and anions. In writing these formulas, *the sum of the total positive charges must be equal to the sum of the total negative charges, that is, the compound must **not** possess an overall charge.* When the charge on the positive ion is not equal to the charge on the negative ion, you must use subscripts to balance the positive charges with the negative charges. Write the positive ion first, followed by the negative ion. For example, iron(II) bromide consists of iron(II) ions, Fe^{2+}, and bromide ions, Br^{1-}. For the total positive charges to be equal to the total negative charges, we need to write one Fe^{2+} and two Br^{1-} as $Fe^{2+}Br^{1-}Br^{1-}$, or using the subscripts, $Fe^{2+}(Br^{1-})_2$. The 2^+ charge of the iron is just balanced by the 2^- charge of the two bromides. Delete the charges and the parentheses and write the formula as $FeBr_2$.

Let us consider more examples to illustrate writing formulas. The names and formulas of the ions in Tables 6–1, 6–2, and 6–4 will be given to you for now, but you should have learned these names and formulas by the time we cover nomenclature (Chapter 7).[6]

EXAMPLE 6–4

Write the correct formula for the compound formed by the combination of the following ions.

ANSWER

(a) sodium (Na^{1+}) and chloride (Cl^{1-})

$(Na^{1+})(Cl^{1-})$, NaCl
$1^+ + 1^- = 0$

(b) aluminum (Al^{3+}) and bromide (Br^{1-})

$(Al^{3+})(Br^{1-})_3$, $AlBr_3$
$3^+ + 3(1^-) = 0$

(c) ferric (Fe^{3+}) and sulfide (S^{2-})

$(Fe^{3+})_2(S^{2-})_3$, Fe_2S_3
$2(3^+) + 3(2^-) = 0$

Note: The least common multiple is 6; hence, 2(3) and 3(2).

[6] After you master nomenclature, you will be asked to write chemical equations (Chapter 9) using these formulas. Using these equations, you will then be asked to determine quantities used or obtained in a given chemical equation (Chapter 10). Therefore, in order to make these calculations and write chemical equations, you must write correct formulas. This means that you must memorize the formulas of ions in Tables 6–1, 6–2, and 6–4 (Task 3). Learn them now. Do not delay!

(d) cupric (Cu^{2+}) and nitrate (NO_3^{1-}) $(Cu^{2+})(NO_3^{1-})_2$, $Cu(NO_3)_2$
$2^+ + 2(1^-) = 0$

Note: There are two nitrate ions; thus, the ( ) must be used. This ( ) means that there are 2 atoms of nitrogen, 6 atoms of oxygen, and 1 atom of copper in *one* formula unit of cupric nitrate.

As you should recognize by now, a knowledge of the ionic charges of the cations and anions and their formulas is mandatory in writing the correct formulas for compounds.

Exercise 6–4

Write the correct formula for the compound formed by the combination of the following ions.

(a) potassium (K^{1+}) and iodide (I^{1-})
(b) sodium (Na^{1+}) and sulfide (S^{2-})
(c) tin(IV) (Sn^{4+}) and oxide (O^{2-})
(d) calcium (Ca^{2+}) and acetate ($C_2H_3O_2^{1-}$)

6–9 Using the Periodic Table to Predict Oxidation Numbers, Properties, Formulas, and Types of Bonding in Compounds

The Periodic Table can be helpful to you in learning the ionic charges on the cations (Table 6–1) and anions (Table 6–2).

Oxidation Numbers

In general, the *Roman group numeral* represents the **maximum *positive* oxidation number** for the elements in that group.[7] For example, aluminum is in group IIIA(13) and hence has a 3^+ oxidation number or ionic charge (see Table 6–1). For the nonmetals, the Roman numeral represents the maximum positive oxidation number. Also, for the nonmetals the **maximum** *negative oxidation number* can be calculated by *subtracting 8* from the Roman group number. For example, chlorine, in group VIIA(17), has a maximum positive oxidation number of 7^+ (group VII) in $KClO_4$ and a maximum negative oxidation number of 1^- (VII –

---

[7] The maximum positive oxidation number is not always the most common oxidation number (see Table 6–1). In all cases, the oxidation number is also the ionic charge on the monatomic ion.

8 = −1) in KCl. See Table 6–2 for the ionic charge on the chloride ion. Sulfur, in group VIA(16), has a maximum positive oxidation number of 6^+ (group VI) in H_2SO_4, and a maximum negative oxidation number of 2^- (VI − 8 = −2) in H_2S. Review Section 6–1 for calculating oxidation numbers of elements. Using the maximum positive and negative oxidation numbers, we can also predict the formulas of some compounds containing two different elements (binary compounds). When barium and iodine form a binary compound, the formula is BaI_2. Barium is in group IIA(2) and has a 2^+ oxidation number or ionic charge, while iodine is in group VIIA (17) and has a 1^- negative oxidation number or ionic charge (VII − 8 = −1). The correct formula (See Section 6–7) is $(Ba^{2+})(I^{1-})_2$, BaI_2. This prediction of formulas applies primarily to the A group elements. Thus, we can determine the positive oxidation numbers or ionic charges on some of the cations given in Table 6–1 by using the Periodic Table. Similarly, the negative oxidation numbers or ionic charges on all of the anions given in Tables 6–2, except the hydride ion (H^{1-}), can be determined from the Periodic Table.

EXAMPLE 6–5

Using the Periodic Table, indicate the maximum positive oxidation number for each of the following elements. For those elements that are non-metals, give *both* the maximum positive oxidation number and the maximum negative oxidation number.

ANSWERS

(a) potassium 1^+. Potassium is in group IA(1); hence it has a 1^+ oxidation number.

(b) strontium 2^+. Strontium is in group IIA(2); therefore, it has a 2^+ oxidation number.

(c) phosphorus 5^+ and 3^-. Phosphorus is a nonmetal and is in group VA(15). It has a maximum positive oxidation number of 5^+ and a maximum negative oxidation number of 3^- (V − 8 = −3).

(d) tellurium (Te) 6^+ and 2^-. Tellurium is a nonmetal and is in group VIA(16). It, therefore, has a maximum positive oxidation number of 6^+ and a maximum negative oxidation number of 2^- (VI − 8 = −2).

Exercise 6–5

Using the Periodic Table, indicate the maximum positivie oxidation number for each of the following elements. For those elements that are nonmetals, give *both* the maximum positive oxidation number and the maximum negative oxidation number.

(a) gallium (Ga)   (b) radium (Ra)
(c) selenium   (d) nitrogen

EXAMPLE 6-6

Using the Periodic Table to determine the oxidation numbers, predict the formulas of the binary compounds formed from the following combinations of elements.

(a) magnesium and oxygen

SOLUTION Magnesium is in group IIA(2) and has a 2^+ oxidation number or ionic charge, while oxygen is in group VIA(16) and has a 2^- oxidation number or ionic charge (VI − 8 = −2). The correct formula is $(Mg^{2+})(O^{2-})$, MgO.   *Answer*

(b) calcium and chlorine

SOLUTION Calcium is in group IIA(2) and has a 2^+ oxidation number or ionic charge, while chlorine is in group VIIA(17) and has a 1^- oxidation number or ionic charge (VII − 8 = −1). The correct formula is $(Ca^{2+})(Cl^{1-})_2$, $CaCl_2$.   *Answer*

(c) potassium and sulfur

SOLUTION Potassium is in group IA(1) and has a 1^+ oxidation number or ionic charge, while sulfur is in group VIA(16) and has a 2^- oxidation number or ionic charge (VI − 8 = −2). The correct formula is $(K^{1+})_2(S^{2-})$, K_2S.   *Answer*

(d) aluminum and sulfur

SOLUTION Aluminum is in group IIIA(3) and has a 3^+ oxidation number or ionic charge, while sulfur is in group VIA(16) and has a 2^- oxidation number or ionic charge. The correct formula is $(Al^{3+})_2(S^{2-})_3$, Al_2S_3.   *Answer*

Exercise 6-6

Using the Periodic Table to determine the oxidation numbers, predict the formulas of the binary compounds formed from the following combination of elements.

(a) calcium and sulfur
(b) sodium and oxygen
(c) sodium and nitrogen
(d) potassium and phosphorus

We can now use the general characteristics outlined in Section 5–3 for predicting properties of elements, formulas of compounds, and types of bonding in compounds.

Properties

We mentioned that there is a somewhat uniform gradation of properties within a given group with increasing atomic number. As an example, let us consider the atomic radii of three elements in group VIA(16) to determine if we can predict the radius of the fourth element in the group, tellurium (Te).

ELEMENT	RADIUS (Å)
O	0.74
S	1.04
Se	1.17
Te	?

The radii increase because of the addition of a new principal energy level; hence, we would expect the radius of tellurium also to increase. We can predict the value of this radius by taking the difference between the radii of sulfur and selenium and adding it to that of selenium. Hence, we would predict that the radius of tellurium would be 1.30 Å [1.17 + (1.17 − 1.04)]. It has been found to be 1.37 Å. We can apply this same general procedure with reasonable reliability to predict many of the properties of elements.[8]

EXAMPLE 6–7

Predict the missing value in the following table.

ELEMENT	RADIUS (Å)
F	0.72
Cl	0.99
Br	1.14
I	?

SOLUTION   The radii increase because of the addition of a new principal energy level; therefore, we would expect the radius of iodine also to increase. We can make a prediction of the radius of the iodine atom by taking the difference between the radii of chlorine and bromine and adding it to the radius of bromine. This procedure predicts the radius of iodine to be 1.29 Å.

$$1.14 + (1.14 − 0.99) = 1.14 + 0.15 = 1.29 \text{ Å} \qquad Answer$$

By indirect measurements it has been found to be 1.33 Å.

---

[8] We can predict properties of the elements more accurately by graphing the values of the property versus the atomic number of the elements.

Exercise 6–7

Predict the missing value in the following table.

ELEMENT	FIRST IONIZATION POTENTIAL[a] (kcal/mol)
F	402
Cl	299
Br	272
I	?

[a] The *ionization potential* of an atom or ion is the amount of energy required to remove the most loosely held electron from the atom or ion.

Formulas

In Section 5–3, paragraph 3, we mentioned that elements in the same group will form compounds with similar formulas, because the electronic configurations of the elements in that group are similar.

EXAMPLE 6–8

Given the following formulas—calcium brom**ide**, $CaBr_2$; water, H_2O; magnesium sulf**ate**, $MgSO_4$—and using the Periodic Table, write formulas for the following compounds.

(a) radium (Ra) bromide

SOLUTION   Radium (Ra) is in the same group (IIA, 2) as calcium; hence, $RaBr_2$.   *Answer*

(b) hydrogen telluride (Te)

SOLUTION   Tellurium is in the same group (VIA, 16) as oxygen; hence, H_2Te. (Note the -ide endings.)   *Answer*

(c) calcium sulfate

SOLUTION   Calcium is in the same group (IIA, 2) as magnesium; hence, $CaSO_4$. (Note the -ate endings.)   *Answer*

(d) strontium selenate

SOLUTION   Strontium is in the same group (IIA, 2) as magnesium, and selenium is in the same group (VIA, 16) as sulfur; hence, $SrSeO_4$.   *Answer*

Exercise 6–8

Given the following formulas—potassium chloride, KCl; sodium bromate, $NaBrO_3$—and using the Periodic Table, write formulas for the following compounds.

(a) rubidium (Rb) bromate
(b) cesium (Cs) chlorate
(c) sodium iodate
(d) rubidium bromide

Bonding

In Section 6–4 we stated that the term "molecule" is reserved for compounds bonded primarily by covalent bonds and the term "formula unit" for compounds bonded primarily by ionic bonds. The greater the difference in electronegativities (see Section 6–5), the greater is the ionic character in a compound. If the *ionic character* is large, then the compound is usually considered to be an *ionic compound*; hence, the smallest unit in this compound would be called a *formula unit* and not a molecule. In compounds consisting of only two *different* elements (binary compounds); the greater the difference in electronegativity of the elements, the greater is the ionic character of the compound. Figure 6–5 (Pauling's electronegativity) shows that the elements in a single group with large electronegativity values are the halogens (group VIIA, 17). Therefore, if the halogens combine with elements having relatively low electronegativities, an ionic compound is formed. The elements with low electronegativity values are found in the alkali metals (group IA, 1, *except* hydrogen) and the alkaline earth metals (group IIA, 2). Therefore we can make a general statement that if **binary compounds** are formed between elements in *group IA* (1, except hydrogen) or *group IIA* (2) with elements in *group VIIA* (17) or *group VIA* (16, *oxygen* and *sulfur* only), **ionic compounds** result. Any compound formed with *fluorine* or *oxygen* and a *metal* is also classified as an ionic compound, because both fluorine and oxygen have high electronegativity values. Note that the smallest unit in these ionic compounds would be a *formula unit*. Consider some examples:

(a) Strontium chloride ($SrCl_2$) is an ionic compound, since strontium is in group IIA(2) and chlorine is in group VIIA(17).

(b) Potassium oxide (K_2O) is an ionic compound, since potassium is in group IA(1) and oxygen is in group VIA(16), and since a compound containing any *metal* with *oxygen* is considered ionic.

(c) Iron(III) fluoride (FeF_3) is an ionic compound, since a compound containing *fluorine* and any *metal* is considered ionic.

Binary compounds containing other combinations of elements are considered to be covalent, with either equal or unequal sharing of electrons, and hence

PERIODIC TABLE OF THE ELEMENTS

	alkali metals	alkaline earth metals															halogens	noble gases
maximum positive ox. no.	1+	2+	3+	4+	5+	6+	7+	8+			1+a	2+	3+	4+	5+	6+	7+	18
maximum negative ox. no.	–	–	–	–	–	–	–							–	3–	2–	3–	
	1	2	3	4	5	6	7	8 9 10			11	12	13	14	15	16	17	
GROUPS	IA	IIA	IIIB	IVB	VB	VIB	VIIB	VIII			IB	IIB	IIIA	IVA	VA	VIA	VIIA	VIIIA

GROUPS

PERIODS

Period																		
1	1.008 H 1 hydrogen																	He 2
2	6.941 Li 3 lithium	9.012 Be 4 beryllium											10.811 B 5 boron	12.011 C 6 carbon	14.007 N 7 nitrogen	15.999 O 8 oxygen	18.998 F 9 fluorine	Ne 10 neon
3	22.990 Na 11 sodium	24.305 Mg 12 magnesium											26.982 Al 13 aluminum	28.0855 Si 14 silicon	30.9738 P 15 phosphorus	32.06 S 16 sulfur	35.453 Cl 17 chlorine	Ar 18 argon
4	39.0983 K 19 potassium	40.08 Ca 20 calcium	44.956 Sc 21 scandium	47.90 Ti 22 titanium	50.9415 V 23 vanadium	51.996 Cr 24 chromium	54.938 Mn 25 manganese	55.847 Fe 26 iron / 58.932 Co 27 cobalt / 58.71 Ni 28 nickel			63.546 Cu 29 copper	65.37 Zn 30 zinc	69.72 Ga 31 gallium	72.59 Ge 32 germanium	74.922 As 33 arsenic	78.96 Se 34 selenium	79.904 Br 35 bromine	Kr 36 krypton
5	85.468 Rb 37 rubidium	87.62 Sr 38 strontium	88.906 Y 39 yttrium	91.22 Zr 40 zirconium	92.9064 Nb 41 niobium	95.94 Mo 42 molybdenum	98.906 Tc 43 technetium	101.07 Ru 44 ruthenium / 102.906 Rh 45 rhodium / 106.4 Pd 46 palladium			107.868 Ag 47 silver	112.41 Cd 48 cadmium	114.82 In 49 indium	118.69 Sn 50 tin	121.75 Sb 51 antimony	127.60 Te 52 tellurium	126.90 I 53 iodine	131.30 Xe 54 xenon
6	132.906 Cs 55 cesium	137.33 Ba 56 barium	138.906 *La 57 lanthanum	178.49 Hf 72 hafnium	180.948 Ta 73 tantalum	183.85 W 74 tungsten	186.2 Re 75 rhenium	190.2 Os 76 osmium / 192.22 Ir 77 iridium / 195.09 Pt 78 platinum			196.967 Au 79 gold	200.59 Hg 80 mercury	204.37 Tl 81 thallium	207.2 Pb 82 lead	208.981 Bi 83 bismuth	(209) Po 84 polonium	(210) At 85 astatine	(222) Rn 86 radon
7	(223) Fr 87 francium	226.025 Ra 88 radium	(227) **Ac 89 actinium	(261) Rf 104 rutherfordium	(262) Ha 105 hahnium	(263) Unh 106 unnilhexium	(262) Uns 107 unnilseptium	(261) Uno 108 unniloctium / (266) Une 109 unnilennium										

TRANSITION ELEMENTS

*Lanthanide series

140.12 Ce 58 cerium	140.908 Pr 59 praseodymium	144.24 Nd 60 neodymium	(145) Pm 61 promethium	150.4 Sm 62 samarium	151.96 Eu 63 europium	157.25 Gd 64 gadolinium	158.925 Tb 65 terbium	162.50 Dy 66 dysprosium	164.930 Ho 67 holmium	167.26 Er 68 erbium	168.934 Tm 69 thulium	173.04 Yb 70 ytterbium	174.967 Lu 71 lutetium

**Actinide series

232.038 Th 90 thorium	231.031 Pa 91 protactinium	238.029 U 92 uranium	237.048 Np 93 neptunium	(244) Pu 94 plutonium	(243) Am 95 americium	(247) Cm 96 curium	(247) Bk 97 berkelium	(251) Cf 98 californium	(254) Es 99 einsteinium	(257) Fm 100 fermium	(256) Md 101 mendelevium	(255) No 102 nobelium	(257) Lr 103 lawrencium

*Certain elements in Group IB also form 2+ and 3+ oxidation numbers.

Numbers below the symbol of the element indicate the atomic numbers. Atomic masses, above the symbol of the element, are based on the assigned relative atomic mass of ^{12}C = exactly 12. ( ) indicates the mass number of the isotope with the longest half-life.

Key: metals; metalloids; nonmetals; noble gases.

1.008 ← Atomic mass
H ← Symbol
1 ← Atomic number
hydrogen ← Name

Figure 6–13 *Periodic Table summarizing most of the generalizations of the elements discussed in this chapter and previous chapters.*

are referred to as molecules. Some examples are carbon dioxide (CO_2—carbon is a nonmetal), sulfur dioxide (SO_2—sulfur is a nonmetal), water (H_2O), and methane (CH_4). As with all general statements, there are exceptions, but knowing that a binary compound is considered ionic if it is formed from elements in certain groups of the Periodic Table will be helpful in further studying the properties of compounds in your general chemistry course.

The preceding general statements about classifying the bonding in compounds apply only to *binary* compounds. We shall now consider the bonding forces in **ternary** and **higher** compounds (three or more different elements) that involve polyatomic ions (see Section 6–7). In general, the combination of *any* element (*hydrogen* being the sole exception) with any polyatomic ion to form a ternary or higher compound results in an ***ionic compound***, because the polyatomic ion can readily accommodate its positive or negative charge over its many atoms. The smallest unit in these compounds is therefore a *formula unit* and not a molecule. Consider some examples:

(a) Sodium sulfate (Na_2SO_4) is an ionic compound, because sulfate (SO_4^{2-}) is a polyatomic ion.

(b) Silver nitrate ($AgNO_3$) is an ionic compound, because nitrate (NO_3^{1-}) is a polyatomic ion.

(c) Ammonium acetate ($NH_4C_2H_3O_2$) is an ionic compound, because both ammonium (NH_4^{1+}) and acetate ($C_2H_3O_2^{1-}$) are polyatomic ions.

Some important differences between ionic compounds and covalent compounds are that ionic compounds have relatively high melting points and conduct an electric current when they are in the liquid state, whereas covalent compounds have relatively low melting points and do not conduct an electric current to any great extent. While there are a few exceptions to these generalizations,[9] they work with almost all simple substances.

EXAMPLE 6–9

Using the Periodic Table, classify the following compounds as essentially ionic or covalent.

(a) NaBr

SOLUTION   Ionic compound, because sodium is in group IA(1) and bromine is in group VIIA(17).   *Answer*

(b) N_2O_5

SOLUTION   Covalent compound, because nitrogen and oxygen are nonmetals.   *Answer*

---

[9] Graphite, a form of elemental carbon, is an important exception. The bonding in graphite is mostly covalent, yet it conducts electricity in the solid state. Some recently developed plastics (polymers, many parts covalently bonded) also conduct electricity to a limited extent in the solid state.

(c) Ag_2O

SOLUTION   Ionic compound, because silver is a metal and any compound formed between a metal and fluorine or oxygen is an ionic compound.   *Answer*

(d) K_2SO_4

SOLUTION   Ionic compound, because sulfate (SO_4^{2-}) is a polyatomic ion and potassium (K^{1+}) is a metal ion.   *Answer*

Exercise 6–9

Using the Periodic Table, classify the following compounds as essentially ionic or covalent.

(a) SO_3 (b) CdF_2

(c) SrO (d) $Sn(C_2H_3O_2)_2$

Figure 6–13 is a Periodic Table summarizing most of the generalizations of the elements discussed in this chapter and the previous chapters.

Silicon: chemistry and the computer revolution

The remarkable progress made in computer technology relies on the development of advanced microprocessors, transistors, and memory chips. Very pure silicon is one of the key ingredients in this rapid development.

Name: Derives from the Latin word *silex* for flint.

Appearance: Dark gray, brittle material with metallic luster.

Occurrence: Silicates and silica (quartz) make up 25 percent of the earth's crust.

Source: Low-grade silicon is obtained by heating quartz $(SiO_2)_n$ and carbon to give silicon and carbon dioxide (CO_2). Very pure silicon for use in the electronics industry is obtained by treating trichlorosilane $(SiHCl_3$, prepared from low-grade silicon) with hydrogen gas (H_2) to produce silicon and hydrogen chloride (HCl).

Its Role in Our World: The largest use of silicon occurs in the steel-making industry. Alloys (mixtures) of iron and silicon are added to iron during the steel-making process to give strength and other desirable properties to the end product. Iron–silicon–boron alloys are used to make bearings for heavy machinery.

Silicate glasses, soluble silicate salts $(SiO_3^{2-}, Si_2O_7^{6-}$, etc.), and gels of variable composition are components of many common products such as soaps, detergents, gel toothpastes, shampoos, antiperspirants, make-up, and paints. They are also used in water purification.

Although the semiconductor industry does not consume large amounts of silicon, it is one of the most visible industrial users. Extremely pure crystals of silicon are required to produce modern transistors, integrated circuits, semiconductors, and other computer chips. The ability of computer manufacturers to make smaller and more powerful computers is dependent on the use of these silicon crystals. A group of cities in the Santa Clara Valley near San Jose, California, where this technology has been developed is called "Silicon Valley."

Unusual Facts: Low-grade silicon is quite pure (95–99%), but not pure enough for the electronics industry. Computer chips and microprocessors require extremely pure silicon that contains less than a few atoms of impurity per billion atoms of silicon!

Problems

1. Calculate the oxidation number for the element indicated in each of the following compounds or ions.

   (a) N in HNO_3
   (b) I in HIO_3
   (c) Cl in HClO
   (d) Mn in MnO_2
   (e) N in HNO_2
   (f) I in IO_4^{1-}
   (g) S in HSO_3^{1-}
   (h) Bi in BiO_3^{1-}
   (j) S in SO_4^{2-}
   (j) As in AsO_4^{3-}

2. Write the Lewis structures and structural formulas for the following molecules.

   (a) HF $(_1^1H, \, _9^{19}F)$
   (b) $CCl_4(_6^{12}C, \, _{17}^{35}Cl)$
   (c) N_2 $(_7^{14}N)$
   (d) C_2H_4 $(_6^{12}C, \, _1^1H)$

3. Write the Lewis structures and structural formulas for the following polyatomic ions.

   (a) SH^{1-} $(_{16}^{32}S, \, _1^1H)$
   (b) SO_3^{2-} $(_{16}^{32}S, \, _8^{16}O)$
   (c) NO_2^{1-} $(_7^{14}N, \, _8^{16}O)$
   (d) ClO_3^{1-} $(_{17}^{35}Cl, \, _8^{16}O)$

4. Write the correct formula for the compound formed by the combination of the following ions.

   (a) lithium (Li^{1+}) and bromide (Br^{1-})

(b) mercury(II) (Hg^{2+}) and iodide (I^{1-})
(c) magnesium (Mg^{2+}) and nitride (N^{3-})
(d) ferric (Fe^{3+}) and chloride (Cl^{1-})
(e) cadmium (Cd^{2+}) and oxide (O^{2-})
(f) calcium (Ca^{2+}) and phosphide (P^{3-})
(g) lithium (Li^{1+}) and hydride (H^{1-})
(h) barium (Ba^{2+}) and nitrate (NO_3^{1-})
(i) aluminum (Al^{3+}) and perchlorate (ClO_4^{1-})
(j) barium (Ba^{2+}) and permanganate (MnO_4^{1-})

5. Using the Periodic Table, indicate the maximum positive oxidation number for each of the following elements. For those elements that are nonmetals, give *both* the maximum positive oxidation number and negative oxidation number.

   (a) barium   (b) selenium
   (c) tin   (d) bromine

6. Using the Periodic Table to determine the oxidation numbers, predict the formulas of the binary compounds formed from the following combination of elements. (If you do not know the symbol for the element, look it up inside the front cover.)

   (a) calcium and iodine   (b) aluminum and oxygen
   (c) magnesium and arsenic   (d) sodium and tellurium

7. Predict the missing value in the following sequences:

   (a) ELEMENT   RADIUS ($\mathring{A}$)

ELEMENT	RADIUS (Å)
K	2.02
Rb	2.16
Cs	?

   (b) ELEMENT   DENSITY (g/mL)

ELEMENT	DENSITY (g/mL)
Ca	1.54
Sr	2.60
Ba	?

8. Given the following formulas—sodium sulfate, Na_2SO_4; magnesium phosphate, $Mg_3(PO_4)_2$; aluminum oxide, Al_2O_3—and using the Periodic Table, write the formulas for the following compounds. (*Hint*: Note the endings of each name for the compound.)

   (a) magnesium arsenate   (b) aluminum sulfide
   (c) sodium selenate   (d) barium arsenate

9. Using the Periodic Table, classify the following compounds as essentially ionic or covalent.

   (a) ZnF_2 (b) SO_2
   (c) $MgSO_4$ (d) RbI

General Problems

10. Consider the undiscovered element atomic number 114.

(a) In what group would it be placed?
(b) How many valence electrons would it have?
(c) Would it be more metallic or more nonmetallic than its precursor in the same group?
(d) What element would it most likely resemble in properties?

11. Consider the undiscovered element atomic number 119.

(a) In what group would it be placed?
(b) How many valence electrons would it have?
(c) What would be its oxidation number?
(d) What element would it most likely resemble in properties?

12. Consider the undiscovered element (Q) with atomic number 115. What is the expected formula of the binary compound Q and hydrogen?

Answers to Exercises

6–1. (a) $+6$ or 6^+; (b) $+4$ or 4^+; (c) $+2$ or 2^+; (d) $+4$ or 4^+

6–2. (a) H^x$\ddot{\text{C}}$l:,   H—Cl;   (b) H^x$\ddot{\text{S}}$:,   H—S
                                                                   $\overset{..}{\underset{x}{\text{H}}}$ |
                                                                                  H

(c)
            H H   H   H
           $^.$x $^.$x | |
        H$_x$C:C$_x$H,   H—C—C—H
           $_.$x $_.$x | |
            H H   H   H

(d) $_x^x$$\overset{xx}{\text{S}}$$_x^x$:C:$_x^x$$\overset{xx}{\text{S}}$$_x^x$,   S=C=S

6–3. (a) :$\ddot{\text{C}}$l$_x$$\overset{xx}{\underset{xx}{\text{O}}}$$_\otimes$-,   [Cl—O]$^{1-}$

(b)
        $_x^{xx}$O$_x^x$
    $_x^x$O:Cl$_x$O$_\otimes$-,
        $_x^{xx}$O$_x^x$

$$\left[\begin{array}{c} \text{O} \\ | \\ \text{O—Cl—O} \\ | \\ \text{O} \end{array}\right]^{1-}$$

(c)
        :$\ddot{\text{O}}$:
    -$_\otimes$$\ddot{\text{O}}^{x}\overset{xx}{\underset{xx}{\text{S}}}^{x}\ddot{\text{O}}$$_\otimes$-,
        :$\ddot{\text{O}}$:

$$\left[\begin{array}{c} \text{O} \\ | \\ \text{O—S—O} \\ | \\ \text{O} \end{array}\right]^{2-}$$

(d) -$_\otimes$^{x}C$_x^{xx}$:N:,   [C≡N]$^{1-}$

6–4. (a) KI; (b) Na_2S; (c) SnO_2; (d) $Ca(C_2H_3O_2)_2$

6–5. (a) 3^+; (b) 2^+; (c) 6^+ and 2^-; (d) 5^+ and 3^-

6–6. (a) CaS; (b) Na_2O; (c) Na_3N; (d) K_3P

6–7. 245 kcal/mol (observed value is also 245 kcal/mol)

6–8. (a) $RbBrO_3$; (b) $CsClO_3$; (c) $NaIO_3$; (d) RbBr (Note the *-ate* and *-ide* endings).

6–9. (a) covalent; (b), (c), and (d) ionic

QUIZ

1. Calculate the oxidation number for the element indicated in each of the following:

(a) I in HIO_2
(b) Cr in $Cr_2O_7^{2-}$

2. Write the Lewis structures and structural formulas for the following compounds:

(a) C_2H_6 ($^{12}_6C$ and ^1_1H)
(b) SO_3^{2-} ($^{32}_{16}S$ and $^{16}_8O$)

3. Write the correct formula for the compound formed by the combination of the following ions:

(a) potassium (K^{1+}) and nitride (N^{3-})
(b) iron(III) (Fe^{3+}) and iodide (I^{1-})
(c) aluminum (Al^{3+}) and sulfate (SO_4^{2-})
(d) lead (IV) (Pb^{4+}) and chromate (CrO_4^{2-})

4. Using the Periodic Table to determine the oxidation numbers, predict the formulas of the binary compounds formed from the following combinations of elements:

(a) calcium and selenium (Se)
(b) aluminum and tellurium (Te)

5. Using the Periodic Table, classify the following compounds as essentially ionic or covalent.

(a) N_2O
(b) CaI_2
(c) CdF_2
(d) $Cu(NO_3)_2$

NOMENCLATURE OF INORGANIC COMPOUNDS

COUNTDOWN You may use the Periodic Table.

5 (1) Write the electronic configuration in sublevels for the following atoms, and (2) give the number of valence electrons for each (Section 4–8).
(a) $^{27}_{13}Al$ $[1s^2, 2s^22p^6, 3s^23p^1 \ (3)]$
(b) $^{35}_{17}Cl$ $[1s^2, 2s^22p^6, 3s^23p^5 \ (7)]$

4 Write the electronic configuration in sublevels for the following ions (Section 6–2 and Section 6–4).
(a) $^{27}_{13}Al^{3+}$ $(1s^2, 2s^22p^6)$
(b) $^{35}_{17}Cl^{1-}$ $(1s^2, 2s^22p^6, 3s^23p^6)$

3 Write the correct formula for the compound formed by the combination of the following ions (Section 6–8).
(a) aluminum (Al^{3+}) and chloride (Cl^{1-}) $(AlCl_3)$
(b) aluminum (Al^{3+}) and sulfide (S^{2-}) (Al_2S_3)
(c) aluminum (Al^{3+}) and phosphide (P^{3-}) (AlP)
(d) magnesium (Mg^{2+}) and nitride (N^{3-}) (Mg_3N_2)

2 Using the Periodic Table to determine the oxidation numbers, predict the formulas of the binary compounds formed from the combination of the elements (Section 6–9, Oxidation Numbers).
(a) potassium and sulfur (K_2S)
(b) barium and nitrogen (Ba_3N_2)

1 Given the following formulas—sodium chloride, NaCl; sodium chlorate, $NaClO_3$—and using the Periodic Table, write the formulas for the following compounds (Section 6–9, Formulas).
(a) potassium iodate (KIO_3)
(b) potassium iodide (KI)

TASK

1. Memorize the names and formulas of ions in Table 6–1 (cations), Table 6–2 (anions), and Table 6–4 (polyatomic ions). You may use the Periodic Table, unless directed otherwise by your instructor.

OBJECTIVES

1. Give the distinguishing characteristics of each of the following terms:
   (a) acid (Section 7–6)
   (b) base (Section 7–6)
   (c) salt (Section 7–6)

2. Determine the name from the formula and the formula from the name for binary compounds containing two nonmetals (Examples 7–1 and 7–2, Exercise 7–1 and 7–2, and Problems 1 and 2).

3. Determine the name from the formula and the formula from the name for binary compounds containing a metal with a fixed oxidation number and a nonmetal (Examples 7–3 and 7–4, Exercises 7–3 and 7–4, and Problems 3 and 4).

4. Determine the name from the formula and the formula from the name for binary compounds containing a metal with a variable oxidation number and a nonmetal (Examples 7–5 and 7–6, Exercises 7–5 and 7–6, and Problems 5 and 6).

5. Determine the name from the formula and the formula from the name for ternary or higher compounds (Examples 7–7 and 7–8, Exercises 7–7 and 7–8, and Problems 7 and 8).

6. Determine the name from the formula and the formula from the name for special ternary compounds (Examples 7–9 and 7–10, Exercises 7–9 and 7–10, and Problems 9 and 10).

7. Given the formulas of hydrogen compounds, write the names as pure compounds and as aqueous (water) solutions (Example 7–11, Exercise 7–11, and Problem 11).

8. Given the names of acids, write the formulas of the acids (Example 7–12, Exercise 7–12, and Problem 12).

9. Determine the name from the formula and the formula from the name for bases (Examples 7–13 and 7–14, Exercises 7–13 and 7–14, and Problems 13 and 14).

10. Given the formulas, classify various compounds as (a) an acid, (b) a base, or (c) a salt (Example 7–15, Exercise 7–15, and Problem 15).

When you first meet a person you learn his or her name. Sometimes you remember it the first time, depending on the impression that person makes on you. And sometimes you forget it and have to be reminded again and again.

In chemistry the compounds have names. You need to be able to (1) name the compound, and (2) given the name of the compound, write the formula of the compound. In this chapter we apply the names and formulas of the cations

(Table 6–1), anions (Table 6–2), and polyatomic ions (Table 6–4) to name compounds and write formulas of compounds.

There are two kinds of names in chemical nomenclature: the systematic chemical name and the common name. The systematic chemical names are used most often, but there are a few compounds whose common names still persist, such as ''water'' (H_2O) and ''ammonia'' (NH_3).

The systematic chemical naming of compounds helps you to give the correct name to the formula of the compound. In this chapter we consider only the systematic chemical names.

7–1   Systematic Chemical Names

Systematic chemical names of inorganic compounds were developed by a group of chemists who were members of the Commission on the Nomenclature of Inorganic Chemistry of the International Union of Pure and Applied Chemistry (IUPAC), which first met in 1921. They developed rules for naming all compounds as well as inorganic compounds, and they meet periodically to revise and update this nomenclature.

Study hint

The name of a compound is like the name of a person. They have a first name (positive portion) and a last name (negative portion).

The names of inorganic compounds are constructed so that every compound can be named from its formula and each formula has a name specific for that formula. The more *positive portion* (the metal, the positive polyatomic ion, the hydrogen ion, or the less electronegative nonmetal) is named and written first. The more *negative portion* (the more electronegative nonmetal or the negative polyatomic ion) is named and written last. In this discussion we divide the compounds into binary (two different elements), ternary and higher compounds (three or more different elements), special ternary compounds, acids, bases, and salts.

7–2   Binary Compounds Containing Two Nonmetals

For all binary compounds, the ending of the second element is **-ide**. When both elements are *nonmetals*, the number of atoms of *each* element is indicated by Greek prefixes, as shown in Table 7–1, except in the case of *mono-* (one), which is rarely used. When no prefix appears, one atom is assumed.

EXAMPLE 7–1

Name the following binary compounds of nonmetals.

ANSWERS

(a) PCl_3 phosphorus *tri*chloride
(b) SO_2 sulfur *dioxide*[1]
(c) CO carbon *mon*oxide[2,3]
                 (*mono-* is used in this case)
(d) N_2O_4 *di*nitrogen *tetr*oxide[4]

TABLE 7–1
GREEK PREFIXES

GREEK PREFIX	NUMBERS
mono-	1
di-	2
tri-	3
tetra-	4
penta-	5
hexa-	6
hepta-	7
octa-	8
nona-	9
(or ennea)[a]	
deca-	10

[a] The IUPAC prefers ennea- to the Latin nona-, but nona- is still widely used.

Exercise 7–1

Name the following binary compounds of nonmetals.

(a) N_2O_5 (b) N_2O_3
(c) PCl_5 (d) SF_6

EXAMPLE 7–2

Write the formulas for the following binary compounds of nonmetals.

ANSWERS

(a) carbon tetrachloride CCl_4
(b) chlorine dioxide ClO_2

---

[1] Sulfur dioxide is found in polluted air and is one of the pollutants most dangerous to human beings. It reacts with water in mucous membranes to form an acid that irritates the mucous membrane. Small amounts of sulfur dioxide in the air appear to increase the rusting of iron products. This decrease in the life of the product requires replacement earlier than expected and, therefore, increases cost. Hence, pollution becomes an economic problem as well as a health problem.

[2] When two vowels appear next to each other, as "oo" in "monooxide," or "ao" in "tetraoxide," "pentaoxide," and "heptaoxide," the vowel from the Greek prefix is dropped for easier pronunciation.

[3] Carbon monoxide, produced primarily from incomplete combustion of gasoline in automobiles is a deadly pollutant.

[4] This compound serves as the oxidizer for the fuel in the two small rocket engines in the space shuttle. These engines place the shuttle in its final orbit and later cause it to leave its orbit and return to earth. The monomer (a single unit) of this compound, NO_2 (nitrogen dioxide), together with other nitrogen oxides, is found in polluted air emitted in the exhausts of trucks and automobiles. These nitrogen oxides are formed primarily from the nitrogen and oxygen in the air as they are sucked through the hot cylinders of the engine. The brown tinge that sometimes appears in polluted air on hot days is probably caused by nitrogen dioxide.

ANSWERS

(c) dichlorine heptoxide Cl_2O_7
(d) dinitrogen oxide N_2O

Exercise 7–2

Write the formulas for the following binary compounds of nonmetals.

(a) tetraphosphorus hexoxide
(b) tetraphosphorus decoxide
(c) sulfur trioxide
(d) iodine monochloride

7–3   Binary Compounds Containing a Metal and a Nonmetal

Metals with Fixed Oxidation Numbers

In these compounds we consider first only metals with fixed oxidation numbers (metals that show only one oxidation state, as 1^+, 2^+, 3^+, etc.). In the names of these compounds, the metal is named first, followed by the stem of the nonmetal with the ending **-ide**, as in all binary compounds. *No Greek prefixes are used.*

EXAMPLE 7–3

Name the following binary compounds consisting of metals with fixed oxidation numbers and nonmetals.

ANSWERS

(a) KCl   potassium chlor*ide*
(b) Na_2S sodium sulf*ide* (no Greek prefixes)
(c) AgBr   silver brom*ide*
(d) MgO   magnesium ox*ide*

Exercise 7–3

Name the following binary compounds consisting of metals with fixed oxidation numbers and nonmetals.

(a) $BaBr_2$
(b) $AlCl_3$
(c) Li_2S
(d) $CaBr_2$

In writing the formulas of compounds, you must know the ionic charges of the metal cations and the nonmetal anions (see Tables 6–1 and 6–2). Use the Periodic Table as a help in writing these formulas.

EXAMPLE 7–4

Write the formulas of binary compounds consisting of metals with fixed oxidation numbers and nonmetals.

ANSWERS

(a) lithium fluoride   LiF (Li is 1^+; F is 1^-; see Tables 6–1 and 6–2)

(b) cadmium phosphide Cd_3P_2 (Cd is 2^+; P is 3^-)

(c) magnesium nitride Mg_3N_2 (Mg is 2^+; N is 3^-)

(d) aluminum sulfide Al_2S_3 (Al is 3^+; S is 2^-)

Exercise 7–4

Write the formulas of binary compounds consisting of metals with fixed oxidation numbers and nonmetals.

(a) aluminum oxide   (b) calcium nitride

(c) strontium chloride   (d) sodium oxide

Metals with Variable Oxidation Numbers

In this group of binary compounds, the metal has a variable oxidation number (metals that show more than one oxidation number when combined, as 1^+ *and* 2^+, 2^+ *and* 3^+, 2^+ *and* 4^+, etc.). In the names of these compounds, the same procedure is followed as with metals having fixed numbers, except that the oxidation number of the metal **must** be specified. There are two methods of specifying oxidation numbers: the newer Stock system[5] and the older *-ous* or *-ic* suffix system. In the Stock system, the oxidation number of the metal is indicated by a Roman numeral in parentheses immediately following the name of the metal. In the *-ous* or *-ic* suffix system, the Latin stem for the metal is used with *-ous* or *-ic* suffix, the *-ous* representing the *lower* oxidation number and the *-ic* the *higher* oxidation number. Table 7–2 lists some common metals with variable oxidation numbers. The names and formulas of these metals with variable oxidation numbers must be memorized.

As you may have noticed from the preceding examples, the *-ous* or *-ic* suffix system can become confusing, because in copper the *-ous* represents a 1^+ oxidation number, whereas in iron it is 2^+. The *-ous* or *ic* suffix system is still used, however, and you should become familiar with both systems.

---

[5] The IUPAC prefers the Stock system.

TABLE 7–2   SOME COMMON METALS WITH VARIABLE OXIDATION
NUMBERS: THE FORMULAS OF THE CATIONS AND THEIR NAMES

VARIABLE OXIDATION NUMBER	METAL (SYMBOL)	CATION	NAME OF CATION
1^+ and 2^+	Copper (Cu)	Cu^{1+} Cu^{2+}	Copper(I) or cuprous Copper(II) or cupric
1^+ and 2^+	Mercury (Hg)	Hg_2^{2+} (a dimer ion) Hg^{2+}	Mercury(I) or mercurous Mercury(II) or mercuric
2^+ and 3^+	Iron (Fe)	Fe^{2+} Fe^{3+}	Iron(II) or ferrous Iron(III) or ferric
2^+ and 4^+	Lead (Pb)	Pb^{2+} Pb^{4+}	Lead(II) or plumbous Lead(IV) or plumbic
2^+ and 4^+	Tin (Sn)	Sn^{2+} Sn^{4+}	Tin(II) or stannous Tin(IV) or stannic

EXAMPLE 7–5

Name the following binary compounds consisting of metals with variable oxidation numbers and nonmetals.

ANSWERS

(a) $CuCl_2$ copper(II) chlor*ide* or cup*ric* chlor*ide*; since chloride is 1^-, then copper must be 2^+.

(b) FeO   iron(II) ox*ide* or ferr*ous* ox*ide*; since oxide is 2^-, then iron must be 2^+.

(c) SnF_4 tin(IV) fluor*ide* or stann*ic* fluor*ide*; since fluoride is 1^-, then tin must be 4^+.

(d) HgO   mercury(II) ox*ide* or mercur*ic* ox*ide*; since oxide is 2^-, then mercury must be 2^+.

Exercise 7–5

Name the following binary compounds consisting of metals with variable oxidation numbers and nonmetals.

(a) CuS   (b) $FeCl_3$

(c) $PbCl_2$ (d) $FeBr_2$

EXAMPLE 7–6

Write the formulas of binary compounds consisting of metals with variable oxidation numbers and nonmetals.

ANSWERS

(a) cupric phosphide Cu_3P_2 (Cu is 2^+; P is 3^-; see Tables 6–1 and 6–2)

(b) iron(III) oxide Fe_2O_3 (Fe is 3^+; O is 2^-)

ANSWERS

(c) mercurous iodide Hg_2I_2 (Hg is Hg_2^{2+}—a dimer ion; I is 1^-)

(d) tin(IV) oxide SnO_2 (Sn is 4^+; O is 2^{-})

Exercise 7-6

Write the formulas of binary compounds consisting of metals with variable oxidation numbers and nonmetals.

(a) ferrous phosphide   (b) mercury(I) chloride

(c) copper(I) iodide   (d) stannic chloride

7-4 Ternary and Higher Compounds

In naming and writing the formulas of ternary and higher compounds, we follow the same procedure as we followed for binary compounds, except that we use the name or formula of the polyatomic ion. Hence, a knowledge of the names and formulas of all the polyatomic ions in Table 6-4 is required. Some of the negative polyatomic ions have suffixes of *-ate*, and *-ite*. The most observable difference in the formulas of the *-ate*, and the *-ite* negative polyatomic ions is that the *-ate* has *one more oxygen* atom than the *ite*.[6] For example, the formula for sulf*ite* is SO_3^{2-}, whereas that of sulf*ate* is SO_4^{2-}. This is true for all the negative polyatomic ions listed in Table 6-4. In this table, there are three polyatomic ions that do not have an *-ate* or *-ite* ending: NH_4^{1+}, ammonium ion—the only positive polyatomic ion in this table; OH^{1-}, hydroxide—an *-ide* ending, the same as binary compounds, considered further in Section 7-6; and CN^{1-}, cyanide—also has an *-ide* ending, the same as binary compounds.

For metals that have a variable oxidation number, either the Stock system or the *-ous* or *-ic* suffix may be used, but the Stock system is preferred.

EXAMPLE 7-7

Name the following ternary or higher compounds.

ANSWERS

(a) $NaNO_2$ sodium nit*rite*[7]

(b) $Cu_3(PO_4)_2$ copper(II) phosph*ate* or cup*ric* phosph*ate* (since phosphate is 3^-, then each copper is 2^+)

(c) $(NH_4)_2SO_3$ ammonium sulf*ite*

(d) $Ba(C_2H_3O_2)_2$ barium acet*ate*

---

[6] A more general method of describing the difference between the *-ate* and the *-ite* is by oxidation numbers of the nonmetal, other than oxygen. The higher oxidation number is the *-ate* and the lower oxidation number is the *-ite*. In the example of sulf*ate* and sulf*ite*, the oxidation numbers of sulfur are 6^+ and 4^+, respectively.

[7] Sodium nitrite and sodium nitrate ($NaNO_3$) are both used as color fixatives and food preserva-

Exercise 7–7

Name the following ternary compounds.

(a) $CaCO_3$

(b) $FePO_4$

(c) $LiNO_3$

(d) $CuSO_4$

EXAMPLE 7–8

Write the formulas of the following ternary compounds.

ANSWERS

(a) iron(II) phosphate $Fe_3(PO_4)_2$ (Fe is 2^+; PO_4 is 3^-; see Tables 6–1 and 6–4)

(b) cupric sulfate $CuSO_4$ (Cu is 2^+; SO_4 is 2^-)

(c) calcium permanganate $Ca(MnO_4)_2$ (Ca is 2^+; MnO_4 is 1^-)

(d) cadmium nitrate $Cd(NO_3)_2$ (Cd is 2^+; NO_3 is 1^-)

Exercise 7–8

Write the formulas of the following ternary or higher compounds.

(a) ammonium carbonate

(b) tin(IV) chromate

(c) silver acetate

(d) barium sulfite

7–5 Special Ternary Compounds

Table 6–4 lists four different polyatomic ions containing chlorine: perchlorate (ClO_4^{1-}), chlorate (ClO_3^{1-}), chlorite (ClO_2^{1-}), and hypochlorite (ClO^{1-}). We have previously mentioned (see Section 7–4) the relationship of chlorite (ClO_2^{1-}) to chlorate (ClO_3^{1-}). Hypochlorite (ClO^{1-}) is related to chlorite (ClO_2^{1-}) by one less oxygen atom.[8] The prefix *hypo-* is a Greek word meaning *under*; hence, *hypo*chlorite has one atom "under" the number of oxygen atoms of chlorite. You

---

tives in various meat products, such as frankfurters, bologna, and poultry. Their value has been questioned. The nitrite ion (NO_2^{1-}) is believed to react with organic compounds in the body to produce new compounds that may produce cancer, but recent studies indicate that this evidence is inconclusive. The nitrate ion (NO_3^{1-}) is not that harmful, but it may be converted to the nitrite ion in the body. The nitrite ion is very effective in preventing the formation of deadly botulism. Sodium hypophosphite $(NaH_2PO_2 \cdot H_2O)$ has been proposed to replace sodium nitrite as a suppressant of botulism formation in smoked meats. Sodium nitrate is used in poultry primarily for its color fixation property, in that it adds a pink color to the meat and not as a preservative. Therefore, a committee of the National Research Council recommended that sodium nitrate be eliminated from all poultry and most meat products, with the possible exceptions of fermented sausages and dry-cured meats. The presence of sodium salts as color fixatives and food preservatives also poses a problem for people who have high blood pressure or heart problems, since they are typically on restricted-sodium diets.

[8] The oxidation number of chlorine in chlorite is 3^+, whereas that of chlorine in hypochlorite is 1^+; hence, the *hypo-* means a lower or "under" oxidation number than that of the *-ite* polyatomic ion.

may remember the term *hypo* by remembering that a hypodermic needle goes under (*hypo*) the skin (*dermis*). Perchlorate (ClO_4^{1-}) is related to chlorate (ClO_3^{1-}) by one more oxygen atom.[9] The prefix *per-* can be used to mean *over*; therefore perchlorate has one atom "over" the number of oxygen atoms of chlorate. *These prefixes can also be applied to other oxyhalogen ions, such as those of bromine and iodine.* Consider some examples:

ClO_4^{1-} is perchlorate, so **BrO_4^{1-}** is *perbromate*, since Br is in the same group as Cl.

ClO_3^{1-} is chlorate, so **BrO_3^{1-}** is *bromate*, since Br is in the same group as Cl.

ClO_2^{1-} is chlorite, so **BrO_2^{1-}** is *bromite*, since Br is in the same group as Cl.

ClO^{1-} is hypochlorite, so **BrO^{1-}** is *hypobromite*, since Br is in the same group as Cl.

The same reasoning can be applied to I. Consider some examples: ClO_3^{1-} is chlorate, so **IO_3^{1-}** is *iodate* and ClO^{1-} is hypochlorite, so **IO^{1-}** is *hypoiodite*, since I is in the same group as Cl. See if you can write the formulas of the polyatomic ions periodate and iodite and relate them to their corresponding polyatomic ion containing chlorine. (Fluorine does not form polyatomic ions with oxygen due to the high electronegativity of both fluorine and oxygen; see Fig. 6–5.)

EXAMPLE 7–9

Name the following ternary or higher compounds.

ANSWERS

(a) NH_4ClO_4 ammonium *per*chlorate[10]
(b) $KBrO_2$ potassium bromite
(c) $NaClO$ sodium *hypo*chlorite[11]
(d) $Ca(IO_3)_2$ calcium iodate

Exercise 7–9

Name the following special ternary compounds.

(a) $Sr(ClO_4)_2$ (b) $LiIO_3$
(c) $Mg(BrO_3)_2$ (d) $NaClO_3$

---

[9] The oxidation number of chlorine in perchlorate is 7^+, whereas that of chlorine in chlorate is 5^+; hence, the *per-* means an "over" or a higher oxidation number than the *-ate* polyatomic ion.

[10] Ammonium perchlorate is the oxidizer in the second stage of the space shuttle. This stage uses a solid-fuel propellant that is 70 percent ammonium perchlorate. A chemical plant in Henderson Nevada, was destroyed by an ammonium perchlorate explosion in the spring of 1988.

[11] Ordinary household "chlorine bleach" is a dilute solution (about 5 percent) of sodium hypochlorite.

EXAMPLE 7–10

Write the formulas for the following special ternary compounds.

ANSWERS

(a) barium hypoiodite $Ba(IO)_2$ (Ba is 2^+: IO is 1^- as is ClO; see Tables 6–1 and 6–4)

(b) calcium perbromate $Ca(BrO_4)_2$ (Ca is 2^+; BrO_4 is 1^- as is ClO_4)

(c) potassium chlorate $KClO_3$ (K is 1^+; ClO_3 is 1^-)

(d) ferric iodate $Fe(IO_3)_3$ (Fe is 3^+; IO_3 is 1^- as is ClO_3)

Exercise 7–10

Write the formulas of the following special ternary compounds.

(a) magnesium perchlorate   (b) mercuric chlorate

(c) potassium hypochlorite   (d) sodium periodate

7–6   Acids, Bases, and Salts

Acids

In our previous discussion, we did not consider the case where the hydrogen ion (H^{1+}) replaced the metal ion or positive polyatomic ion. This is a special case. Hydrogen compounds have completely different properties if they are in the gaseous or liquid state (pure substances) than if they are dissolved in water (aqueous solution) and hence may be named differently.

In the gaseous or liquid state, the hydrogen compounds are sometimes named as hydrogen derivatives; for example, HCl is hydrogen chloride, HCN is hydrogen cyanide, and HBr is hydrogen bromide.

acids Substances that yield hydrogen ions [$H^{1+}{}_{(aq)}$] when dissolved in water.

In aqueous (water) solution, these compounds are called acids. **Acids** are hydrogen compounds that yield hydrogen ions (H^{1+}) in aqueous (water) solution. For **binary compounds**, the prefix **hydro-**, meaning hydrogen or in water, is added; the **-ide** of the anion name is replaced by **-ic acid**. Therefore, hydrogen chloride in aqueous solution is *hydrochloric acid*. The same procedure is applied to other binary compounds and also to hydrogen cyanide (HCN), which is called *hydro*cyan*ic acid* in aqueous solution.

Study hint

-ate = -ic acid; -ite = -ous acid. If you "ate" too much, you may h-*ic*-cup!

For **ternary and higher compounds**, the word "hydrogen" is dropped, and the name of the polyatomic ion is used; the **-ate** or **ite** is dropped and **-ic** or **-ous acid**, respectively, is added. Therefore, "hydrogen phosph*ate*" (H_3PO_4) in aqueous solution is phosphor*ic acid* and "hydrogen phosph*ite*" (H_3PO_3) is phosphor*ous acid*.[12] Table 7–3 summarizes these changes.

---

[12] In each of the ternary acids involving phosphorus, "or" from phosph*or*us, is reinserted in the acid name.

TABLE 7–3   SUMMARY OF THE NAMING OF BINARY AND TERNARY COMPOUNDS OF HYDROGEN IN THE GAS OR LIQUID STATE AND IN AQUEOUS SOLUTION

| GENERAL | | EXAMPLE | | |
GAS OR LIQUID	AS AQUEOUS SOLUTION	FORMULA	NAME OF GAS OR LIQUID	NAME OF AQUEOUS SOLUTION
Binary and *-ide* endings Hydrogen ____ -ide	Hydro ____ -ic acid	HCl	Hydrogen chloride	Hydrochloric acid
Ternary and *higher* Hydrogen ____ -ate	____ -ic acid	H_3PO_4	Hydrogen phosphate	Phosphoric acid
Hydrogen ____ -ite	____ -ous acid	H_3PO_3	Hydrogen phosphite	Phosphorous acid

EXAMPLE 7–11

Name the following hydrogen compounds as pure compounds and in aqueous (water) solution.

ANSWERS

	PURE COMPOUND	AQUEOUS SOLUTION
(a) HI	hydrogen iod*ide*	*hydr*iod*ic acid*[13]
(b) H_2S	hydrogen sulf*ide*	*hydro*sulfuric *acid*[14]
(c) H_2SO_4	hydrogen sulf*ate*	sulfur*ic acid*
(d) $HBrO_4$	hydrogen *per*brom*ate*	*per*brom*ic acid*[15]

Exercise 7–11

Name the following hydrogen compounds as pure compounds and in aqueous (water) solution.

(a) HBr
(c) H_2SO_3

(b) HNO_3
(d) HClO

EXAMPLE 7–12

Write the formulas of the following acids.

ANSWERS

(a) hydrofluoric acid   HF (H is 1^+; F is 1^-; see Tables 6–1 and 6–2)

---

[13] For pronunciation, the "o" in hydr*o*- is dropped when followed by a vowel.

[14] In each of the acids involving sulfur, "ur" from sulf*ur* is reinserted in the acid name.

[15] Note that the prefix of the negative polyatomic ion is carried over to the name of the aqueous solution.

ANSWERS

(b) sulfurous acid	H_2SO_3 (H is 1^+; SO_3 is 2^-; see Tables 6–1 and 6–4)
(c) chloric acid	$HClO_3$ (H is 1^+; ClO_3 is 1^-)
(d) chromic acid	H_2CrO_4 (H is 1^+; CrO_4 is 2^-)

Exercise 7–12

Write the formulas of the following acids.

(a) acetic acid

(b) perchloric acid

(c) hydriodic acid

(d) nitric acid

Bases

base A substance that yields hydroxide ions (OH^{1-}) when dissolved in water.

The compound formed with a hydroxide polyatomic ion (OH^{1-}) and a metal ion can be defined as a **base**. Even though bases are not binary compounds, they have the ending *-ide*.

EXAMPLE 7–13

Name the following bases.

ANSWERS

(a) LiOH   lithium hydroxide

(b) KOH   potassium hydroxide

(c) $Ca(OH)_2$ calcium hydroxide

(d) $Al(OH)_3$ aluminum hydroxide

Exercise 7–13

Name the following bases.

(a) NaOH

(b) $Cd(OH)_2$

(c) $Ba(OH)_2$

(d) $Zn(OH)_2$

EXAMPLE 7–14

Write the formulas of the following bases

ANSWERS

(a) ferric hydroxide $Fe(OH)_3$ (Fe is 3^+; OH is 1^-; see Tables 6–1 and 6–4)

(b) barium hydroxide $Ba(OH)_2$ (Ba is 2^+; OH is 1^-)

(c) magnesium hydroxide[16] $Mg(OH)_2$ (Mg is 2^+; OH is 1^-)

(d) lead(II) hydroxide $Pb(OH)_2$ (Pb is 2^+; OH is 1^-)

---

[16] This compound is found in the antacid and laxative, milk of magnesia.

Exercise 7–14

Write the formulas of the following bases.

(a) strontium hydroxide   (b) copper(II) hydroxide

(c) iron(II) hydroxide   (d) potassium hydroxide

Salts

A *salt* is a compound formed when *one or more* of the hydrogen ions of an acid is replaced by a cation (metal or positive polyatomic ion) *or* when one or more of the hydroxide ions of a base is replaced by an anion (nonmetal or negative polyatomic ion). Therefore, a **salt** is an ionic compound made of a positively charged ion (cation) and a negatively charged ion (anion). The binary compounds of metal cations with nonmetal anions (Section 7–3) and the ternary compounds of metal cations or ammonium ions with negative polyatomic ions (Sections 7–4 and 7–5) are examples of salts. Potassium bromide (KBr), sodium nitrate ($NaNO_3$), and ammonium sulfate [$(NH_4)_2SO_4$] are examples of salts.

salt Ionic compound made up of a positively charged ion (cation) and a negatively charged ion (anion).

Now since we have developed simplified definitions of acids, bases, and salts, you should be able to identify them, given their formulas. The following are generalized formulas for acids, bases, and salts:

HX = ACID   [H^{1+} is a hydrogen ion and X^{1-} is an anion (nonmetal ion or negative polyatomic ion) in aqueous solution.]

MOH = BASE   (M^{1+} is a metal ion and OH^{1-} is a hydroxide ion.)

MX = SALT   [M^{1+} is a cation (metal ion or positive polyatomic ion) and X^{1-} is an anion (nonmetal ion or negative polyatomic ion).]

EXAMPLE 7–15

Classify each of the following compounds as an acid, a base, or a salt. Assume that all soluble compounds are in aqueous (water) solution.

ANSWERS

(a) $HC_2H_3O_2$ acid
(b) $Ca(C_2H_3O_2)_2$ salt
(c) $Mg(OH)_2$ base
(d) MgS   salt

Exercise 7–15

Classify each of the following compounds as (1) an acid, (2) a base, or (3) a salt. Assume that all soluble compounds are in aqueous (water) solution.

(a) H_2CrO_4 (b) $K_2Cr_2O_7$

(c) $Ca(OH)_2$ (d) $NiSO_4$

Oxygen: chemistry and life on earth

The launch of the space shuttle Discovery. The combustion of liquid hydrogen and liquid oxygen to form water and energy is used to launch rockets into space (NASA).

Name: From Greek *oxy-* and French *-gen*, meaning acid producing. Many simple acids (Chapters 7 and 14) contain oxygen in the combined form.

Appearance: Colorless, odorless, and tasteless gas. Can be liquified to produce a pale blue liquid, and solidified to produce a light blue solid.

Occurrence: Oxygen is found as a diatomic gas (O_2) and also as ozone (O_3). Oxygen makes up 21 percent of dry atmospheric air as O_2, and 46 percent of the earth's crust, mostly in the form of oxides and silicates. The concentration of ozone in the atmosphere varies with location, altitude, and season, but it is typically quite small—about 1×10^{-6} percent.

Source: Obtained commercially by the large-scale liquefaction and distillation of air.

Its Role in Our World: Over half of all commercially produced oxygen is used in the steel industry. The production of steel-grade iron for various steel alloys and stainless steel requires large amounts of oxygen gas.

Oxygen is used in the production of many important industrial chemicals such as sulfuric acid (battery acid).

Medical uses of oxygen include treating people with breathing problems and surgery patients under anesthesia.

Oxygen is used as a fuel or an oxidizer for the propulsion of space vehicles. America's space shuttle is a prime example of this application.

Ozone (O_3) is used extensively in the purification of water.

Unusual Facts: Although oxygen is an absolute necessity for life on earth, there are hazards associated with it. Combustion of flammable materials proceeds explosively in the presence of pure oxygen. One such tragic consequence was the fire that destroyed an Apollo space capsule and killed three astronauts during a test years ago (January 1967).

Problems

1. Name the following binary compounds of nonmetals.

    (a) SO_3 (b) PBr_3
    (c) SiO_2 (d) N_2S_5

2. Write the formulas for the following binary compounds of nonmetals.

    (a) carbon dioxide   (b) sulfur dioxide
    (c) dichlorine monoxide   (d) disulfur decafluoride

3. Name the following binary compounds of metals with fixed oxidation numbers and nonmetals.

    (a) KI   (b) Al_2O_3
    (c) $SrCl_2$ (d) AgBr

4. Write the formulas of the following binary compounds of metals with fixed oxidation numbers and nonmetals.

    (a) potassium oxide   (b) calcium sulfide
    (c) bismuth chloride   (d) magnesium nitride

5. Name the following binary compounds of metals with variable oxidation numbers and nonmetals.

    (a) SnI_4 (b) PbS
    (c) Fe_2S_3 (d) CuBr

6. Write the formulas of the following binary compounds of metals with variable oxidation numbers and nonmetals.

    (a) iron(II) chloride   (b) stannous sulfide
    (c) tin(IV) oxide   (d) gold(III) oxide

7. Name the following ternary or higher compounds.

    (a) CuCN   (b) $Ca(HSO_4)_2$
    (c) $KMnO_4$ (d) Li_2SO_3

8. Write the formulas of the following ternary or higher compounds.

   (a) silver phosphate

   (b) copper(I) carbonate

   (c) zinc hydrogen carbonate

   (d) aluminum sulfate

9. Name the following special ternary compounds.

   (a) $NaBrO_2$

   (b) $KClO$

   (c) $KClO_3$

   (d) $Ca(IO_3)_2$

10. Write the formulas of the following special ternary compounds.

    (a) cadmium iodite

    (b) mercury(II) chlorate

    (c) copper(II) perchlorate

    (d) calcium hypochlorite

11. Name the following hydrogen compounds, as pure compounds and in aqueous solution.

    (a) HBr

    (b) $HClO_4$

    (c) $HMnO_4$

    (d) HCN

12. Write the formulas of the following acids.

    (a) nitric acid

    (b) iodic acid

    (c) sulfuric acid

    (d) dichromic acid

13. Name the following bases.

    (a) $Bi(OH)_3$

    (b) KOH

    (c) $Pb(OH)_2$

    (d) $Zn(OH)_2$

14. Write the formulas of the following bases.

    (a) cadmium hydroxide

    (b) calcium hydroxide

    (c) iron(III) hydroxide

    (d) silver hydroxide

15. Classify each of the following compounds as (1) an acid, (2) a base, or (3) a salt. Assume that all soluble compounds are in aqueous (water) solution.

    (a) $AlCl_3$

    (b) $H_2C_2O_4$

    (c) $Ca(ClO_4)_2$

    (d) $Ba(OH)_2$

General Problems

16. Write the correct formula for each of the following compounds.

    (a) stannous phosphate

    (b) barium permanganate

    (c) calcium hypoiodite

    (d) phosphorus trichloride

    (e) magnesium nitrite

    (f) iron(II) phosphate

17. Write the correct name for each of the following compounds.

    (a) $Fe(OH)_2$

    (b) B_2O_3

    (c) $Na_2C_2O_4$

    (d) $HClO_3$ in aqueous solution

    (e) $Mg(HCO_3)_2$

    (f) $Ca(CN)_2$

18. Calcium fluoride is sometimes used to produce "fluoridated" drinking water to improve the dental health of the public.

   (a) Give the formula of calcium fluoride.
   (b) Write the electronic configuration in sublevels for the calcium and fluoride ions ($^{40}_{20}Ca$ and $^{19}_{9}F$).
   (c) Classify calcium fluoride as (1) an acid, (2) a base, or (3) a salt.

Answers to Exercises

7-1. (a) dinitrogen pentoxide; (b) dinitrogen trioxide; (c) phosphorus pentachloride; (d) sulfur hexafluoride

7-2. (a) P_4O_6; (b) P_4O_{10}; (c) SO_3; (d) ICl

7-3. (a) barium bromide; (b) aluminum chloride; (c) lithium sulfide; (d) calcium bromide

7-4. (a) Al_2O_3; (b) Ca_3N_2; (c) $SrCl_2$; (d) Na_2O

7-5. (a) copper(II) sulfide or cupric sulfide; (b) iron(III) chloride or ferric chloride; (c) lead(II) chloride or plumbous chloride; (d) iron(II) bromide or ferrous bromide

7-6. (a) Fe_3P_2; (b) Hg_2Cl_2 [mercury(I) exists as a dimer ion, Hg_2^{2+}; see Table 6-1]; (c) CuI; (d) $SnCl_4$

7-7. (a) calcium carbonate: (b) iron(III) phosphate or ferric phosphate; (c) lithium nitrate; (d) copper(II) sulfate or cupric sulfate

7-8. $(NH_4)_2CO_3$; (b) $Sn(CrO_4)_2$; (c) $AgC_2H_3O_2$; (d) $BaSO_3$

7-9. (a) strontium perchlorate; (b) lithium iodate; (c) magnesium bromate; (d) sodium chlorate

7-10. (a) $Mg(ClO_4)_2$; (b) $Hg(ClO_3)_2$; (c) KClO; (d) $NaIO_4$

7-11. (a) hydrogen bromide, hydrobromic acid; (b) hydrogen nitrate, nitric acid; (c) hydrogen sulfite, sulfurous acid; (d) hydrogen hypochlorite, hypochlorous acid

7-12. (a) $HC_2H_3O_2$; (b) $HClO_4$; (c) HI; (d) HNO_3

7-13. (a) sodium hydroxide; (b) cadmium hydroxide; (c) barium hydroxide; (d) zinc hydroxide

7-14. (a) $Sr(OH)_2$; (b) $Cu(OH)_2$; (c) $Fe(OH)_2$; (d) KOH

7-15. (a) (1), acid; (b) (3), salt; (c) (2), base; (d) (3), salt

QUIZ   You may use the Periodic Table.

1. Write the correct name for each of the following compounds:

   (a) LiBr   (b) AlI_3

    (c) PbS_2 (d) $SnSO_4$

    (e) P_2S_5 (f) $Ca(ClO_2)_2$

    (g) $Sn(BrO_3)_2$ (h) NH_4NO_3

2. Write the correct formula for each of the following compounds:

    (a) barium oxide   (b) calcium phosphide

    (c) iron(III) sulfide   (d) stannous fluoride

    (e) phosphoric acid   (f) aluminum hydroxide

    (g) tin(II) chromate   (h) calcium oxalate

3. Classify each of the following compounds as (1) an acid, (2) a base, or (3) a salt. Assume that all soluble compounds are in aqueous (water) solution.

    (a) $SnCO_3$ (b) $H_2C_2O_4$

    (c) H_3BO_3 (d) $Fe(OH)_3$

8

CHEMICAL CALCULATIONS

COUNTDOWN

5 Determine the number of significant digits in the following numbers (Section 1–1).

(a) 0.00765   (3);   (b) $76\overline{0}$ (3)

4 Round off the following numbers to three significant digits (Section 1–2).

(a) 27.48   (27.5)   (b) 27.45   (27.4)

3 Perform the indicated mathematical operations and express your answer to the proper number of significant digits (Section 1–2).

(a) 4.74
2.752
1.657
                     (b) $\dfrac{4.85 \times 3.278}{6.721} =$

(9.15)   (2.37)

2 Carry out the operations indicated on the following exponential numbers. Express your answer to three significant digits in scientific notation (Sections 1–1, 1–2, 1–3, and 1–4).

(a) $3.25 \times 10^{23} \times 1.45 \times 10^{-6} =$

(4.71×10^{17})

(b) $\dfrac{16.02 \times 10^{12}}{1.21 \times 10^{3}} =$

(1.32×10^{10})

1 Convert 86.5 mg to kilograms (kg). Express your answer in scientific notation (Sections 1–4 and 2–3).

$(8.65 \times 10^{-5} \text{ kg})$

TASKS

1 Memorize Avogadro's number, 6.02×10^{23} (Section 8–2).

2 Memorize the molar volume of a gas: 1 mol (6.02×10^{23} molecules) of an ideal gas occupies a volume of **22.4 liters** at 0°C and $76\overline{0}$ torr (**STP** conditions, Section 8–3).

OBJECTIVES

1 Give the distinguishing characteristics of each of the following terms:
   (a) mole (Section 8–2)
   (b) Avogadro's number (Section 8–2)
   (c) molar mass (Section 8–2)
   (d) molar volume of a gas (Section 8–3)
   (e) empirical formula (Section 8–5)
   (f) molecular formula (Section 8–5)

2 Given the formula of a compound and the Table of Approximate Atomic Masses (inside the back cover), calculate the formula mass or molecular mass of the compound (Example 8–1, Exercise 8–1, Problem 1).

3 Given the Table of Approximate Atomic Masses and the formula of a substance, interconvert among the mass of the substance, the number of particles of a substance, and the number of moles of a substance (Examples 8–2, 8–3, 8–4, 8–5, and 8–6, Exercises 8–2, 8–3, and 8–4, Problems 2, 3, and 4).

4 Given the number of moles of a compound, its formula, and the Table of Approximate Atomic Masses, calculate the number of moles of a given element in the compound and the mass of the element in the compound (Examples 8–7, Exercise 8–5, Problem 5).

5 Given the Table of Approximate Atomic Masses and the formula of a gas, interconvert among the mass of the gas, the volume of the gas, and the number of moles of the gas (Examples 8–8, and 8–9, Exercises 8–6 and 8–7, Problems 6 and 7).

6 Given the volume of a gas at STP and its mass, calculate the molecular mass of the gas (Example 8–10. Exercise 8–8, Problem 8).

7 Given the formula of a gas and the Table of Approximate Atomic Masses, calculate the density of the gas at STP (Example 8–11, Exercise 8–9, Problem 9).

8 Given the formula of compound and the Table of Approximate Atomic Masses or the experimental analysis of a compound, calculate the percent composition of the compound (Examples 8–12 and 8–13, Exercises 8–10 and 8–11, Problems 10 and 11).

9 Given the percent composition of a compound and the Table of Approximate Atomic Masses, calculate the empirical formula for the compound (Examples 8–14 and 8–15, Exercise 8–12, Problem 12).

10 Given the percent composition of a compound, the Table of Approximate Atomic Masses, and the molecular mass, calculate the molecular formula for the compound (Example 8–16, Exercise 8–13, Problem 13).

Chemists are assigned various tasks. Examples of some of these tasks might be (1) determining the substance(s) present in a natural product that slows down the growth of cancer cells, (2) preparing substances similar to the cancer-inhibiting substance, (3) developing the formulation for a new antibiotic, (4) identifying the minute amounts of compounds found in the upper atmosphere, and many others. In the process of doing these tasks chemists need to make calculations. In this chapter we will consider some of the basic calculations made by chemists.

Previously, we considered a general description of elements and compounds with few quantitative (how much?) calculations. We will now consider quantitative calculations involving elements and compounds. In our calculations we will use the *factor-unit* method in problem solving introduced in Section 2–3. We suggest that you review this method.

> **Study hint**
>
> Review Section 2–3 in Chapter 2 before you read further.

8–1 Calculation of Formula or Molecular Masses

In Section 3–7 we identified the subscripts in a formula of a compound as representing the number of atoms of the respective elements in a molecule or formula unit of the compound. For example, in a molecule of glucose (dextrose $C_6H_{12}O_6$), there are 6 atoms of carbon, 12 atoms of hydrogen, and 6 atoms of oxygen. In Section 4–1 we introduced the atomic mass scale based on an arbitrarily assigned value of exactly 12 amu for ^{12}C. The atomic masses of all the elements are found inside the front cover and approximate atomic masses of the elements are found inside the back cover. The formula masses or molecular masses of compounds can be calculated from the atomic masses of the elements.

The term *formula mass* is used for compounds that are represented as *formula units*—that is, the compound exists as ions and has primarily ionic bonding (see Section 6–4). The term *molecular mass* is applied to compounds that exist as *molecules* and have primarily covalent bonding (see Section 6–5). *The methods for calculating formula masses and molecular masses are the same.*

EXAMPLE 8–1

Calculate the formula mass of Na_2SO_4.[1]

SOLUTION In this formula unit, there are 2 atoms of Na, 1 atom of S, and 4 atoms of O. Hence, the formula mass is calculated as follows:

---

[1] In all calculations involving atomic masses, the Table of Approximate Atomic Masses inside the back cover will be used, unless otherwise stated. Refer to this table for the atomic masses of the elements in solving problems.

$$2 \; \text{atoms Na} \times \frac{23.0 \; \text{amu}}{1 \; \text{atom Na}} = 46.0 \; \text{amu}$$

$$1 \; \text{atom S} \times \frac{32.1 \; \text{amu}}{1 \; \text{atom S}} = 32.1 \; \text{amu}$$

$$4 \; \text{atoms O} \times \frac{16.0 \; \text{amu}}{1 \; \text{atom O}} = 64.0 \; \text{amu}$$

Formula mass Na_2SO_4 = $\overline{142.1 \; \text{amu}}$ *Answer*

The answer is expressed to the smallest place present in all the numbers that are added (see Section 1–2), which, in this example, is the tenths decimal place. The calculation can be simplified as follows:

$$2 \times 23.0 \; \text{amu} = 46.0 \; \text{amu}$$
$$1 \times 32.1 \; \text{amu} = 32.1 \; \text{amu}$$
$$4 \times 16.0 \; \text{amu} = 64.0 \; \text{amu}$$

Formula mass Na_2SO_4 = $\overline{142.1 \; \text{amu}}$ *Answer*

Exercise 8–1

Calculate the molecular mass of water (H_2O). *Hint:* The same method is used in solving for molecular masses as is used in solving for formula masses.

8–2 Calculation of Moles of Units. Avogadro's Number

mole (mol) The amount of a substance that contains the same number of atoms, formula units, molecules, or ions as there are atoms in exactly 12 grams of carbon-12 (approximately 6.02×10^{23} atoms).

Avogadro's (a′vō·ga′drō) **number** (*N*) The number of atoms in exactly 12 grams of carbon-12 (approximately 6.02×10^{23}); it is equivalent to 1 mol of a substance.

In our discussion of atomic masses, the standard used was the ^{12}C isotope. We also use ^{12}C to define a new term–the mole. The **mole (mol)** is the *amount of a substance containing the *same number of* atoms, formula units, molecules, or ions as there are atoms in *exactly 12 g of carbon-12*. Now this poses another question: How many atoms are there in exactly 12 g of ^{12}C? Experimentally, by X-ray diffraction and other methods, the number of atoms in exactly 12 g of ^{12}C has been found to be 6.02×10^{23} atoms. This number,[2] 6.02×10^{23} or 602,000,000,000,000,000,000,000, is called **Avogadro's** (ä′vō·gä′drō) **number** (*N*) and is named in honor of the Italian physicist and chemist Amedeo Avogadro (1776–1856).

Figure 8–1 may be of some help in understanding the meaning of this extremely large number. *There are 6.02×10^{23} atoms of ^{12}C in 1 mol of ^{12}C atoms, and this number of atoms of ^{12}C has a mass of exactly 12 g, the atomic mass for carbon expressed in grams* (Fig. 8–2). For 1 mol or 6.02×10^{23} oxygen atoms (note the same number of oxygen atoms as of carbon), there is a mass of

---

[2] Avogadro's number has been measured to seven significant digits, 6.022045×10^{23}, but we will use only three significant digits in this text.

| 0.1 million | 1.0 million | 1.5 million | 1.7 million |
| (100,000) years | (1,000,000) years | (1,500,000) years | (1,700,000) years |

Figure 8-1 *Counting Avogadro's number,* N, *of peas. If all the people now alive on the earth (5.6 billion) started counting Avogadro's number,* N, *of peas at a rate of two peas per second, it would take approximately 1.7 million (1,700,000) years. That is a lot of peas!*

16.0 g (to three significant digits), the atomic mass for oxygen expressed in grams. You should note that for the same number of atoms, oxygen has a greater mass since an oxygen atom is heavier than a carbon atom. The same statement can be made for any element: 1 *mol of atoms of any element contains 6.02 ×* 10^{23} *atoms of the element and is equal to the atomic mass of the element expressed in grams.*

| 1 mol of atoms | | 6.02 × 10^{23} atoms | | atomic mass of the |
| of an element | = | of element | = | element in grams |

The mole has been referred to as "the chemist's dozen." It is a quantity of matter based on a certain number (6.02 × 10^{23}) of elementary units per mole, just as a dozen is defined as 12 units per dozen or a gross as 144 units per gross. The total mass of a substance in a mole of the substance will be related to the mass of each individual elementary unit of the substance and could be

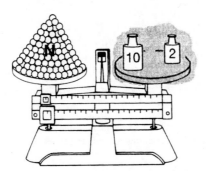

Figure 8-2 *One mole of* ^{12}C *atoms (6.02 × 10^{23} atoms,* N) *has a mass of exactly 12 g*

expressed as *amu* for 1 atom or molecule *or* more conveniently as *grams* for 1 mol of the substance (see Table 8–1).

The reasoning we just applied to atoms of an element can also be applied to formula units, and to molecules of a compound. Therefore, in *1 mol of a compound there are 6.02 × 10²³ formula units or molecules, and the number of formula units or molecules has a mass equal to the molecular or formula mass expressed in grams* (see Table 8–1).

1 mol of a compound	=	6.02 × 10²³ formula units or molecules of the compound	=	formula or molecular mass in grams

For 1 mol (6.02 × 10²³ molecules) of water (H_2O), there is a mass of 18.0 g ($2 \times 1.0 + 1 \times 16.0 = 18.0$ amu, to the tenths place), the molecular mass of water expressed in grams. For 1 mol (6.02 × 10²³ formula units) of sodium sulfate (Na_2SO_4), there is a mass of 142.1 g ($2 \times 23.0 + 1 \times 32.1 + 4 \times 16.0 = 142.1$ amu, to the tenths place), the formula mass of sodium sulfate expressed in grams.

molar mass The mass in grams of one mole of a substance.

The formula mass or molecular mass expressed in grams has a special name, the molar mass. The **molar mass** is the mass in *grams* of *one* mole of a substance. For sodium sulfate (Na_2SO_4), the molar mass is 142.1 g (see Example 8–1).

We can apply the same reasoning to *ions* or *any* particle. Therefore, in *1 mol of ions there are 6.02 × 10¹² ions, and this number of ions has a mass equal to the atomic or formula mass of the ion expressed in grams.* For 1 mol or 6.02 × 10²³ sodium ions (Na^{1+}), there is a mass of 23.0 g (to three significant digits), the atomic mass of sodium expressed in grams. For 1 mol or 6.02 × 10²³ sulfate ions (SO_4^{2-}), there is a mass of 96.1 g ($1 \times 32.1 + 4 \times 16.0 = 96.1$ amu to the tenths place), the formula mass of sulfate ion expressed in grams.

1 mol of ions	=	6.02 × 10²³ ions	=	atomic or formula mass of the ions in grams

TABLE 8–1   THE MOLE AND ITS EQUIVALENT IN MASS

UNIT	MASS OF UNIT (amu)	UNITS IN 1 MOL	GRAMS IN 1 MOL
H (atom)	1.0	6.02 × 10²³	1.0
H_2 (molecule)	2.0	6.02 × 10²³	2.0
N (atom)	14.0	6.02 × 10²³	14.0
N_2 (molecule)	28.0	6.02 × 10²³	28.0
H_2SO_4	98.1	6.02 × 10²³	98.1
NaCl	58.5	6.02 × 10²³	58.5

EXAMPLE 8–2

Calculate the number of moles in 47.0 g of sodium sulfate (Na_2SO_4).

SOLUTION The formula mass of sodium sulfate is 142.1 amu (see Example 8–1). Therefore, 1 mol of sodium sulfate = 142.1 g (molar mass), and the number of moles of sodium sulfate is calculated as

$$47.0 \ \text{g Na}_2\text{SO}_4 \times \frac{1 \ \text{mol Na}_2\text{SO}_4}{142.1 \ \text{g Na}_2\text{SO}_4} = 0.331 \ \text{mol Na}_2\text{SO}_4 \qquad Answer$$

EXAMPLE 8–3

Calculate the number of moles in 1.25×10^{23} formula units of sodium sulfate (Na_2SO_4).

SOLUTION There are 6.02×10^{23} formula units in 1 mol of formula units of a compound; hence, the number of moles of sodium sulfate is calculated as

$$1.25 \times 10^{23} \ \text{formula units Na}_2\text{SO}_4 \times \frac{1 \ \text{mol Na}_2\text{SO}_4}{6.02 \times 10^{23} \ \text{formula units Na}_2\text{SO}_4}$$

$$= 0.208 \ \text{mol Na}_2\text{SO}_4 \qquad Answer$$

Exercise 8–2

Calculate the number of moles of water (H_2O) in the following.

(a) 35.0 g of H_2O (b) 3.56×10^{23} molecules of H_2O

EXAMPLE 8–4

Calculate the number of grams in 0.425 mol of sodium sulfate (Na_2SO_4).

SOLUTION

$$0.425 \ \text{mol Na}_2\text{SO}_4 \times \frac{142.1 \ \text{g Na}_2\text{SO}_4}{1 \ \text{mol Na}_2\text{SO}_4} = 60.4 \ \text{g Na}_2\text{SO}_4 \qquad Answer$$

EXAMPLE 8–5

Calculate the number of formula units in 0.275 mol of sodium sulfate (Na_2SO_4).

SOLUTION

$$0.275 \ \text{mol Na}_2\text{SO}_4 \times \frac{6.02 \times 10^{23} \ \text{formula units Na}_2\text{SO}_4}{1 \ \text{mol Na}_2\text{SO}_4}$$

$$= 1.66 \times 10^{23} \ \text{formula units Na}_2\text{SO}_4 \qquad Answer$$

Exercise 8–3

Given 0.745 mol of water (H_2O), calculate the following.

(a) the number of grams of water
(b) the number of molecules of water

EXAMPLE 8–6

Calculate the number of formula units of sodium sulfate (Na_2SO_4) in 24.5 g of sodium sulfate.

SOLUTION   First, convert the grams to moles since 1 mol of Na_2SO_4 = 142.1 g (molar mass). Then convert the moles to formula units since there are 6.02×10^{23} formula units in 1 mol of any compound. Therefore,

$$24.5 \text{ g } Na_2SO_4 \times \frac{1 \text{ mol } Na_2SO_4}{142.1 \text{ g } Na_2SO_4} \times \frac{6.02 \times 10^{23} \text{ formula units } Na_2SO_4}{1 \text{ mol } Na_2SO_4}$$

$$= 1.04 \times 101^{23} \text{ formula units } Na_2SO_4 \quad \textit{Answer}$$

Exercise 8–4

Calculate the number of molecules of water (H_2O) in 42.5 g of water.

We stated that in the formulas of compounds, the subscripts represent the number of atoms of each element in a formula unit or molecule of the compound (see Section 3–7). These subscripts also represent the ***number of moles of atoms*** of the elements in ***1 mol of molecules or formula units*** of the compound. Let us consider the case of water (H_2O), consisting of 2 atoms of hydrogen and 1 atom of oxygen. Water has a molecular mass of 18.0 amu, consisting of 2.0 amu of hydrogen atoms and 16.0 amu of oxygen atoms. One mole of water molecules has a mass of 18.0 g (molar mass), consisting of 2.0 g of hydrogen atoms and 16.0 g of oxygen atoms. Hence, the moles of atoms of each element in ***1 mol*** of water molecule is

$$2.0 \text{ g } H \text{ atoms} \times \frac{1 \text{ mol H atoms}}{1.0 \text{ g } H \text{ atoms}} = 2 \text{ mol H atoms}$$

$$16.0 \text{ g } O \text{ atoms} \times \frac{1 \text{ mol O atoms}}{16.0 \text{ g } O \text{ atoms}} = 1 \text{ mol O atoms}$$

Therefore, in ***1 mol*** of molecules or formula units of a compound, the subscripts represent the number of moles of atoms of elements present.

A simple analogy may be helpful. Each person normally has 2 arms and 1 head. If we have a dozen people, we would have 2 dozen arms and 1 dozen heads. The ratio of arms to head (2 to 1) is the same for one person as it is for a group of people—a dozen. Therefore, *one person* is analogous to *one molecule* or *formula unit*, and *one dozen* people is analogous to *one mole* of molecules or formula units.

EXAMPLE 8–7

Given 0.825 mol of sodium sulfate (Na_2SO_4), calculate the following.

(a) the number of moles of sodium ions

SOLUTION   Since there are two Na^{1+} ions in one formula unit of Na_2SO_4, there will be 2 mol of Na^{1+} ions in 1 mol of Na_2SO_4. Therefore,

$$0.825 \; \cancel{\text{mol Na}_2\text{SO}_4} \times \frac{2 \; \text{mol Na}^{1+}}{1 \; \cancel{\text{mol Na}_2\text{SO}_4}} = 1.65 \; \text{mol Na}^{1+} \qquad \textit{Answer}$$

(The moles of atoms or ions in 1 mol of formula units or molecules are regarded as exact values and are not considered in computing significant digits.)

(b) the number of grams of sodium ions

> **SOLUTION** In 1 mol of Na^{1+}, there are 23.0 g. Therefore,

$$0.825 \; \cancel{\text{mol Na}_2\text{SO}_4} \times \frac{2 \; \cancel{\text{mol Na}^{+}}}{1 \; \cancel{\text{mol Na}_2\text{SO}_4}} \times \frac{23.0 \; \text{g Na}^{1+}}{1 \; \cancel{\text{mol Na}^{+}}} = 38.0 \; \text{g Na}^{1+}$$

Answer

Exercise 8–5

Given 0.125 mol of water (H_2O), calculate the following.

(a) the number of moles of hydrogen atoms
(b) the number of grams of hydrogen atoms

8–3 Molar Volume of a Gas and Related Calculations

For a gas, under conditions where it behaves ideally, it has been experimentally determined that 6.02×10^{23} molecules (Avogadro's number, N) of a gas or 1 mol of *gas molecules* occupies a volume of 22.4 liters at a temperature of 0°C (273 K) and a pressure of $76\bar{0}$ torr (mm Hg).[3] The conditions of 0°C and $76\bar{0}$ torr are defined as *standard temperature and pressure* (STP). This volume of **22.4 L** occupied by 1 mol of any **gas molecules** at *0°C* and $76\bar{0}$ torr is called the **molar volume of a gas** and is approximately the volume occupied by three standard basketballs (see Fig. 8–3). This molar volume of a gas relates the mass of a gas to its volume at STP and can be used in various types of calculations.

molar volume of a gas The volume occupied by 1 mol of any gas, 22.4 L of gas molecules at 0°C and $76\bar{0}$ torr.

Moles or Mass and Volume

EXAMPLE 8–8

Given 5.60 liters of oxygen gas (O_2) at STP, calculate the following.

(a) the number of moles of oxygen

> **SOLUTION** One mole of O_2 at STP occupies a volume of 22.4 L. Therefore, the number of moles of O_2 molecules in 5.60 L is

---

[3] In Chapter 12 we consider the units of pressure and evaluate the effect of temperature and pressure on the volume of a gas.

Figure 8–3 *Molar volume of a gas. 6.02 × 10²³ molecules of any gas (1 mol of gas molecules) occupies a volume of 22.4 liters at 0°C and 76̄0 torr (STP conditions). This volume is approximately the volume occupied by* **three** *standard basketballs.*

calculated as

$$5.60 \text{ L } \cancel{O_2} \times \frac{1 \text{ mol } O_2}{22.4 \text{ L } \cancel{O_2}} = 0.250 \text{ mol } O_2 \qquad Answer$$

(b) the number of grams of oxygen

SOLUTION One mole of O_2 has a mass of 32.0 g (2 × 16.0 = 32.0 amu); therefore, the mass in grams in 5.60 L of oxygen at STP can be calculated as

$$5.60 \text{ L } \cancel{O_2} \times \frac{1 \text{ mol } \cancel{O_2}}{22.4 \text{ L } \cancel{O_2}} \times \frac{32.0 \text{ g } O_2}{1 \text{ mol } \cancel{O_2}} = 8.00 \text{ g } O_2 \qquad Answer$$

Exercise 8–6

Given 4.65 L of nitrogen gas (N_2) at STP, calculate the following.

(a) the number of moles of nitrogen
(b) the number of grams of nitrogen

EXAMPLE 8–9

Calculate the volume in liters at STP that 12.0 g of oxygen gas (O_2) would occupy.

SOLUTION One mole of O_2 has a mass of 32.0 g (molar mass); therefore, the volume in liters of 12.0 g oxygen gas at STP can be calculated as

$$12.0 \text{ g } \cancel{O_2} \times \frac{1 \text{ mol } \cancel{O_2}}{32.0 \text{ g } \cancel{O_2}} \times \frac{22.4 \text{ L } O_2 \text{ STP}}{1 \text{ mol } \cancel{O_2}} = 8.40 \text{ L } O_2 \text{ STP} \qquad Answer$$

Exercise 8–7

Calculate the volume at STP that 22.8 g of nitrogen gas (N_2) would occupy.

Molecular Mass (Molar Mass)

The molecular mass or molar mass of a gas can be calculated by solving for *grams per mole* of the gas, which is numerically equal to the molecular mass in *amu*.

EXAMPLE 8-10

Calculate the molecular mass of a gas if 5.00 L measured at STP has a mass of 9.85 g.

SOLUTION Solving for grams per mole, we calculate the molecular mass as

$$\frac{9.85 \text{ g}}{5.00 \text{ L STP}} \times \frac{22.4 \text{ L STP}}{1 \text{ mol}} = 44.1 \text{ g/mol}$$

Molecular mass = 44.1 amu   *Answer*

Exercise 8-8

Calculate the molecular mass of a gas if 7.25 L measured at STP has a mass of 8.75 g.

Density

The density of a gas is another property that can be determined from the molar volume of a gas. The density of any gas at STP can be directly calculated from the molecular mass of a gas by solving for the number of grams per liter of the gas.

EXAMPLE 8-11

Calculate the density of oxygen gas (O_2) at STP.

SOLUTION The units of density for a gas are grams per liter. Hence, from the molecular mass of O_2 (32.0 amu), the density can be calculated as

$$\frac{32.0 \text{ g } O_2}{1 \text{ mol } O_2} \times \frac{1 \text{ mol } O_2}{22.4 \text{ L STP}} = 1.43 \text{ g/L at STP}   \textit{Answer}$$

Exercise 8-9

Calculate the density of nitrogen gas (N_2) at STP.

8-4 Calculation of Percent Composition of Compounds

Percent means parts per hundred. For example, if your college has a student enrollment of 1000 and there are 400 men students, the percent of men students is 40 ($\frac{400}{1000} \times 100 = 40$ percent) or 40 (men students) *per hundred* (students). In the same manner, the percent composition of each element in a compound can be calculated. The exact numbers, such as 400 and 1000, may or may not be given. If not, then the *formula* will be given and from it the molecular or formula mass can be calculated; thus, the percent composition of each element in the compound can be calculated. Any units such as amu, g, and lb may be assigned to the molecular or formula mass, as long as the same units are used throughout the entire calculation.

EXAMPLE 8–12

Calculate the percent composition of ethyl chloride (C_2H_5Cl).

SOLUTION   The molecular mass of C_2H_5Cl is calculated as 64.5 amu.

$$2 \times 12.0 = 24.0 \text{ amu}$$
$$5 \times 1.0 = 5.0 \text{ amu}$$
$$1 \times 35.5 = \underline{35.5 \text{ amu}}$$
$$\text{Molecular mass of } C_2H_5Cl = \overline{64.5 \text{ amu}}$$

The percentage of each element in the compound is calculated by dividing the contribution of each element (amu) by the molecular mass (amu) and multiplying by 100:

Percent carbon: $\dfrac{24.0 \text{ amu}}{64.5 \text{ amu}} \times 100 = 37.2\%\text{C}$

Percent hydrogen: $\dfrac{5.0 \text{ amu}}{64.5 \text{ amu}} \times 100 = 7.8\%\text{H}$

Percent chlorine: $\dfrac{35.5 \text{ amu}}{64.5 \text{ amu}} \times 100 = 55.0\%\text{Cl}$ *Answers*

Exercise 8–10

Calculate the percent composition of glucose ($C_6H_{12}O_6$).

EXAMPLE 8–13

Calculate the percent of metal in an oxide if 0.400 g of metal combines with 0.365 g of oxygen to form the oxide.

SOLUTION   The total mass of the oxide is 0.765 g (0.400 g + 0.365 g). The percent of the metal is calculated by dividing the mass of the metal in grams by the total mass of the oxide and multiplying by 100 to find percent:

$$\frac{0.400 \text{ g}}{0.765 \text{ g}} \times 100 = 52.3\% \text{ metal}   \textit{Answer}$$

Exercise 8–11

Calculate the percent of metal in an oxide if 1.64 g of metal combines with 0.450 g of oxygen to form the oxide.

8–5 Calculation of Empirical (Simplest) and Molecular Formulas

The examples and exercises in Section 8–4 give the formulas of the compounds and ask you to calculate the percent composition. The examples and exercises in this section provide the percent composition of the compounds and ask you to determine the formulas.

empirical formula The formula of a compound that contains the smallest whole number ratio of atoms present in a molecule or formula unit of a compound.

The **empirical formula** (simplest formula) of a compound is the formula containing the *smallest whole number ratio of the atoms* that are present in a molecule or formula unit of the compound. This empirical formula is found from the percent composition of the compound, which is determined *experimentally* from analysis of the compound in the laboratory. The empirical formula gives only the ratio of the atoms present expressed as *small whole numbers*.

molecular formula A formula composed of an appropriate number of symbols of elements representing *one* molecule of the given compound. Also defined as the formula containing the *actual* number of atoms of each element in *one* molecule of the compound.

The **molecular formula** of the compound is the formula containing the *actual* number of atoms of each element present in one molecule of the compound. The molecular formula is a *whole number multiple* of the empirical formula. The empirical formula of glucose (a sugar) is CH_2O and the molecular formula of glucose is $6 \times (CH_2O) = C_6H_{12}O_6$. Thus, the *ratio* of carbon : hydrogen : oxygen in glucose is $1:2:1$, but an actual glucose molecule contains 6 carbons, 12 hydrogens, and 6 oxygens. The molecular formula is determined from the empirical formula *and* the molecular mass of the compound, which may be experimentally measured by various methods (see Example 8–16).

A simple analogy may help to illustrate these two types of formulas. In your college, the ratio of men to women may be $1:2$ (empirical formula), but the actual number of men to women may be $400:800$ (molecular formula). In the case of hydrogen peroxide, the empirical formula is HO (1 atom H : 1 atom O), but the molecular formula is H_2O_2 (2 atoms H : 2 atoms O).

In some cases, both the empirical and molecular formulas are the same, as in the case of H_2O. The formulas of compounds existing as *molecules* (covalent compounds) are always referred to as *molecular formulas*. For those compounds existing as *formula units* (ionic compounds), there are no molecular formulas, because these compounds do not exist as molecules. Hence, their formulas are called *empirical formulas*.

EXAMPLE 8–14

Calculate the empirical formula for the compound containing 32.4% sodium, 22.6% sulfur, and 45.1% oxygen.[4]

---

[4] The difference of 0.1% between 100.1% (32.4 + 22.6 + 45.1) and exactly 100% emphasizes the experimental nature of the data for this type of problem and is due to experimental error.

SOLUTION   Remember that percent means *parts per hundred*. Thus, in 100 g of the compound there would be 32.4 g of Na, 22.6 g of S, and 45.1 g of O. The first step is to calculate the **moles** *of atoms* of each element present, as follows:

$$32.4 \text{ g Na} \times \frac{1 \text{ mol Na atoms}}{23.0 \text{ g Na}} = 1.41 \text{ mol Na atoms}$$

$$22.6 \text{ g S} \times \frac{1 \text{ mol S atoms}}{32.1 \text{ g S}} = 0.704 \text{ mol S atoms}$$

$$45.1 \text{ g O} \times \frac{1 \text{ mol O atoms}}{16.0 \text{ g O}} = 2.82 \text{ mol O atoms}$$

The elements are combined in a ratio of 1.41 mol of Na atoms to 0.704 mol of S atoms to 2.82 mol of O atoms as $Na_{1.41 \text{ mol of atoms}}$, $S_{0.704 \text{ mol of atoms}}$, $O_{2.82 \text{ mol of atoms}}$. The empirical formula must express these relationships in terms of *small whole numbers*.

The second step, then, is to express these relationships in **small whole numbers** by dividing each value by the *smallest* one, as follows:

$$\text{For Na: } \frac{1.41}{0.704} = 2.00 = 2$$

$$\text{For S: } \frac{0.704}{0.704} = 1.00 = 1$$

$$\text{For O: } \frac{2.82}{0.704} = 4.01 \approx 4$$

Hence, the elements are combined in a ratio of 2 mol of Na atoms to 1 mol of S atoms to 4 mol of O atoms, and the empirical formula is Na_2SO_4.

Answer

EXAMPLE 8–15

Calculate the empirical formula for a compound with a composition of 26.6% potassium, 35.4% chromium, and 38.1% oxygen.

SOLUTION   In 100 g of the compound there would be 26.6 g of K, 35.4 g of Cr, and 38.1 g of O. The first step is to calculate the **moles** *of atoms* of each element present, as follows:

$$26.6 \text{ g K} \times \frac{1 \text{ mol K atoms}}{39.1 \text{ g K}} = 0.680 \text{ mol K atoms}$$

$$35.4 \text{ g Cr} \times \frac{1 \text{ mol Cr atoms}}{52.0 \text{ g Cr}} = 0.681 \text{ mol Cr atoms}$$

$$38.1 \text{ g O} \times \frac{1 \text{ mol O atoms}}{16.0 \text{ g O}} = 2.38 \text{ mol O atoms}$$

Second, reduce these values to **simpler numbers** by dividing each one by the *smallest* value, as follows:

$$\text{For K:} \frac{0.680}{0.680} = 1$$

$$\text{For Cr:} \frac{0.681}{0.680} = 1.00 \approx 1$$

$$\text{For O:} \frac{2.38}{0.680} = 3.5$$

These relative ratios may be converted to small whole numbers by **multiplying by 2**; the empirical formula is then $K_2Cr_2O_7$.   *Answer*

If the ratio of these numers ends in 0.5, then *all* the numbers must be multiplied by **2** to obtain whole numbers. If the ratio of these numbers ends in 0.33 (0.33 . . . , the fraction $\frac{1}{3}$), then *all* the numbers must be multiplied by *3* to obtain whole numbers.

Exercise 8–12

Calculate the empirical formula for a compound with a composition of 52.9% aluminum and 47.1% oxygen.

EXAMPLE 8–16

Calculate the molecular formula of a compound from the following experimental data: 30.4% nitrogen, 69.6% oxygen, and molecular mass of 91.0 amu.

SOLUTION   The empirical formula is calculated from the percent composition by first calculating the ***moles** of atoms* of nitrogen and oxygen in $10\overline{0}$ g of the compound:

$$30.4 \cancel{\text{ g N}} \times \frac{1 \text{ mol N atoms}}{14.0 \cancel{\text{ g N}}} = 2.17 \text{ mol N atoms}$$

$$69.6 \cancel{\text{ g O}} \times \frac{1 \text{ mol O atoms}}{16.0 \cancel{\text{ g O}}} = 4.35 \text{ mol O atoms}$$

Second, these values are reduced to small whole numbers by dividing by the *smallest* value, as follows.

$$\frac{2.17}{2.17} = 1$$

$$\frac{4.35}{2.17} = 2.00 \approx 2$$

Therefore, the empirical formula is NO_2. The molecular formula will be *equal* either to the empirical formula or to some *multiple* (2, 3, 4, etc.)

of it. The empirical formula mass of NO_2 is calculated as

$$1 \times 14.0 = 14.0 \text{ amu}$$
$$2 \times 16.0 = \underline{32.0 \text{ amu}}$$

Empirical formula mass = 46.0 amu

The molecular mass as given in the problem was 91.0 amu.[5] The multiple of the empirical formula is found to be approximately 2.

$$\frac{\text{Molecular mass}}{\text{Empirical formula mass}} = \frac{91.0 \text{ amu}}{46.0 \text{ amu}} = 1.98 \approx 2$$

Therefore, the molecular formula is

$$(NO_2)_2 = N_2O_4 \quad \textit{Answer}$$

Exercise 8–13

Calculate the molecular formula of a compound from the following experimental data: 56.4% phosphorus, 43.6% oxygen, and molecular mass of 220 amu.

Carbon: from jewelry to golf clubs

Name: The name carbon derives from the Latin *carbo*, meaning coal or charcoal. The name for graphite comes from the Greek *graphein*, meaning "to write."

Forms: Pure carbon occurs in two forms, graphite and diamond. Impure carbon takes many other forms (see Occurrence).

Appearance: Graphite is a soft black substance in which the carbons link to form large sheets. Diamond, which can be colorless single crystals, is the hardest natural substance known.

Occurrence: Carbon occurs in many different circumstances. Deposits of coal, graphite, and diamond are mostly carbon, while oil and natural gas contain carbon in combination with hydrogen. Carbon occurs as $CaCO_3$ (limestone, marble, chalk) and $CaMg(CO_3)_2$ (dolomite). Living organisms contain significant amounts of carbon. Carbon is found in air as CO_2 to the extent of about $\approx 0.035\%$ by volume.

Source: Some forms of carbon (natural graphite and diamonds) are obtained from natural sources. Other forms of carbon (activated carbon, artificial graphite, artificial diamonds, carbon black, graphitized carbon) are prepared from coal, oil, wood, or natural gas by charring and then heating with various additives (fillers and binders). The additives and binders in artificial graphite give it a variety of important properties (heat resistant, inert to acids and bases).

Its Role in Our World: Diamonds are of course used as gems, but they also find use as drill tips, abrasives, and cutting tools due to their hardness.

The biggest use of carbon is in the steel industry. Coke (a form of carbon), oxygen, and iron oxides are heated in a blast furnace, where

---

[5] The difference between the actual value of 92.0 amu and the experimental value of 91.0 amu results from experimental error in the determination of the molecular mass.

The two stable forms of pure carbon are quite different in appearance and are probably familiar to you: graphite (upper right) and diamond (center).

coke and oxygen form carbon monoxide (CO), which then changes iron(II) and iron(III) ions in the ore to iron metal.

Graphite is relatively inert and has a high electrical conductivity, which renders it suitable for making electrodes. Graphite electrodes are used to heat furnaces to very high temperatures in the production of stainless steels and aluminum metal. Graphite electrodes are also used in the electrolytic production of fluorine (F_2), sodium (Na), and lithium (Li).

Synthetic graphite is used to line blast furnaces and rocket nozzles because it is a good heat insulator and retains its strength at high temperature.

Carbon fibers made from rayon are very strong but relatively light. They are used to give strength and flexibility to golf club shafts, tennis rackets, skis, bows, and fishing rods.

Carbon black is used to add strength to tires, and as a pigment in inks and paints. Other uses of carbon include water and gas purification (activated charcoal), dry cell electrodes (graphite), and pencil lead (graphite).

Unusual Facts: Passing large amounts of current between two graphite electrodes creates a brilliant, high intensity light. Such carbon arc lamps are used in searchlights, spotlights, and as a light source in movie theater projectors.

Problems

1. Calculate the formula or molecular masses of each of the following compounds.

   (a) $C_2H_6O_2$

   (b) CO_2

   (c) $Ca(OH)_2$

   (d) $Ba(NO_3)_2$

2. Calculate the number of moles in each of the following quantities of compounds.

   (a) 87.5 g $C_2H_6O_2$
   (b) 36.4 g of CO_2
   (c) 1.50×10^{23} molecules of $C_2H_6O_2$
   (d) 9.84×10^{24} molecules of CO_2

3. Given 1.25 mol of carbon dioxide (CO_2), calculate the following.

   (a) the number of grams of carbon dioxide
   (b) the number of molecules of carbon dioxide

4. Calculate the number of molecules in each of the following quantities.

   (a) 75.0 g of $C_2H_6O_2$

   (b) 83.0 g of CO_2

5. Given 2.45 mol of ethylene glycol ($C_2H_6O_2$), calculate the following.

   (a) the number of moles of oxygen atoms
   (b) the number of grams of oxygen atoms

6. Given 12.6 L of carbon dioxide gas (CO_2) at STP, calculate the following.

   (a) the number of moles of carbon dioxide
   (b) the number of grams of carbon dioxide

7. Calculate the volume in liters at STP that the following gases would occupy.

   (a) 7.00 g of nitrogen gas (N_2)
   (b) 0.140 g of carbon monoxide gas (CO)

8. Calculate the molecular mass of the following gases, given the following data.

   (a) 3.20 L at STP has a mass of 0.572 g.
   (b) 4.00 L at STP has a mass of 5.00 g.

9. Calculate the density of the following gases at STP.

   (a) carbon dioxide (CO_2)

   (b) methane (CH_4)

10. Calculate the percent composition of the following compounds.

    (a) $CaBr_2$

    (b) $Ca(ClO_3)_2$

11. Calculate the percent of metal in the following compounds from the experimental data.

    (a) 0.500 g of a metal combines with 0.400 g of oxygen to form an oxide.
    (b) 0.350 g of a metal combines with 0.255 g of oxygen to form an oxide.

12. Calculate the empirical formula for each of the following compounds from the experimental data.

   (a) 48.0% zinc and 52.0% chlorine
   (b) 18.9% iron and 81.1% bromine

13. Calculate the molecular formula for each of the following compounds from the experimental data.

   (a) 80.0% carbon, 20.0% hydrogen, and molecular mass of 30.0 amu
   (b) 83.7% carbon, 16.3% hydrogen, and molecular mass of 86.0 amu

General Problems

14. Cyclopropane, a gaseous hydrocarbon once used as an anesthetic, has an experimental composition of 85.6% carbon and 14.4% hydrogen. At STP, 7.52 L of cyclopropane has a mass of 14.1 g. Calculate the molecular formula of cyclopropane.

15. In most states, a person operating a motor vehicle is considered to be driving while intoxicated (DWI) when the person has a blood alcohol level of 100 mg of alcohol (C_2H_6O) per 100 mL of blood. How many alcohol molecules per mL of blood must be present for the person to be considered DWI?

16. How many milliliters of concentrated nitric acid will be needed to supply 4.20 mol of HNO_3? Concentrated nitric acid is 72.0% HNO_3 and has a specific gravity of 1.42. (*Hint*: A 72.0% HNO_3 solution means 72.0 g of HNO_3 in 100 g of solution.)

17. Calculate the number of molecules in a 50.0 mg daily dose of the steroidal antiinflammatory medication prednisone (molecular mass = 358.4 amu).

18. Sulfur hexafluoride is one of the densest known gases. Calculate the density of sulfur hexafluoride at STP.

Answers to Exercises

8–1. 18.0 amu

8–3. (a) 13.4 g;
(b) 4.48 × 10²³ molecules

8–5. (a) 0.250 mol; (b) 0.250 g

8–7. 18.2 L STP

8–9. 1.25 g/L at STP

8–11. 78.5% metal

8–13. P_4O_6

8–2. (a) 1.94 mol; (b) 0.591 mol

8–4. 1.42 × 10²⁴ molecules

8–6. (a) 0.208 mol; (b) 5.81 g

8–8. 27.0 amu

8–10. 40.0% C, 6.7% H, 53.3% O

8–12. Al_2O_3

QUIZ

1. Calculate the number of molecules in 25.0 g of nitrogen dioxide (NO_2).

2. Calculate the molecular mass of a gas sample that has a volume of 4.50 L at STP and a mass of 4.10 g.

3. Calculate the percent composition of phosgene ($COCl_2$), a colorless, toxic gas.

ELEMENT	ATOMIC MASS UNITS (amu)
H	1.0
C	12.0
N	14.0
O	16.0
Cl	35.5

4. Quinine, one of the most important antimalarial agents, is isolated from the bark of the cinchona tree. The elemental composition of quinine is 74.04% carbon, 7.46% hydrogen, 8.64% nitrogen, and 9.86% oxygen. Determine the empirical formula for quinine.

5. The elemental composition of lactic acid is 40.00% carbon 6.71% hydrogen, and 53.29% oxygen, and it has a molecular mass of 90.0 amu. What is the molecular formula of lactic acid? Lactic acid is an organic compound that can be found in sour milk. It is also a product of muscle metabolism and causes your muscles to "ache" after vigorous exercise.

9

CHEMICAL EQUATIONS

COUNTDOWN You may use the Periodic Table.

5 Classify the following changes as physical or chemical (Section 3–5).
(a) sawing wood   (physical)
(b) burning wood   (chemical)
(c) rusting of your car's left rear fender, producing a
    big hole in the fender   (chemical)
(d) smashing your car's left rear fender against a fire hydrant (physical)

4 Write the correct formula for the compound formed by the combination of the following ions (Section 6–8).
(a) iron(III) (Fe^{3+}) and chloride (Cl^{1-})   ($FeCl_3$)
(b) iron(III) (Fe^{3+}) and sulfide (S^{2-})   (Fe_2S_3)
(c) stannic (Sn^{4+}) and oxide (O^{2-})   (SnO_2)
(d) stannic (Sn^{4+}) and acetate ($C_2H_3O_2^{1-}$)   [$Sn(C_2H_3O_2)_4$]

3 Write the correct formula for each of the following binary compounds (Sections 7–2 and 7–3):
(a) potassium oxide   (K_2O)
(b) lead(II) iodide   (PbI_2)
(c) tetraphosphorus trisulfide   (P_4S_3)
(d) barium phosphide   (Ba_3P_2)

2 Write the correct formula for each of the following ternary and higher compounds (Sections 7–4, 7–5, and 7–6):
(a) magnesium acetate   [$Mg(C_2H_3O_2)_2$]
(b) aluminum sulfate   [$Al_2(SO_4)_3$]
(c) chromic acid   (H_2CrO_4)
(d) mercury(II) bromate   [$Hg(BrO_3)_2$]

1 Classify each of the following compounds as (1) an acid, (2) a base, or (3) a salt. Assume that all soluble compounds are in aqueous (water) solution (Section 7–6).

(a) $HMnO_4$ (1, acid);   (b) $NaMnO_4$ (3, salt)

(c) $Ca(MnO_4)_2$ (3, salt);   (d) $Mn(OH)_2$ (2, base)

TASKS

1 Learn the terms, symbols, and their meanings used in chemical equations (Table 9–1).

2 Be able to interpret and use the order of the electromotive or activity series (Section 9–8), as directed by your instructor.

3 Be able to interpret and use the rules for the solubility of inorganic substances in water (Section 9–9), as directed by your instructor.

OBJECTIVES

1 Give the distinguishing characteristics of each of the following terms:

(a) chemical equation (Section 9–1)
(b) law of conservation of mass (Section 9–1)
(c) catalyst (Section 9–2)
(d) word equation (Section 9–4)
(e) combination reactions (Section 9–6)
(f) basic oxide (Section 9–6)
(g) acid oxide (Section 9–6)
(h) decomposition reactions (Section 9–7)
(i) hydrate (Section 9–7)
(j) single-replacement reactions (Section 9–8)
(k) double-replacement reactions (Section 9–9)
(l) neutralization reactions (Section 9–10)

2 Given the formulas of reactants and products, balance various chemical equations (Example 9–1, Exercise 9–1, Problem 1).

3 Given the names of reactants and products, write the formulas for the reactants and products and balance word equations (Example 9–2, Exercise 9–2, Problem 2).

4 Given the formulas of the reactants and the Periodic Table, complete and balance various *combination* reaction equations (Example 9–3, Exercise 9–3, Problem 3).

5 Given the formula of the reactant and the Periodic Table, complete and balance various *decomposition* reaction equations (Example 9–4, Exercise 9–4, Problem 4).

6 Given the formulas of the reactants, the Periodic Table, and the electromotive or activity series of metals, complete and balance various *single replacement* reaction equations (Examples 9–5 and 9–6, Exercise 9–5, Problem 5).

7 Given the formulas of the reactants, the Periodic Table, and the rules for the solubility of inorganic substances in water, complete and balance various *double-replacement* reaction equations. Indicate any precipitate by (s) and any gas by (g) (Example 9–7, Exercise 9–6, Problem 6).

8 Given the formulas of the reactants, the Periodic Table, and the rules for the solubility of inorganic substances in water, complete and balance various *neutralization* reaction equations. Indicate any precipitate by (s) (Example 9–8, Exercise 9–7, Problem 7).

You may have wondered why your eyes burn on smoggy days or how antacids neutralize acid in the stomach. In both cases, a chemical reaction has taken place.

In this chapter we consider the chemical properties (see Section 3–4) and chemical changes (see Section 3–5) of elements and compounds. We will learn to write chemical equations for these chemical reactions. Before we can do this, we need to become proficient with a new tool—balancing chemical equations.

9–1 Definition of a Chemical Equation. The Law of Conservation of Mass

chemical equation A short-hand way of expressing a chemical change (reaction) in terms of symbols and formulas.

A **chemical equation** is a shorthand way of expressing a chemical change (reaction) in terms of symbols and formulas. We cannot write an equation for a reaction unless we know the substances that are reacting and the substances that are formed. For an equation to be considered correct, *it must be balanced*. We balance equations because the law of conservation of mass requires that we do so. The **law of conservation of mass,** experimentally determined by Antoine Laurent Lavoisier, states that mass is neither created nor destroyed in ordinary chemical changes and the total mass involved in a physical or chemical change remains constant. As a consequence, *the number of atoms or moles of atoms of each element must be the same on both sides of the equation.* This is the reason we balance equations.

law of conservation of mass Mass can be neither created nor destroyed, and the total mass of the sub-stances involved in a physi-cal or chemical change remains constant.

Chemical equations may be written in two general ways: (1) as "molecular equations" or (2) as "ionic equations." In this chapter we consider only molecu-lar equations;[1] after we master the skill of balancing molecular equations, we will consider ionic equations (Chapter 10).

9–2 Terms, Symbols, and Their Meanings

A chemical equation is a shorthand way of expressing a chemical change, so various terms and symbols are used, just as in shorthand. In an equation the substances that combine with one another, the *reactants*, are written on the

---

[1] The term "molecular equation" is used to include elements and compounds that exist not only as molecules but also as formula units.

left. The substances that are formed, the *products*, are written on the right. A single arrow ($\rightarrow$) or an equals sign ($=$) or a double arrow ($\rightleftarrows$) separates the reactants from the products, depending on the nature of the reaction. A plus sign ($+$) separates the reactants from one another and the products from each other. The three physical states of substances involved in the reaction are sometimes indicated as subscripts (in parenthesis) following the formula of the substance, as follows:

1. *Gas* by (g), or if a gas is a product, by an arrow point upward ($\uparrow$): $H_{2(g)}$ or $H_2\uparrow$.

2. *Liquid* by (ℓ): $H_2O_{(\ell)}$.

3. *Solid* by (s), or if a solid is a product precipitating or coming out of a solution, by an arrow pointing downward ($\downarrow$) or by underscoring: $AgCl_{(s)}$, $AgCl\downarrow$, or $\underline{AgCl}$. The use of (s) is the preferred method.

| catalyst A substance that speeds up a chemical reaction but is recovered relatively unchanged at the end of the reaction. |

Since water is often used to dissolve solids, a substance dissolved in water is indicated by the subscript (aq), meaning *aq*ueous solution, such as $NaCl_{(aq)}$. A capital Greek delta (Δ) may appear above or below the arrow that separates the reactants and products, meaning that heat is necessary to make the reaction go, such as $\xrightarrow{\Delta}$. Also above or below the arrow may appear the symbols for an element or a compound, such as $\xrightarrow{Pt}$. These symbols denote a catalyst. A **catalyst**[2] is a substance that speeds up a chemical reaction, but is recovered relatively unchanged at the end of the reaction. The various enzymes used in the digestion of foods are catalysts. One example is ptyalin in saliva, which catalyzes the breakdown of large molecules, such as starch, to smaller molecules, such as maltose. Another catalyst, chlorophyll, is used in photosynthesis to form glucose (a sugar) from carbon dioxide, water, and sunlight.

These symbols may or may not appear in the equation depending on the emphasis placed on the reactants and products in the equation. Hence, in some equations you may see many of these symbols, and in other equations you may see none. Table 9–1 summarizes these terms, symbols, and their meanings.

9–3 Guidelines for Balancing Chemical Equations

Just as not all symbols appear in all equations, so there are no absolute rules for how to balance equations. (Remember, though, that it *is* a rule that you *must* balance equations.) You should, however, find the following guidelines generally applicable for most of the simple equations you will encounter in this chapter. Also bear in mind that you must balance the number of atoms or moles of atoms

---

[2] A substance that slows down a chemical reaction is called an *inhibitor*. An inhibitor may be thought of as a "negative catalyst."

TABLE 9–1   TERMS, SYMBOLS, AND THEIR MEANINGS USED IN
CHEMICAL EQUATIONS

TERM OR SYMBOL	MEANING
Reactants	Left side of equation
Products	Right side of equation
→, ⇄, =	Separates the products from the reactants
Subscript (g) or ↑	Gas or gas as a product
Subscript (ℓ)	Liquid
Subscript (s), ↓ , or underscoring formula	Solid or solid as a product precipitating or coming out of a solution
Subscript (aq)	Aqueous solution (dissolved in water)
Δ above or below arrow ($\overset{\Delta}{\rightarrow}$ or $\underset{\Delta}{\rightarrow}$)	Heat needed for reaction to start or go to completion
Symbol above or below arrow ($\overset{Pt}{\rightarrow}$ or $\underset{Pt}{\rightarrow}$)	Catalyst

of each element. Hence, there must be the *same number* of atoms or moles of atoms of each element on each side of the equation. We call this process "balancing by inspection." This expression refers to the fact that no mathematical process is involved. Rather, we look at (inspect) the equation, work with the equation according to the guidelines, and balance it.

<table>
<tr><td>

Study hint

Review Chapter 7 on nomenclature to be sure you can write the correct formulas. This is a *MUST*.

</td></tr>
</table>

1. Write the correct formulas for the reactants and the products, with the reactants on the left and the products on the right separated by →, ⇄, or =. Each reactant and each product is separated from each other by a + sign. *Once you have written the correct formulas, do not change them during the subsequent balancing operation.* (Review Chapter 7 on nomenclature to be able to write the correct formulas.)

2. Choose the compound that contains the *greatest number of atoms*, whether it is a reactant or a product. Start with the *element* in the compound that also has the greatest number of atoms. This element as a rule should *not* be hydrogen, oxygen, or a polyatomic ion. Balance the number of atoms in this compound with the corresponding atom on the other side of the equation by placing a coefficient *in front* of the formula of the element or compound. If you place a 2 in front of H_2O, as $2H_2O$, then there are 4 atoms of H and 2 atoms of O, and the same number of atoms must appear on the other side of the equation. If you place no number in front of a formula, the coefficient is assumed to be 1. *Under no circumstances do you change the correct formula of a compound to balance the equation.*

3. Next balance the polyatomic ions that remain the *same* on both sides of the equation. You may balance these polyatomic ions as a single unit. In some cases you may need to go back to the coefficient you placed before the compound in guideline 2 and change it to balance the polyatomic ions. If this is the case, remember to adjust the coefficient on the other side of the equation accordingly.

4. Balance the H atoms and then the O atoms. If they appear in the polyatomic ion, and you balanced them before, you need not consider them again.

5. Check all coefficients to see that they are *whole numbers* and in the *lowest possible ratio*. If some of the coefficients are fractions, then all coefficients must be multiplied by some number so as to make them all, including the fractions, whole numbers.[3] If a coefficient such as $\frac{5}{2}$ or $2\frac{1}{2}$ exists, then you must multiply all the coefficients by 2. The $\frac{5}{2}$ or $2\frac{1}{2}$ is then 5, a whole number. You must reduce the coefficients to the lowest possible ratios. If the coefficients are 6, 9 → 3, 12, you can reduce them by dividing *each one* by 3 to give the lowest possible ratio of whole numbers, 2, 3 → 1, 4.

6. Check each atom or polyatomic ion with a $\sqrt{}$ above the atom or ion on both sides of the equation to ensure that the equation is balanced. As you become proficient in balancing equations, this may not be necessary. These $\sqrt{}$s are not part of the final form of the equation and are only used as a teaching device to make sure that you balance each atom or ion.

9–4   Examples Involving the Balancing of Equations. Word Equations

Now we will apply these guidelines in balancing the following equations by inspection.

EXAMPLE 9–1

Balance each of the following equations by inspection.

(a) $Ca(OH)_{2(aq)} + H_3PO_{4(aq)} \rightarrow Ca_3(PO_4)_{2(s)} + H_2O_{(\ell)}$ (unbalanced)

SOLUTION   Guideline 1 need not be considered, since the formulas are given. Considering guideline 2, the compound with the greatest number of atoms is $Ca_3(PO_4)_2$ and the element we start with is Ca, which has 3 atoms. (The polyatomic PO_4^{3-} ion is balanced in guideline 3.) To balance the Ca atoms, place a 3 in front of the $Ca(OH)_2$, as $3Ca(OH)_2$. The formula of $Ca(OH)_2$ is not changed to balance the Ca atoms. The equation now appears as

$3Ca(OH)_{2(aq)} + H_3PO_{4(aq)} \longrightarrow Ca_3(PO_4)_{2(s)} + H_2O_{(\ell)}$ (unbalanced)

Guideline 3: the polyatomic ion PO_4^{3-} appears on both sides of the equation; hence, place a 2 in front of H_3PO_4, as $2H_3PO_4$ to balance the $2PO_4^{3-}$ in $Ca_3(PO_4)_2$. Balance the H atoms as in guideline 4 by placing a 6 in front of the H_2O, as $6H_2O$, since there are 12H atoms

---

[3] In balancing equations for some purposes, it is sometimes convenient to leave one of the coefficients as a fraction. In this book we will consider an equation to be balanced only if the coefficients are whole numbers in the lowest possible ratios.

on the left [3 × 2 = 6 from 3Ca(OH)$_2$ and 2 × 3 = 6 from 2H$_3$PO$_4$]. The O atoms are balanced by 6H$_2$O because there are 6O atoms in 3 Ca(OH)$_2$. [The O atoms in the PO$_4^{3-}$ are not included because they were balanced with the Ca$_3$(PO$_4$)$_2$.] The equation is now

$$3Ca(OH)_{2(aq)} + 2H_3PO_{4(aq)} \longrightarrow Ca_3(PO_4)_{2(s)} + 6H_2O_{(\ell)} \text{ (balanced)}$$

The coefficients are all whole numbers, and the lowest possible ratios as suggested in guideline 5. Check off each atom or polyatomic ion as in guideline 6. The final balanced equation is

$$3\overset{\checkmark}{C}a\overset{\checkmark}{(O}\overset{\checkmark}{H})_{2(aq)} + 2\overset{\checkmark}{H}_3\overset{\checkmark}{P}O_{4(aq)} \longrightarrow \overset{\checkmark}{C}a_3\overset{\checkmark}{(P}O_4)_{2(s)} + 6\overset{\checkmark}{H}_2\overset{\checkmark}{O}_{(\ell)} \text{ (balanced)}$$

Answer

(b) $C_4H_{10(g)} + O_{2(g)} \overset{\Delta}{\rightarrow} CO_{2(g)} + H_2O_{(g)}$ (unbalanced)

butane

SOLUTION Considering guideline 2, the compound with the greatest number of atoms is C$_4$H$_{10}$, and we start with C, which has 4 atoms. (H atoms are balanced later—guideline 4.) To balance the C atoms, place a 4 in front of the CO$_2$, as 4CO$_2$. For guideline 4 (3 is not applicable since there are no polyatomic ions), balance the H atoms with a 5 in front of the H$_2$O, as 5H$_2$O, which gives a total of 13O atoms in the products (8O from 4CO$_2$, and 5O from 5H$_2$O). To balance the O atoms in the reactants, you must use a fraction, $\frac{13}{2}$ or $6\frac{1}{2}$, to obtain 13O atoms in the reactants. The equation now appears as

$$C_4H_{10(g)} + \tfrac{13}{2}O_{2(g)} \overset{\Delta}{\longrightarrow} 4CO_{2(g)} + 5H_2O_{(g)}$$

For guideline 5, the coefficients must be whole numbers. To obtain a whole number from $\frac{13}{2}$, multiply it by 2; then multiply *all* the coefficients by 2. The coefficients are in the lowest possible ratio, and each atom is checked off as in guideline 6. The final balanced equation is

$$2\overset{\checkmark}{C}_4\overset{\checkmark}{H}_{10(g)} + 13\overset{\checkmark}{O}_{2(g)} \overset{\Delta}{\longrightarrow} 8\overset{\checkmark}{C}\overset{\checkmark}{O}_{2(g)} + 10\overset{\checkmark}{H}_2\overset{\checkmark}{O}_{(g)} \text{ (balanced)}$$

butane

Answer

(Checking off each atom provides a double check, and assures you that you multiplied *each* coefficient by 2.)

Exercise 9–1

Balance each of the following equations by inspection.

(a) $MnO_2 + Al \overset{\Delta}{\rightarrow} Al_2O_3 + Mn$

(b) $PCl_5 + H_2O \rightarrow H_3PO_4 + HCl$

word equation Express a chemical equation in words instead of symbols and formulas.

Word equations are another form of chemical equations. A **word equation** expresses the chemical equation in words instead of symbols and formulas. To write and balance word equations, we need only to apply our guidelines in Section 9–3 with special emphasis on guideline 1: The correct formulas for the elements or compounds must be written from the names. Here we apply the nomenclature you learned in Chapter 7.

EXAMPLE 9–2

Change the following word equations into chemical equations and balance by inspection.

(a) *Word Equation:*

Calcium bromide + sulfuric acid $\longrightarrow$

hydrogen bromide + calcium sulfate

SOLUTION   By guideline 1 we must first write the correct formulas from each of the names of the compounds. See Table 6–1 (cations), 6–2 (anions), and 6–4 (polyatomic ions) for the formulas of ions to refresh your memory.

Chemical Equation:

$$CaBr_2 + H_2SO_4 \longrightarrow HBr + CaSO_4 \quad \text{(unbalanced)}$$

Once the correct formula has been written, it must not be changed to balance the equation. The starting point is the $CaBr_2$ (guideline 2). Place a 2 in front of the HBr to balance the Br atoms. The $SO_4{}^{2-}$ is balanced (guideline 3) and so are the H atoms (guideline 4). The coefficients are whole numbers and in the lowest possible ratio (guideline 5). Check each atom or polyatomic ion (guideline 6), and the balanced equation appears as follows:

$$\overset{\checkmark}{Ca}\overset{\checkmark}{Br_2} + \overset{\checkmark}{H_2}\overset{\checkmark}{SO_4} \longrightarrow 2\overset{\checkmark\checkmark}{HBr} + \overset{\checkmark}{Ca}\overset{\checkmark}{SO_4} \quad \text{(balanced)} \quad \textit{Answer}$$

(b) *Word Equation:*

$$\text{Oxygen} + C_8H_{18} \xrightarrow{\Delta} \text{carbon monoxide}_{(g)} + \text{water vapor}$$

a component
of gasoline

SOLUTION   We now write the correct formulas from the names (guideline 1).

Chemical Equation:

$$O_2 + C_8H_{18} \xrightarrow{\Delta} CO_{(g)} + H_2O_{(g)} \quad \text{(unbalanced)}$$

Balance the C atoms (guideline 2) by placing an 8 in front of the CO, as 8CO, and then balance the H atoms (guideline 4) with a 9 in front of the H_2O, as $9H_2O$. You now need 17 O atoms on the reactant side (8 from 8CO plus 9 from $9H_2O$). Place $\frac{17}{2}O_2$ or $8\frac{1}{2}$ in front of O_2, as $\frac{17}{2}O_2$ (guideline 4), to obtain the needed 17O atoms. The following equation results:

$$\tfrac{17}{2}O_2 + C_8H_{18} \xrightarrow{\Delta} 8CO_{(g)} + 9H_2O_{(g)}$$

The coefficients are not whole numbers (guideline 5); hence, *all* coefficients must be multiplied by 2. Check each atom (guideline 6), and obtain the following balanced equation:[4]

$$17\overset{\checkmark}{O}_2 + 2\overset{\checkmark}{C}_8\overset{\checkmark}{H}_{18} \xrightarrow{\Delta} 16\overset{\checkmark}{C}\overset{\checkmark}{O}_{(g)} + 18\overset{\checkmark}{H}_2\overset{\checkmark}{O} \quad \text{(balanced)} \quad Answer$$

Exercise 9-2

Change the following word equations into chemical equations and balance by inspection.

(a) Potassium nitrate $\xrightarrow{\Delta}$ potassium nitrite + oxygen

(b) Strontium sulfite + acetic acid →

strontium acetate + water + sulfur dioxide

9-5 Completing Chemical Equations. The Five Simple Types of Chemical Reactions

Now we shall not only balance the equation but also complete it by writing the products. To write the products in a chemical equation, we must consider a few generalizations about ordinary chemical reactions. Therefore, we divide the ordinary chemical reactions that we consider in this chapter into five simple types for which we can write equations.

1. Combination reactions

2. Decomposition reactions

3. Single-replacement reactions

4. Double-replacement reactions

5. Neutralization reactions

---

[4] Gasoline is a mixture of carbon-containing compounds, one of which has the formula C_8H_{18}. In footnote 3, Chapter 7, we mentioned that carbon monoxide is a deadly air pollutant. Carbon monoxide is released into the air by incomplete combustion of gasoline from automobiles and is harmful to human beings in that the carbon monoxide has a greater affinity for the hemoglobin in the red blood cells than does oxygen. Thus, the hemoglobin is "tied up" by the carbon monoxide and is not able to carry oxygen. Carbon monoxide hence robs the tissues of the oxygen required for survival.

combination reaction A process in which two or more substances (either elements or compounds) react to produce one substance:

$$A + Z \longrightarrow AZ$$

where A and Z are elements or compounds.

Another type of reaction is oxidation-reduction. Special techniques are required to write balanced equations for complicated oxidation-reduction reactions (see Sections 14–1 and 14–2). In general, these equations are very difficult to balance "by inspection." However, most combination, decomposition, and replacement reactions are simple cases of oxidation-reduction reactions that can be balanced "by inspection."

9–6   Combination Reactions

In a **combination reaction**, two or more substances (either elements or compounds) react to produce *one* substance (always a compound). Combination reactions are also called *synthesis reactions*. This reaction can be shown by a general equation,

$$A + Z \longrightarrow AZ$$

where A and Z are elements or compounds. Six general classes of combination reactions are shown below and in Example 9–3:

Study hint

Some students find it helpful to make flash cards of these reactions with the reactants on one side and the product on the other. For those of you who plan to do this, we will mark the reaction by writing "card" and then its number next to the equation. To be able to complete and balance equations, you will need to apply the general principles you learned in Chapter 6 on oxidation numbers and formulas of compounds. You should refer to the Periodic Table as you study this material.

	A	$+$	Z	$\rightarrow$	AZ
(a)	metal		oxygen		metal oxide
(b)	nonmetal		oxygen		nonmetal oxide
(c)	metal		nonmetal		salt
(d)	water		metal oxide		base
(e)	water		nonmetal oxide		oxyacid
(f)	metal oxide		nonmetal oxide		salt

EXAMPLE 9–3

Study hint

Do *not* balance the equation by changing the formula of MgO:

$$Mg_{(s)} + O_{2(g)}$$

$$\xrightarrow{\Delta} MgO_{2(s)} \text{ (wrong)}$$

The equation is wrong because the formula of magnesium oxide is incorrect.

Complete and balance the following combination reaction equations.

(a) Metal + oxygen $\xrightarrow{\Delta}$ metal oxide[5]   (Card 1)

$$Mg_{(s)} + O_{2(g)} \xrightarrow{\Delta}$$

SOLUTION   The formula of the product is determined from a knowledge of the ionic charges of Mg and O in the combined state. Magnesium in the combined state has an ionic charge of 2^+ and oxygen 2^-. Hence, the correct formula of the metal oxide is MgO.

$$2Mg_{(s)} + O_{2(g)} \xrightarrow{\Delta} 2MgO_{(s)} \quad Answer$$

---

[5] The reactions involving oxygen gas, such as a metal plus oxygen and a nonmetal plus oxygen, are sometimes called *combustion reactions* because combustion (burning) is the reaction between oxygen and many substances. Besides oxygen plus a metal or nonmetal, organic (carbon-based) compounds can also burn in oxygen to form water and carbon dioxide on complete combustion. A common example is the metabolism of food given by the complete combustion of sucrose (sugar) to yield carbon dioxide, water, and energy.

$$C_{12}H_{22}O_{11(s)} + 12O_{2(g)} \longrightarrow 12CO_{2(g)} + 11H_2O_{(g)} + energy$$

This energy keeps our bodies warm and maintains our body temperature. One tablespoonful of sugar (12 g) yields 45 kilocalories of energy.

(b) Nonmetal + oxygen → nonmetal oxide[5]   (Card 2)

(i) $S_{(s)} + O_{2(g)} \xrightarrow{\Delta}$
    limited

(ii) $S_{(s)} + O_{2(g)} \xrightarrow{\Delta}$
    excess

SOLUTION   We can determine the formula of the product from a knowledge of the oxides of the nonmetal, that is, SO_2 and SO_3. Therefore, these two oxides should be included on the product side of your card. With *limited* oxygen the oxide with the smaller amount of oxygen in its formula, SO_2, is formed. With *excess* oxygen the oxide with the greater amount of oxygen in its formula, SO_3, is formed.

$$S_{(s)} + O_{2(g)} \xrightarrow{\Delta} SO_{2(g)} \qquad Answer$$
limited

$$2S_{(s)} + 3O_{2(g)} \xrightarrow{\Delta} 2SO_{3(g)} \qquad Answer$$
excess

(iii) $C_{(s)} + O_{2(g)} \xrightarrow{\Delta}$
    limited

(iv) $C_{(s)} + O_{2(g)} \xrightarrow{\Delta}$
    excess

SOLUTION   Again, we use our knowledge of the oxides of carbon to determine the formulas of the products, CO and CO_2. Place these oxides on your card. Limited oxygen gives CO; excess oxygen affords CO_2.

$$2C_{(s)} + O_{2(g)} \xrightarrow{\Delta} 2CO_{(g)} \qquad Answer$$
limited

$$C_{(s)} + O_{2(g)} \xrightarrow{\Delta} CO_{2(g)} \qquad Answer$$
excess

(c) Metal + nonmetal → salt   (Card 3)

$$Na_{(s)} + Cl_{2(g)} \longrightarrow$$

SOLUTION   Determine the formula of the product (NaCl) from your knowledge of the ionic charges on Na and Cl in the combined state, Na^{1+} and Cl^{1-}.

$$2Na_{(s)} + Cl_{2(g)} \longrightarrow 2NaCl_{(s)} \qquad Answer$$

(d) Water + metal oxides → base (metal hydroxide)   (Card 4)

$$H_2O_{(\ell)} + MgO_{(s)} \longrightarrow$$

SOLUTION   Determine the formula of the product from your knowledge of the ionic charge on Mg in the combined state (Mg^{2+}) and the ionic charge on the hyroxide ion (OH^{1-}): $Mg(OH)_2$. Metal oxides,

basic oxides Metal oxides.

like MgO and others, that react with water to form a base (metal hydroxide) are called **basic oxides**.

$$H_2O_{(\ell)} + MgO_{(s)} \longrightarrow Mg(OH)_{2(s)} \quad \textit{Answer}$$

(e) Water + nonmetal oxide → oxyacid[6]   (Card 5)

(i) $H_2O_{(\ell)} + SO_{2(g)} \longrightarrow$

SOLUTION Determine the formula of the product from the oxidation number (see Section 6-3) of S in SO_2, which is 4^+. This S atom forms an acid with the *same* oxidation number in the product; hence, the formula of the acid is H_2SO_3 (oxidation number of S = 4^+) and *not* H_2SO_4 (oxidation number of S = 6^+). The oxidation number is the *same* in the reactants as in the products for all reactions of nonmetal oxides with water that we consider in this chapter. Nonmetal oxides, like SO_2 and others, that react with water to form an acid are called **acid oxides**. Sulfur dioxide, an air pollutant, is harmful to people partly because it combines with moisture in the eyes, throat, and lungs to form sulfurous acid (H_2SO_3), as shown by the following balanced equation.

acid oxides Nonmetal oxides.

$$H_2O_{(\ell)} + SO_{2(g)} \longrightarrow H_2SO_{3(aq)} \quad \textit{Answer}$$

(ii) $H_2O_{(\ell)} + SO_{3(g)} \longrightarrow$

SOLUTION Determine the formula of the product from the oxidation number of S in SO_3, which is 6^+. The S atom forms an acid with the same oxidation number, so the formula of the acid is H_2SO_4.[7]

$$H_2O_{(\ell)} + SO_{3(g)} \longrightarrow H_2SO_{4(aq)} \quad \textit{Answer}$$

On your card you should write "SO_2 forms SO_3^{2-}" (oxidation number of S = 4^+) and SO_3 forms "SO_4^{2-}" (oxidation number of S = 6^+).

(f) Metal oxide (basic oxide) + nonmetal oxide (acid oxide) → salt (Card 6)

$$MgO_{(s)} + SO_{3(g)} \longrightarrow$$

---

[6] Sulfur and nitrogen oxides react with water to form their respective acids. During precipitation as rain, snow, sleet, or fog, these acids fall back on the earth and are referred to as "acid rain."

[7] Monsanto and many other chemical companies have developed processes to control sulfur oxides as air pollutants. The Monsanto process oxidizes the sulfur oxides to SO_3 and then converts this gas to dilute sulfuric acid, as shown in the equation. The oxygen in the air can also oxidize sulfur dioxide to sulfur trioxide. Therefore, on foggy days, with polluted air, you may be inhaling a dilute solution of sulfuric acid. High-sulfur coal can also be cleaned by a similar process developed by General Motors. When the coal is burned, sulfur dioxide is produced. The sulfur dioxide gas is trapped and converted by various reactions to calcium sulfate and calcium sulfite. These salts are used as landfills. The process removes about 90% of the sulfur dioxide from the gas.

SOLUTION The SO_3 will form the SO_4^{2-} polyatomic ion, as mentioned, and thus the correct formula for the salt based on the ionic charge of Mg^{2+} in the combined state and on the ionic charge on the SO_4^{2-} ion is $MgSO_4$.

$$MgO_{(s)} + SO_{3(g)} \longrightarrow MgSO_{4(s)} \quad \textit{Answer}$$

If the gas were SO_2, what would be the product? (See Card 5.)

Exercise 9–3

Complete and balance the following combination reaction equations.

(a) $Al_{(s)} + O_{2(g)} \xrightarrow{\Delta}$
(b) $C_{(s)} + O_{2(g)} \xrightarrow[\text{limited}]{\Delta}$
(c) $SrO_{(s)} + H_2O_{(\ell)} \rightarrow$
(d) $N_2O_{5(s)} + H_2O_{(\ell)} \rightarrow$

9–7 Decomposition Reactions

decomposition reaction A process in which one substance undergoes a reaction to form two or more substances:

$$AZ \longrightarrow A + Z$$

where A and Z are elements or compounds

In a **decomposition reaction**, one substance undergoes a reaction to form two or more substances. The substance broken down is always a compound, and the products may be *elements* or *compounds*. Heat is often necessary for this process. This reaction can be represented by a general equation,

$$AZ \longrightarrow A + Z$$

where A and Z are elements or compounds. In general, a prediction of the products in a decomposition reaction can only be determined by a knowledge of each individual reaction; therefore, you will need to make cards for each individual reaction. Five general classes of decomposition reactions are illustrated below and in Example 9–4:

AZ	→	A	+	Z
(a) compound		compound or element		oxygen
(b) metal carbonate		metal oxide		carbon dioxide
(c) hydrated salt		salt		water
(d) compound		element or compound		water
(e) metal hydrogen carbonate		salt		water + carbon dioxide

EXAMPLE 9–4

Complete and balance the following decomposition reaction equations.

(a) Some compounds decompose to yield oxygen:

(i) $HgO_{(s)} \xrightarrow{\Delta}$

SOLUTION The red mercury(II) oxide, when heated, forms droplets of mercury along the edge of the tube, and oxygen is evolved. The presence of oxygen, which supports combustion (see Section 3–3), is demonstrated by inserting a glowing wooden splint into the test tube. The piece of wood catches fire and burns upon contact with high levels of oxygen.

$$2HgO_{(s)} \xrightarrow{\Delta} 2Hg_{(\ell)} + O_{2(g)} \qquad Answer \quad (Card\ 7)$$

For all equations the order for writing the products makes no difference. Either O_2 or Hg may be written first. Make sure that the equation is balanced.

(ii) $KNO_{3(s)} \xrightarrow{\Delta}$

SOLUTION This reaction is a method for the production of oxygen in the laboratory. The products are potassium nitrite and oxygen. The sodium salt would also show a similar reaction, since sodium is in the same group (IA, 1) as potassium.

$$2KNO_{3(s)} \xrightarrow{\Delta} 2KNO_{2(s)} + O_{2(g)} \qquad Answer \quad (Card\ 8)$$

Once you write the product, try balancing the equation. Do *not* memorize the coefficients.

(iii) $KClO_{3(s)} \xrightarrow[\Delta]{MnO_2}$

SOLUTION This reaction is the usual method for the production of oxygen in the laboratory. The products are potassium chloride and oxygen. The MnO_2 is the catalyst.

$$2KClO_{3(s)} \xrightarrow[\Delta]{MnO_2} 2KCl_{(s)} + 3O_{2(g)} \qquad Answer \quad (Card\ 9)$$

(iv) $H_2O_{(\ell)} \xrightarrow[\substack{electric \\ current}]{direct}$

SOLUTION Electrolysis decomposes water into hydrogen and oxygen (see Section 3–3) if the water contains a trace of an ionic compound, such as sodium sulfate.

$$2H_2O_{(\ell)} \xrightarrow[\substack{electric \\ current}]{direct} 2H_{2(g)} + O_{2(g)} \qquad Answer \quad (Card\ 10)$$

(v) $H_2O_{2(aq)} \xrightarrow[light]{\Delta\ or}$

SOLUTION Hydrogen peroxide decomposes when heated or in the presence of light to yield water and oxygen.[8]

$$2H_2O_{2(aq)} \xrightarrow[light]{\Delta\ or} 2H_2O_{(\ell)} + O_{2(g)} \qquad Answer \quad (Card\ 11)$$

---

[8] Hydrogen peroxide is used to bleach cloth and hair. It has also been used as an antiseptic. The release of oxygen can be observed when a dilute solution (3%) used as an antiseptic bubbles when it comes into contact with a wound.

(b) Some metal carbonates, when heated, decompose to yield carbon dioxide and a metal oxide:

$$CaCO_{3(s)} \xrightarrow{\Delta}$$

SOLUTION   When limestone (calcium carbonate) is heated, carbon dioxide is one of the products, along with calcium oxide. The presence of carbon dioxide, which does *not* support combustion,[9] is demonstrated by inserting a glowing wooden splint into the test tube. The glowing piece of wood goes out upon contact with carbon dioxide. Carbonates of magnesium and strontium would give similar products, since they are in the same group (IIA, 2) as calcium.

$$CaCO_{3(s)} \xrightarrow{\Delta} CaO_{(s)} + CO_{2(g)} \qquad Answer \quad (Card\ 12)$$

(c) **Hydrates** decompose to liberate water when heated:

$$MgSO_4 \cdot 7H_2O_{(s)} \xrightarrow{\Delta}$$

hydrate A crystalline substance that contains chemically bound water molecules in definite proportions. An example is Epsom salts, magnesium sulfate *hepta*hydrate ($MgSO_4 \cdot 7H_2O$).

SOLUTION   When a hydrate such as Epsom salt crystals (magnesium sulfate heptahydrate) is heated, magnesium sulfate and water are produced.

$$MgSO_4 \cdot 7H_2O_{(s)} \xrightarrow{\Delta} MgSO_{4(s)} + 7H_2O_{(g)} \qquad Answer \quad (Card\ 13)$$

If the hydrate were $BaCl_2 \cdot 2H_2O$, or any other hydrate, what would be the products?

(d) Some compounds (not hydrates) decompose when heated to yield water:

$$C_{12}H_{22}O_{11(s)} \xrightarrow{\Delta}$$

SOLUTION   Sugar ($C_{12}H_{22}O_{11}$) decomposes when heated to form a caramel brown or black solid (carbon) and water (see Sections 3–1 and 3–3) as a vapor (see Fig. 9–1).

$$C_{12}H_{22}O_{11(s)} \xrightarrow{\Delta} 12C_{(s)} + 11H_2O_{(g)} \qquad Answer \quad (Card\ 14)$$

(e) Some compounds, such as metal hydrogen carbonates (metal bicarbonates), decompose to yield water, carbon dioxide, and a carbonate salt when heated:

$$NaHCO_{3(s)} \xrightarrow{\Delta}$$

SOLUTION   Baking soda (sodium hydrogen carbonate or sodium bicarbonate) decomposes to give water, carbon dioxide, and sodium

---

[9] Carbon dioxide does not support combustion and it is more dense than oxygen or air. Many types of fire extinguishers take advantage of these properties. By directing a stream of CO_2 at a fire, the denser CO_2 forms a blanket of gas over the fire and pushes the air (oxygen) up and away and extinguishes the fire.

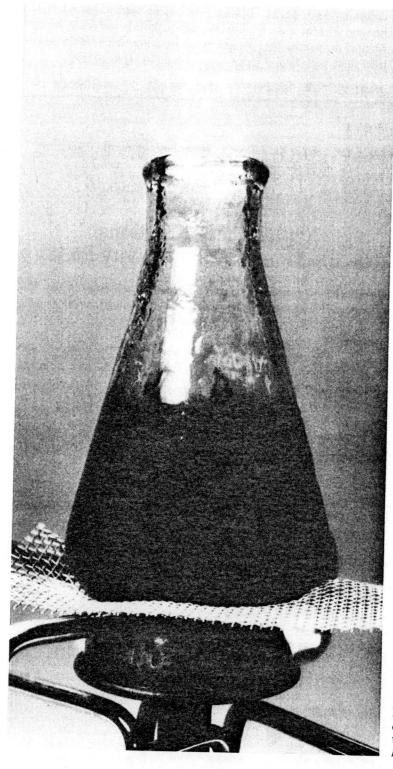

Figure 9–1 *The decomposition of sugar to produce carbon and water vapor (Courtesy of Dr. E. R. Degginger.)*

carbonate when heated. Baking soda can be used to put out fires because the heat of the fire liberates carbon dioxide and water, which smother the flames. The potassium salt would exhibit a similar reaction, since potassium is in the same group (IA, 1) as sodium.

$$2NaHCO_{3(s)} \xrightarrow{\Delta} Na_2CO_{3(s)} + H_2O_{(g)} + CO_{2(g)} \; Answer \; (Card \; 15)$$

Exercise 9–4

Complete and balance the following decomposition reaction equations.

(a) $NaNO_{3(a)} \xrightarrow{\Delta}$ (b) $MgCO_{3(s)} \xrightarrow{\Delta}$

(c) $KHCO_{3(s)} \xrightarrow{\Delta}$ (d) $CaCl_2 \cdot 2H_2O_{(s)} \xrightarrow{\Delta}$

single-replacement reaction A process in which an element (A or X) replaces another element (B or Z) in a compound:

1. A metal replacing a metal ion in its salt or hydrogen ion in an acid:

$$A + BZ \longrightarrow AZ + B$$

2. A nonmetal replacing a nonmetal ion in its salt or acid:

$$X + BZ \longrightarrow BX + Z$$

Study hint

This type of reaction is similar to a woman "cutting in" on a couple at a dance. The first woman *replaces* the woman from the couple, just as a metal *replaces* a metal ion from a salt.

Li
K
Ba
Ca
Na
Mg
Al
Zn
Fe
Cd
Ni
Sn
Pb
(H)
Cu
Hg
Ag
Au

9–8 Single-Replacement Reactions. The Electromotive or Activity Series

In a **single-replacement reaction**, one element reacts by replacing another element in a compound. Single-replacement reactions are also called *replacement, substitution,* or *displacement reactions.* This reaction can be represented by two general equations:

1. A metal (A) replacing a metal ion (B = metal) in its salt or a hydrogen ion (B = H) in an acid.

$$A + BZ \longrightarrow AZ + B \quad (Card \; 16)$$

2. A nonmetal (X) replacing another nonmetal (Z) in its salt (B = metal) or acid (B = H).

$$X + BZ \longrightarrow BX + Z \quad (Card \; 17)$$

In the *first* case, the replacement depends on the two metals involved, A and B. The metals can be arranged in a series called the *electromotive or activity series,* so that each element in the series will displace any element following it from an aqueous solution of its salt. Although hydrogen is not a metal, it is *included* in this series: **Li, K, Ba, Ca, Na, Mg, Al, Zn, Fe, Cd, Ni, Sn, Pb, (H), Cu, Hg, Ag, Au.** You should be able to interpret and use this series so that you can complete and balance chemical equations involving single-replacement reactions. (This series is listed inside the back cover of this book.)

EXAMPLE 9–5

Complete and balance the following single-replacement reaction equations.

(a) $Fe_{(s)} + CuSO_{4(aq)} \longrightarrow$

SOLUTION Iron is higher in the electromotive or activity series than copper. Hence, iron will replace the copper(II) from its salt. For metals existing in variable oxidation numbers, the *lower* oxidation number is often formed. Thus, the new salt is the iron(II) salt, $FeSO_4$, rather than the iron(III) salt, $Fe_2(SO_4)_3$.

$$Fe_{(s)} + CuSO_{4(aq)} \longrightarrow FeSO_{4(aq)} + Cu_{(s)} \qquad Answer$$

(b) $Sn_{(s)} + HCl_{(aq)} \longrightarrow$

SOLUTION   Tin is higher in the electromotive or activity series than hydrogen, and hence tin will replace the hydrogen ion from the acid. The salt with the *lower* oxidation number for the metal is formed, and the new salt is the tin(II) salt $SnCl_2$, not the tin(IV) product $SnCl_4$. Remember to write hydrogen gas as a diatomic molecule, H_2.

$$Sn_{(s)} + 2HCl_{(aq)} \longrightarrow SnCl_{2(aq)} + H_{2(g)} \qquad Answer$$

(c) $Cu_{(s)} + MgCl_{2(aq)} \longrightarrow$

SOLUTION   Copper is below magnesium in the electromotive or activity series; therefore, no reaction will occur.

$$Cu_{(s)} + MgCl_{2(aq)} \longrightarrow \text{no reaction} \qquad Answer$$

(d) $K_{(s)} + H_2O_{(\ell)} \longrightarrow$

SOLUTION   Potassium is high in the electromotive or activity series and can replace *one* hydrogen atom from water to form the hydroxide ion and hydrogen gas. The first five metals in the series exhibit similar properties. Thus, lithium, barium, calcium, and sodium behave much as potassium does. If we write water as H—OH, it helps to remind us that potassium replaces only one hydrogen atom from each water molecule. Only very active metals high in the electromotive or activity series react with water because water has strong covalent bonds and is neither a strong acid nor a salt. The formula of the metal hydroxide is KOH because K has a 1^+ ionic charge in the combined state and OH has a 1^- ionic charge. The balanced equation is (see Fig. 9–2)

$$2K_{(s)} + 2H\text{—}OH_{(\ell)} \longrightarrow 2KOH_{(aq)} + H_{2(g)} \qquad Answer$$

In the *second case*, when a nonmetal replaces a nonmetal ion from its salt or acid, the replacement depends on the two nonmetals involved, X and Z. A series similar to the electromotive or activity series exists for the halogen nonmetals: F_2, Cl_2, Br_2, I_2. Fluorine will replace chloride ion from an aqueous solution of its salt; chlorine will replace bromide; and bromine will replace iodide. Notice that this series follows the nonmetallic properties as given in the Periodic Table (see Section 5–3, number 4).

F_2
Cl_2
Br_2
I_2

EXAMPLE 9–6

Complete and balance the following single-replacement reaction equations.

(a) $Cl_{2(g)} + NaBr_{(aq)} \longrightarrow$

Figure 9–2 *A single-replacement reaction. The production of hydrogen gas from calcium metal and water. See if you can write the balanced equation for this reaction (Courtesy of Fundamental Photographs, all rights reserved.)*

SOLUTION Chlorine gas replaces bromide from an aqueous solution of its salt to yield the chloride salt and bromine. Recall that bromine is a diatomic molecule (Br_2; see Section 6–4).

$$Cl_{2(g)} + 2NaBr_{(aq)} \longrightarrow 2NaCl_{(aq)} + Br_{2(aq)} \qquad Answer$$

(b) $Br_{2(aq)} + NaI_{(aq)} \longrightarrow$

SOLUTION Bromine dissolved in water replaces iodide from an aqueous solution of its salt to yield the bromide salt and iodine dissolved in water. Recall that iodine is also a diatomic molecule (I_2; see Section 6–5).

$$Br_{2(aq)} + 2NaI_{(aq)} \longrightarrow 2NaBr_{(aq)} + I_{2(aq)} \qquad Answer$$

Exercise 9–5

Complete and balance the following single-replacement reaction equations.

(a) $Al_{(s)} + SnCl_{2(aq)} \rightarrow$

(b) $Hg_{(\ell)} + SnCl_{2(aq)} \rightarrow$

(c) $Ba_{(s)} + H{-}OH_{(\ell)} \rightarrow$

(d) $Br_{2(aq)} + KI_{(aq)} \rightarrow$

9–9 Double-Replacement Reactions. Rules for the Solubility of Inorganic Substances in Water

double-replacement reaction A process in which two compounds undergo a reaction where the positive ion (cation) of one compound exchanges with the positive ion (cation) of the second compound:

$$\longrightarrow AZ + BX$$

Study hint

A double-replacement reaction is similar to two couples at a dance *exchanging partners* and dancing some more.

Study hint

In double-replacement reactions there are *four* separate particles, A, X, B, and Z, whereas in single-replacement reactions for *metals* there are only *three*, A, B, and Z. Single-replacement reactions *depend* on the electromotive or activity series, but double-replacement reactions *do not.*

In **double-replacement reactions**, *two* compounds are involved in a reaction, with the positive ion (cation) of one compound *exchanging* with the positive ion (cation) of another compound. In other words, the two positive ions *exchange* negative ions (anions) or partners. Double-replacement reactions are also called *metathesis* (meaning a change of state, substance, or form) or *double-decomposition reactions.* This reaction can be represented by the general equation

$$AX + BZ \longrightarrow AZ + BX \qquad \text{(Card 18)}$$

Double-replacement reactions will generally proceed if one of the following conditions is satisfied:

1. An insoluble or slightly soluble product is formed.

2. A weakly ionized species is produced as a product. The most common species of this type is water.

3. A gas is produced as a product.

The most common type of double-replacement reaction we will discuss belongs to the first of these three classes. A precipitate appears during the reaction because one of the products is insoluble (or slightly soluble) in water (see Fig. 9–3). This precipitate can be indicated in an equation by underscoring, as AgCl, or by a downward arrow, as AgCl $\downarrow$. A better method is to indicate it by an (s), as $AgCl_{(s)}$, which we will use in this book. To recognize that a precipitate will form, you must be able to interpret and use the following rules for the solubility of inorganic substances in water (there are exceptions).

Figure 9-3 *The precipitation of barium sulfate from aqueous solution. The beaker on the left contains a solution of barium nitrate in water. In the center beaker a few drops of dilute sulfuric acid have been added and a cloud of solid barium sulfate can be observed. At the right the precipitation has been completed and the solid barium sulfate has settled to the bottom of the beaker. (Courtesy of Katharine P. Daub.)*

1. Nearly all *nitrates* (NO_3^{1-}) and *acetates* ($C_2H_3O_2^{1-}$) are *soluble*.

2. All *chlorides* (Cl^{1-}) are *soluble*, except $AgCl$, Hg_2Cl_2, and $PbCl_2$. ($PbCl_2$ is soluble in hot water.)

3. All *sulfates* (SO_4^{2-}) are *soluble*, except $BaSO_4$, $SrSO_4$, and $PbSO_4$. ($CaSO_4$ and Ag_2SO_4 are only slightly soluble.)

4. Most of the *alkali metal* [group IA (1), Li, Na, K, etc.] salts, and *ammonium* (NH_4^{1+}) salts are *soluble*.

5. All the common acids are *soluble*.

6. All *oxides* (O^{2-}) and *hydroxides* (OH^{1-}) are **insoluble**, except those of the alkali metals and certain alkaline earth metals [group IIA (2), Ca, Sr, Ba, Ra]. [$Ca(OH)_2$ is only moderately soluble.]

7. All *sulfides* (S^{2-}) are **insoluble**, except those of the alkali metals, alkaline earth metals, and ammonium sulfide.

8. All *phosphates* (PO_4^{3-}) and *carbonates* (CO_3^{2-}) are **insoluble**, except those of the alkali metals and the ammonium salts.

These generalizations will be quite useful in writing double-replacement equations. They are given inside the back cover.

EXAMPLE 9–7

Complete and balance the following double-replacement reaction equations. Indicate any precipitate by (s) and any gas by (g).

(a) A salt and an acid to form a precipitate:

$$AgNO_{3(aq)} + HCl_{(aq)} \longrightarrow$$

SOLUTION   The silver ion changes places with the hydrogen ion to form the insoluble silver chloride (AgCl; see solubility rule 2), and the hydrogen reacts with the nitrate ion to form a new acid, nitric acid (HNO_3). The formation of an insoluble compound acts as the driving force behind these reactions. The precipitate is AgCl and is determined from the solubility rules (see solubility rule 2).

$$AgNO_{3(aq)} + HCl_{(aq)} \longrightarrow AgCl_{(s)} + HNO_{3(aq)} \quad Answer$$

(b) A salt and a base to form a precipitate:

$$Ni(NO_3)_{2(aq)} + NaOH_{(aq)} \longrightarrow$$

SOLUTION   In the exchange of ions, a new salt, $NaNO_3$, and a new base, $Ni(OH)_2$, which is insoluble in water (see solubility rule 6) are formed:

$$Ni(NO_3)_{2(aq)} + 2NaOH_{(aq)} \longrightarrow Ni(OH)_{2(s)} + 2NaNO_3 \quad Answer$$

(c) Two salts to form a precipitate:

$$NaCl_{(aq)} + AgNO_{3(aq)} \longrightarrow$$

SOLUTION   In the exchange of ions, two new salts are formed: silver chloride (AgCl), which is insoluble in water (see solubility rule 2) and sodium nitrate ($NaNO_3$), which is soluble in water (see solubility rule 1).

$$NaCl_{(aq)} + AgNO_{3(aq)} \longrightarrow AgCl_{(s)} + NaNO_{3(aq)} \quad Answer$$

The reaction above can be summarized by two general statements:

$$\text{Salt}_1 + \text{acid}_1 \text{ or base}_1 \longrightarrow \text{salt}_2 + \text{acid}_2 \text{ or base}_2$$

$$\text{Salt}_1 + \text{salt}_2 \longrightarrow \text{salt}_3 + \text{salt}_4$$

In both cases, one of the two products is usually insoluble in water or weakly ionized.

(d) Metal carbonate and an acid:

$$CaCO_{3(s)} + HCl_{(aq)} \longrightarrow$$

SOLUTION The calcium ion changes places with the hydrogen ion to form the salt $CaCl_2$. The hydrogen ion reacts with the carbonate ion to form carbonic acid (H_2CO_3), which is *unstable* and decomposes to form water and carbon dioxide. Calcium carbonate is one of the ingredients in a popular antacid (Tums®), and this reaction is the basis of the action of the antacid in neutralizing some acid (HCl) in the stomach.

$$CaCO_{3(s)} + 2HCl_{(aq)} \longrightarrow CaCl_{2(aq)} + H_2O_{(\ell)} + CO_{2(g)} \qquad Answer$$

Note that the formation of a gas and a weakly ionized species (H_2O) provide the driving force for this reaction.

This type of reaction can be summarized as

$$\text{Metal carbonate} + \text{acid} \longrightarrow \text{salt} + \text{water} + \text{carbon dioxide}$$

Exercise 9–6

Complete and balance the following double-replacement reaction equations. Indicate any precipitate by (s) and any gas by (g).

(a) $AgNO_{3(aq)} + H_2S_{(g)} \rightarrow$

(b) $CdSO_{4(aq)} + NaOH_{(aq)} \rightarrow$

(c) $Na_2CO_{3(aq)} + HCl_{(aq)} \rightarrow$

(d) $BaCl_{2(aq)} + (NH_4)_2CO_{3(aq)} \rightarrow$

neutralization reaction A process in which an acid (HX) or an acid oxide reacts with a base (MOH) or a basic oxide. In most of these reactions, water is one the the products:

$$HX + MOH$$

$$\longrightarrow MX + HOH$$

Study hint

As you may have noticed, neutralization reactions are just a special type of double-replacement reaction.

9–10 Neutralization Reactions

A **neutralization reaction** is one in which an acid or an acid oxide reacts with a base or basic oxide. In most of these reactions, water is one of the products. The formation of water acts as the driving force behind the neutralization, since water is only slightly ionized, and heat is also given off in its formation.

This reaction can be represented by a general equation,

$$HX + MOH \longrightarrow MX + HOH \qquad (Card\ 19)$$

where HX is an acid or acid oxide and MOH is a base or basic oxide, and water is usually one of the products. Four general classes of neutralization reactions are shown below and in Example 9–8.

ACIDIC REACTANT + BASIC REACTANT → SALT + WATER

(a) acid	base	salt	yes
(b) acid	basic oxide (metal oxide)	salt	yes
(c) acid oxide (nonmetal oxide)	base	salt	yes
(d) acid oxide (nonmetal oxide)	basic oxide (metal oxide)	salt	no

The differences between neutralization reactions and ordinary double-replacement reactions are (1) an acid or an acid oxide reacts with a base or a basic oxide in a neutralization reaction and (2) water is usually one of the products of a neutralization reaction.

EXAMPLE 9–8

Complete and balance the following neutralization reaction equations. Indicate any precipitate by (s).

(a) An acid and a base:

(i) $HCl_{(aq)} + NaOH_{(aq)} \longrightarrow$

SOLUTION   The hydrogen ion reacts with the hydroxide ion to form a weakly ionized water molecule. This leaves the sodium ion (Na^{1+}) and the chloride ion (Cl^{1-}) to form an aqueous solution of sodium chloride ($NaCl$; see solubility rule 2).

$$HCl_{(aq)} + NaOH_{(aq)} \longrightarrow NaCl_{(aq)} + H_2O_{(\ell)} \quad Answer$$

(ii) $HCl_{(aq)} + Mg(OH)_{2(s)} \longrightarrow$

SOLUTION   Again, an acid and a base react to form the salt, $MgCl_2$, and water.[10]

$$2HCl_{(aq)} + Mg(OH)_{2(s)} \longrightarrow MgCl_{2(aq)} + 2H_2O_{(\ell)} \quad Answer$$

This type of neutralization reaction can be summarized as

$$Acid + base \longrightarrow salt + water$$

(b) An acid and a basic oxide (metal oxide):

$$HCl_{(aq)} + ZnO_{(s)} \longrightarrow$$

SOLUTION   The zinc ion changes places with the hydrogen ion to form the salt zinc chloride ($ZnCl_2$; Zn^{2+} and Cl^{1-}). The hydrogen ion reacts with the oxide ion to form water, which is weakly ionized.

$$2HCl_{(aq)} + ZnO_{(s)} \longrightarrow ZnCl_{2(aq)} + H_2O_{(\ell)} \quad Answer$$

---

[10] This reaction occurs in the stomach when we use milk of magnesia [$Mg(OH)_2$] as an antacid. The milk of magnesia neutralizes some acid in the stomach to form a salt and water, which are less irritating to stomach tissue than the acid.

This type of neutralization reaction can be summarized as follows:

Acid + basic oxide (metal oxide) $\longrightarrow$ salt + water

(c) An acid oxide (nonmetal oxide) and a base:

$$CO_{2(g)} + LiOH_{(g)} \longrightarrow$$

SOLUTION The carbon dioxide reacts with lithium hydroxide to form a salt, lithium carbonate (Li_2CO_3), and water. The oxidation number of C in CO_2 is 4^+, the same as it is in the salt, Li_2CO_3. The correct formula of the salt is written from a knowledge of the ionic charges on lithium and the carbonate ion—that is, Li^{1+} and CO_3^{2-}—hence, Li_2CO_3.

$$CO_{2(g)} + 2LiOH_{(s)} \longrightarrow Li_2CO_{3(s)} + H_2O_{(\ell)} \quad \textit{Answer}$$

On card 19 you should write "CO_2 forms CO_3^{2-}" (oxidation number of C = 4^+).

This type of neutralization reaction can be summarized as follows:

Acid oxide (nonmetal oxide) + base $\longrightarrow$ salt + water

(d) An acid oxide (nonmetal oxide) and a basic oxide (metal oxide):

$$SO_{3(g)} + BaO_{(s)} \longrightarrow$$

SOLUTION We discussed this type of reaction previously under combination reactions [see Section 9–6, Example 9–3(f)] because only *one* substance was formed. This reaction is also a neutralization reaction, in that a basic oxide reacts with an acid oxide to form a salt. Note that no water is formed in the reaction.

$$SO_{3(g)} + BaO_{(s)} \longrightarrow BaSO_{4(s)} \quad \textit{Answer}$$

This type of neutralization reaction can be summarized as follows:

Acid oxide (nonmetal oxide) + basic oxide (metal oxide) $\longrightarrow$ salt

TABLE 9–2 SUMMARY OF THE FIVE SIMPLE
TYPES OF CHEMICAL REACTIONS

TYPE OF REACTION	EXAMPLE OF REACTION
Combination reactions:	$A + Z \rightarrow AZ$
Decomposition reactions:	$AZ \rightarrow A + Z$
Single-replacement reactions:	$A + BZ \rightarrow AZ + B$
	$X + BZ \rightarrow BX + Z$
Double-replacement reactions:	$AX + BZ \rightarrow AZ + BX$
Neutralization reactions:[a]	$HX + MOH \rightarrow MX + HOH$

[a] HX is an acid and MOH is a base. An acid oxide (nonmetal oxide) may be substituted for the acid, and a basic oxide (metal oxide) may be substituted for the base in a neutralization reaction. A nonmetal oxide plus a metal oxide forms a salt, but no water.

Exercise 9–7

Complete and balance the following neutralization reaction equations. Indicate any precipitate by (s).

(a) $Zn(OH)_{2(s)} + HNO_{3(aq)} \rightarrow$

(b) $Fe_2O_{3(s)} + HCl_{(aq)} \rightarrow$

(c) $SO_{2(g)} + KOH_{(aq)} \rightarrow$

(d) $CaO_{(s)} + CO_{2(g)} \rightarrow$

Table 9–2 summarizes the general reactions for the five simple types of chemical reactions discussed in this chapter.

Sucrose: how sweet it is! ───────────

For better or worse, sucrose (sugar) is part of our lives.

Name: The chemical name for sucrose is α-D-glucopyranosyl-β-D-fructofuranoside. More common names besides sugar include saccharose, cane sugar, and beet sugar. Its chemical formula is $C_{12}H_{22}O_{11}$. Sucrose is a *carbohydrate* compound made up of carbon (carbo-) and water (-hydrate). A true carbohydrate has a chemical formula of $(CH_2O)_n$. Sucrose is made up of two smaller carbohydrates, fructose [$(CH_2O)_6$] and glucose [dextrose, also $(CH_2O)_6$]. In combining the two smaller units into sucrose, a molecule of water is lost; thus sucrose has a formula of $(C)_{12}(H_2O)_{11}$ rather than $(CH_2O)_{12}$.

Appearance: Sugar is a white crystalline solid that turns brown (caramelizes) and then black on heating. This is the basis for candy making. The slow cooking of sucrose turns the candy a light brown (soft caramels), and then darker brown (toffee or peanut brittle) as the heating is prolonged. If you don't stir the mixture well enough, it gets too hot and scorches (turns black!).

Occurrence: Sucrose is found universally throughout the plant kingdom in fruits, seeds, flowers, and roots. The primary sources of sucrose are sugarcane (15–20% of stalks) and sugar beet (10–17% of beet). Lesser amounts of sucrose (as syrups) are obtained from sugar maple and sorghum.

Source: Sucrose is the highest volume organic compound produced worldwide. The cane is crushed and washed with water to remove most of the sucrose. The resulting solution is evaporated to give a *syrup*. Further concentration of the brownish-black syrup and crystallization provides raw sucrose. More purification by crystallization yields the white solid with which we are familiar.

Its Role in Our World: Sucrose has been known as a sweetener throughout history. Sugarcane is believed to have been native to either India or New Guinea, and by 1000 B.C. it had spread throughout the South Pacific. By 400 B.C. sucrose could be found in the Middle East and by the twelfth century it was used throughout most of Europe. Venice became a center for the sugar trade, and Marco Polo recorded advanced sugar refining techniques used in China during his travels. Columbus brought sugar to the New World and by 1750, sugar was used worldwide.

The consumption of sucrose in the United States is over 45 kg (100 pounds) per person per year. Sucrose is used not only as a sweetener,

but as a preservative, a bulking agent, a flavor enhancer, and a texturizer. Another important characteristic of sucrose is that it can serve as a food for yeast. Yeast converts sucrose and water into ethyl alcohol (C_2H_6O) and carbon dioxide (see box on carbon dioxide in Chapter 10).

Many forms of sucrose are available. Granulated sugar is the typical form that is produced. Further grinding produces powdered sugar, to which must be added a little cornstarch to keep the powdered sugar from "caking." Sugar cubes are prepared by mixing sucrose and a sucrose syrup, pouring the mixture into molds, and allowing the cubes to harden by draining and evaporation. Sugar cubes are more expensive than granulated sugar because of the extra steps involved. Brown sugar can be made in two ways: (1) by mixing sucrose and an appropriate syrup to impart the desired taste and color, or (2) by concentrating a syrup and allowing it to crystallize rapidly to trap compounds that impart the color and flavor of brown sugar. A true *molasses* is really a thick, black syrup with a strong taste that is used as cattle feed or for fermentation. What you buy in the store as a "molasses" is a concentrated syrup with a more pleasing taste than a true molasses.

Unusual Facts: The most important variety of sugarcane for the commercial production of sucrose prior to 1920 was the *noble cane*. In the 1920s, the mosaic virus decimated the commercial cane fields and new strains had to be developed that were resistant to mosaic disease. There are many such strains now available which are used commercially. Another interesting aspect to the growing of sugarcane is that sucrose does not spoil in the canes like fruit on the tree or vine does. Thus, sugar does not "ripen" and can be harvested at almost any time, weather and season permitting. The canes are cut off at ground level, and another crop will usually grow without replanting. Eventually, the yields decrease, however, and the old plants are replaced by new ones.

Problems

1. Balance each of the following equations by inspection.

(a) $CaCl_{2(aq)} + (NH_4)_2CO_{3(aq)} \rightarrow CaCO_{3(s)} + NH_4Cl_{(aq)}$

(b) $KClO_{3(s)} \overset{\Delta}{\rightarrow} KCl_{(s)} + O_{2(g)}$

(c) $Al(OH)_{3(s)} + NaOH_{(aq)} \rightarrow NaAlO_{2(aq)} + H_2O_{(\ell)}$

(d) $Fe(OH)_{3(s)} + H_2SO_{4(aq)} \rightarrow Fe_2(SO_4)_{3(aq)} + H_2O_{(\ell)}$

(e) $Na_{(s)} + H_2O_{(\ell)} \rightarrow NaOH_{(aq)} + H_{2(g)}$

(f) $Mg_{(s)} + N_{2(g)} \overset{\Delta}{\rightarrow} Mg_3N_{2(s)}$

(g) $Al_{(s)} + O_{2(g)} \overset{\Delta}{\rightarrow} Al_2O_{3(s)}$

(h) $AgNO_{3(aq)} + CuCl_{2(aq)} \rightarrow AgCl_{(s)} + Cu(NO_3)_{2(aq)}$

2. Change the following word equations into chemical equations and balance by inspection.

(a) Iron + chlorine $\rightarrow$ iron(III) chloride

(b) Potassium nitrate $\overset{\Delta}{\rightarrow}$ potassium nitrite + oxygen

(c) Calcium + water $\rightarrow$ calcium hydroxide + hydrogen

(d) Sodium hydroxide + sulfuric acid $\rightarrow$ sodium hydrogen sulfate + water

3. Complete and balance the following combination reaction equations.

(a) $Ca_{(s)} + O_{2(g)} \overset{\Delta}{\rightarrow}$

(b) $S_{(s)} + excess\ O_{2(g)} \overset{\Delta}{\rightarrow}$

(c) $CaO_{(s)} + H_2O_{(\ell)} \rightarrow$

(d) $Al_{(s)} + Cl_{2(g)} \overset{\Delta}{\rightarrow}$

4. Complete and balance the following decomposition reaction equations.

(a) $HgO_{(s)} \overset{\Delta}{\rightarrow}$

(b) $H_2O_{(\ell)} \xrightarrow[\text{electric}]{\text{direct}}$ current

(c) $SrCO_{3(s)} \overset{\Delta}{\rightarrow}$

(d) $CaSO_4 \cdot 2H_2O_{(s)} \overset{\Delta}{\rightarrow}$

5. Complete and balance the following single-replacement reaction equations.

(a) $Zn_{(s)} + NiCl_{2(aq)} \rightarrow$

(b) $Pb_{(s)} + HCl_{(aq)} \overset{\Delta}{\rightarrow}$

(c) $Na_{(s)} + H_2O_{(\ell)} \rightarrow$

(d) $Cl_{2(g)} + NaBr_{(aq)} \rightarrow$

6. Complete and balance the following double-replacement reaction equations. Indicate any precipitate by (s) and any gas by (g).

(a) $Pb(NO_3)_{2(aq)} + HCl_{(aq)} \xrightarrow{\text{cold}}$

(b) $Bi(NO_3)_{3(aq)} + NaOH_{(aq)} \rightarrow$

(c) $Pb(C_2H_3O_2)_{2(aq)} + K_2SO_{4(aq)} \rightarrow$

(d) $CaCO_{3(s)} + HCl_{(aq)} \rightarrow$

7. Complete and balance the following neutralization reaction equations. Indicate any precipitate by (s).

(a) $Ca(OH)_{2(aq)} + HC_2H_3O_{2(aq)} \rightarrow$
(b) $Fe_2O_{3(s)} + H_3PO_{4(aq)} \rightarrow$
(c) $CO_{2(g)} + KOH_{(aq)} \rightarrow$
(d) $BaO_{(s)} + HCl_{(aq)} \rightarrow$

General Problems

8. Change the following words into chemical formulas, complete, and balance. Indicate any precipitate by (s) and any gas by (g).

(a) iron(II) sulfide + hydrochloric acid $\rightarrow$
(b) calcium carbonate + sulfuric acid $\rightarrow$
(c) cadmium oxide + hydrochloric acid $\rightarrow$
(d) aluminum + lead(II) chloride $\rightarrow$

9. Bathtub scum typically results from the precipitation of insoluble calcium salts from hard water. One of the major components of this scum is calcium carbonate. One way to remove such deposits is with dilute aqueous hydrochloric acid. Write a balanced chemical equation for this process.

10. Titanium diboride has been developed as an extremely hard, wear-resistant coating for use on materials that must survive extremely erosive (eroding) environments. It is expected to be used as a coating on valves in coal liquefaction reactors. Titanium diboride is prepared by allowing titanium tetrachloride, boron trichloride, and hydrogen to react with each other at atmospheric pressure to yield the diboride and hydrogen chloride. Write a balanced chemical equation for this reaction.

11. In the refining of magnesium, waste material is produced that contains 2% magnesium nitride (Mg_3N_2). This waste material reacts with the water in the atmosphere to produce ammonia gas and magnesium hydroxide. Complete and balance the equation for the reaction described. Recently, this waste has been used as a fertilizer. It has been applied to peas, beans, and corn in Oregon and to winter wheat, barley, and oats in Idaho with no toxicity and an increase in yields of crops over controlled plots.

Answers to Exercises

(The numbers represent the coefficients in front of the formulas in the balanced equation.)

9–1. (a) $3 + 4 \overset{\Delta}{\rightarrow} 2 + 3$
(b) $1 + 4 \rightarrow 1 + 5$

9–2. (a) $2KNO_3 \overset{\Delta}{\rightarrow} 2KNO_2 + O_2$
(b) $SrSO_3 + 2HC_2H_3O_2 \rightarrow Sr(C_2H_3O_2)_2 + H_2O + SO_2$

9–3. (a) $4Al_{(s)} + 3O_{2(g)} \overset{\Delta}{\rightarrow} 2Al_2O_3$

(b) $2C_{(s)} + O_{2(g)} \xrightarrow{\Delta} 2CO$
(c) $SrO_{(s)} + H_2O_{(\ell)} \rightarrow Sr(OH)_2$
(d) $N_2O_{5(s)} + H_2O_{(\ell)} \rightarrow 2HNO_3$
    (N has an ox no of 5^+ in both N_2O_5 and HNO_3)

9–4. (a) $2NaNO_{3(s)} \xrightarrow{\Delta} 2NaNO_2 + O_2$
(b) $MgCO_{3(s)} \xrightarrow{\Delta} MgO + CO_2$
(c) $2KHCO_{3(s)} \xrightarrow{\Delta} K_2CO_3 + CO_2 + H_2O$
(d) $CaCl_2 \cdot 2H_2O_{(s)} \xrightarrow{\Delta} CaCl_2 + 2H_2O$

9–5. (a) $2Al_{(s)} + 3SnCl_{2(aq)} \rightarrow 2AlCl_3 + 3Sn$
(b) $Hg_{(\ell)} + SnCl_{2(aq)} \rightarrow$ no reaction
(c) $Ba_{(s)} + 2H—OH_{(\ell)} \rightarrow Ba(OH)_2 + H_2$
(d) $Br_{2(aq)} + 2KI_{(aq)} \rightarrow 2KBr + I_2$

9–6. (a) $2AgNO_{3(aq)} + H_2S_{(g)} \rightarrow Ag_2S_{(s)} + 2HNO_3$
(b) $CdSO_{4(aq)} + 2NaOH_{(aq)} \rightarrow Cd(OH)_{2(s)} + Na_2SO_4$
(c) $Na_2CO_{3(aq)} + 2HCl_{(aq)} \rightarrow 2NaCl + H_2O + CO_{2(g)}$
(d) $BaCl_{2(aq)} + (NH_4)_2CO_{3(aq)} \rightarrow BaCO_{3(s)} + 2NH_4Cl$

9–7. (a) $Zn(OH)_{2(s)} + 2HNO_{3(aq)} \rightarrow Zn(NO_3)_2 + 2H_2O$
(b) $Fe_2O_{3(s)} + 6HCl_{(aq)} \rightarrow 2FeCl_3 + 3H_2O$
(c) $SO_{2(g)} + 2KOH_{(aq)} \rightarrow K_2SO_3 + H_2O$
(d) $CaO_{(s)} + CO_{2(g)} \rightarrow CaCO_{3(s)}$

QUIZ You may use the Periodic Table, the Electromotive or Activity Series, and the Rules for the Solubility of Inorganic Substances in Water.

1. Balance the following equations:

(a) $MnO_{2(s)} + Al_{(s)} \xrightarrow{\Delta} Al_2O_{3(s)} + Mn_{(s)}$
(b) $C_4H_{10(g)} + O_{2(g)} \xrightarrow{\Delta} CO_{2(g)} + H_2O_{(g)}$

2. Vinegar (acetic acid) and baking soda (sodium hydrogen carbonate) are two ingredients in many recipes. Complete and balance the chemical equation that describes the reaction between these two common substances.

3. Complete and balance the following equations. Indicate the precipitate in (f) by (s).

(a) $Ba + O_2 \xrightarrow{\Delta}$
(b) $NaNO_3 \xrightarrow{\Delta}$
(c) $BaCl_2 \cdot 2H_2O \xrightarrow{\Delta}$
(d) $Cl_2 + NaI \rightarrow$
(e) $Al + SnCl_2 \rightarrow$
(f) $AgNO_3 + H_2S \rightarrow$

10

IONIC EQUATIONS

COUNTDOWN You may use the Periodic Table and the Rules for the Solubility of Inorganic Substances in Water.

5 Write the correct formula for each of the following compounds (Sections 7–2, 7–3, 7–4, 7–5, and 7–6):

(a) disulfur decafluoride (S_2F_{10})

(b) tin(IV) chromate $[Sn(CrO_4)_2]$

(c) strontium bromate $[Sr(BrO_3)_2]$

(d) phosphoric acid (H_3PO_4)

4 Classify each of the following compounds as (1) an acid, (2) a base, or (3) a salt. Assume that all soluble compounds are in aqueous (water) solution (Section 7–6).

(a) $Cr(OH)_2$ (2, base);   (b) $Cr(C_2H_3O_2)_2$ (3, salt)

(c) H_2CrO_4 (1, acid);   (d) $Cr_2(C_2O_4)_3$ (3, salt)

3 Balance each of the following equations by inspection (Sections 9–3 and 9–4).

(a) $(NH_4)_2Cr_2O_{7(s)} \xrightarrow{\Delta} N_{2(g)} + H_2O_{(g)} + Cr_2O_{3(s)}$

$$(1 \longrightarrow 1 + 4 + 1)$$

(b) $BCl_{3(g)} + H_{2(g)} \xrightarrow{\Delta} B_{(s)} + HCl_{(g)}$

$$(2 + 3 \longrightarrow 2 + 6)$$

2 Complete and balance the following double-replacement reaction equations. Indicate any precipitate by (s) and any gas by (g) (Section 9–9).

(a) $MgCO_{3(s)} + HCl_{(aq)} \longrightarrow$

$$[MgCO_{3(s)} + 2HCl_{(aq)} \longrightarrow MgCl_2 + H_2O + CO_{2(g)}]$$

(b) $AsCl_{3(aq)} + H_2S_{(g)} \longrightarrow$

$$[2AsCl_{3(aq)} + 3H_2S_{(g)} \longrightarrow As_2S_{3(s)} + 6HCl]$$

1 Complete and balance the following neutralization reaction equations. Indicate any precipitate by (s) (Section 9-10).

(a) $Sr(OH)_{2(aq)} + H_2SO_{4(aq)} \longrightarrow$

$$[Sr(OH)_{2(aq)} + H_2SO_{4(aq)} \longrightarrow SrSO_{4(s)} + 2H_2O]$$

(b) $PbO_{(s)} + HCl_{(aq)} \xrightarrow{\text{cold}} [PbO_{(s)} + 2HCl_{(aq)} \xrightarrow{\text{cold}} PbCl_{2(s)} + H_2O]$

TASK   Memorize the list of strong electrolytes, weak electrolytes, and nonelectrolytes (Table 10-1).

OBJECTIVES

1 Give the distinguishing characteristics of each of the following terms:
(a) electrolytes (Section 10-1)
(b) nonelectrolytes (Section 10-1)
(c) strong electrolytes (Section 10-1)
(d) weak electrolytes (Section 10-1)
(e) ionic equations (Section 10-2)
(f) net ionic equations (Section 10-2)

2 Given the formulas of reactants and their states, complete and balance the equations, expressing them as total ionic equations and net ionic equations (Example 10-1, Exercise 10-1, Problems 1 and 2).

You may have wondered why you should not touch a radio, TV, or anything electrical while taking a bath. The reason is that salts are present in the bath water and on your body. These salts are composed of ions. Should there be a short in the appliance, the ions in the water could conduct the electrical current from the broken appliance through your body and seriously shock or even kill you.

Ions in solution may undergo reactions as well. In Chapter 9 we stated that chemical equations may be written as molecular equations or as ionic equations. We discussed how to balance and complete molecular equations. In this chapter we consider how to write ionic compounds as ions and their electrical properties. We also consider how to write molecular equations in ionic form.

10-1 Electrolytes versus Nonelectrolytes

electrolytes Substances whose aqueous solution can conduct an electric current.

nonelectrolyte Substances whose aqueous solution do *not* conduct an electric current.

Before we can consider ionic equations, we must understand the meaning of the terms "electrolytes" and "nonelectrolytes." Substances whose aqueous solutions conduct an electric current are called **electrolytes**, and those substances whose aqueous solutions do *not* conduct an electric current are referred to as **nonelectrolytes**. To determine whether a substance is an electrolyte or a nonelectrolyte, prepare an aqueous solution of the substance and test the solution with two

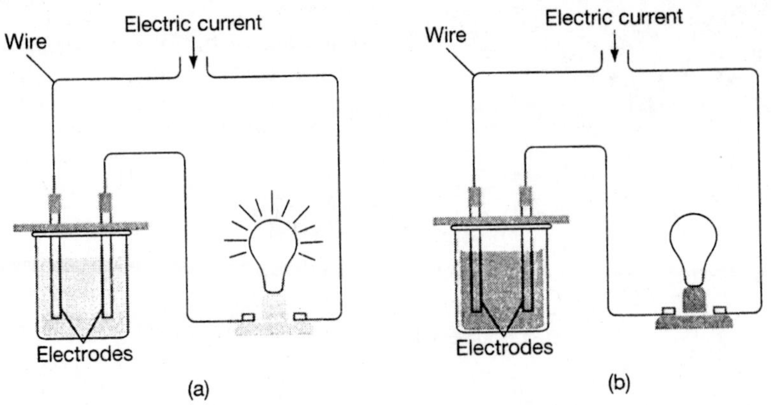

Figure 10–1 *An apparatus for determining conduction of an electric current in an aqeous solution of a substance. (a) An electrolyte. (b) A nonelectrolyte.*

electrodes connected to a source of electric current (direct or alternating)[1] with a standard light bulb in the circuit, as shown in Fig. 10–1. If the *bulb glows*, the substance is an *electrolyte*; and if it does *not glow*, it is a *non*electrolyte.

The reason for the conduction of electric current by electrolytes was explained in 1884 by the Swedish physicist and chemist Svante August Arrhenius.[2] His explanation was that ions exist in aqueous solutions of electrolytes and that no ions are present in aqueous solutions of nonelectrolytes. If a *direct* electric current (battery) is used, one of the electrodes becomes positively charged and the other negatively charged. The negative electrode is called the *cathode*, and the positive electrode is the *anode*. The positive ions in the solution migrate to the cathode (negative electrode) and are called *cations*, whereas the negative ions in the solution migrate to the anode (positive electrode) and hence are called *anions*.

Study hint

Noting that the "t" in "cation" looks like a "+" will help you remember that cations are attracted to the cathode and unlike charges attract, so the cathode must be negative.

Common *electrolytes* include salts, acid, and bases. Some common *nonelectrolytes* are sugar (sucrose, $C_{12}H_{22}O_{11}$), ethyl alcohol (C_2H_6O), and glycerine ($C_3H_8O_3$). Solutions of these nonelectrolytes do *not* conduct an electric current and exist as molecules, not ions, in solution. Pure water is shown to be essentially a nonelectrolyte because the standard light bulb *does not* glow in the electrode test.[3] There do not appear to be sufficient ions present in pure water to conduct an electric current. Tap water, however, contains dissolved salts and will conduct

---

[1] An electric current must be supplied. The electrolyte *does not produce* a current; it conducts an electric current. A source of direct current is an automobile battery; a source of alternating current is the electric outlet in your home. ***Do not attempt this experiment unless you are supervised by a qualified individual.***

[2] At the time Arrhenius proposed his explanation, he was a graduate student and only 25 years old. At first, his explanation was not widely accepted by his more seasoned colleagues, but as time passed it received wide acclaim. Arrhenius received the Nobel Prize in chemistry for his work in 1903.

[3] More sensitive tests indicate that pure water does contain a very small number of ions (H^{1+} and OH^{1-}). There are not enough of them, however, to conduct enough current to light the standard light bulb in this test.

electricity, which is why it is dangerous to touch electrical objects when taking a bath.

Electrolytes may be further subdivided into *strong* and *weak* electrolytes based on how many ions are formed in the solution. **Strong electrolytes** are substances whose aqueous solutions conduct sufficient electrical current to produce a *bright glow* in a standard light bulb. In such solutions there are many ions present because the substances dissolve easily and form many ions. These strong electrolytes are *soluble* salts, *strong* acids and bases. Examples of soluble salts are sodium acetate ($NaC_2H_3O_2$), sodium chloride ($NaCl$), and potassium sulfate (K_2SO_4), plus many others. The solubility rules given in Section 9–9 and on the inside back cover of the text help you to determine the solubility of the salt in water. Examples of strong acids are sulfuric acid (H_2SO_4), hydrochloric acid (HCl), hydrobromic acid (HBr), hydriodic acid (HI), nitric acid (HNO_3), and perchloric acid ($HClO_4$). Examples of strong bases are group IA(1) hydroxides [sodium hydroxide ($NaOH$), potassium hydroxide (KOH)], barium hydroxide [$Ba(OH)_2$], strontium hydroxide [$Sr(OH)_2$], and calcium hydroxide [$Ca(OH)_2$]. **Weak electrolytes** are substances whose aqueous solutions conduct sufficient current to produce a *dull glow* in a standard light bulb, because only a few ions exist in solution. In such solutions there are only a few ions present, because the substances either dissolve poorly or do not tend to form ions when they do dissolve. These *weak* electrolytes are slightly soluble salts and weak acids and bases. Examples of slightly soluble salts are lead(II) or plumbous acetate [$Pb(C_2H_3O_2)_2$] and mercury(II) or mercuric chloride ($HgCl_2$). Most acids such as acetic acid ($HC_2H_3O_2$) are examples of weak acids. Most bases such as ammonia (NH_3) water are examples of weak bases.

Figure 10–2 summarizes the process for determining whether a substance is nonelectrolyte, strong electrolyte, or a weak electrolyte. Table 10–1 summa-

> **strong electrolytes** Substances whose aqueous solutions conduct sufficient electric current to produce a *bright glow* in a standard light bulb because there are many ions in solution. These substances are soluble salts and strong acids and bases. You must memorize the strong acids and bases.

> **weak electrolytes** Substances whose aqueous solutions conduct sufficient electric current to produce a *dull glow* in the standard light bulb becuse only a few ions exist in solution. These substances are slightly soluble salts and weak acids and bases.

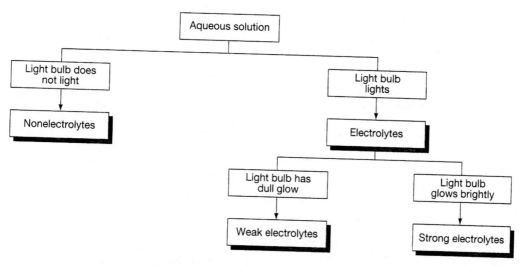

Figure 10–2 *A substance can be classed as a nonelectrolyte, a strong electrolyte, or a weak electrolyte by testing an aqueous solution of the substance with the apparatus described in Fig. 10–1.*

TABLE 10–1   SUMMARY OF STRONG ELECTROLYTES, WEAK
ELECTROLYTES, AND NONELECTROLYTES

STRONG ELECTROLYTES	WEAK ELECTROLYTES	NONELECTROLYTES
Strong acids H_2SO_4 (sulfuric acid) HCl (hydrochloric acid) HBr (hydrobromic acid) HI (hydroiodic acid) HNO_3 (nitric acid) $HClO_4$ (perchloric acid)	*Weak acids* $HC_2H_3O_2$ (acetic acid) Most other acids	$C_{12}H_{22}O_{11}$ (sugar or sucrose) C_2H_6O (ethyl alcohol) $C_3H_8O_3$ (glycerine) H_2O (water)
Strong bases Group IA (1) hydroxides like: NaOH (sodium hydroxide) KOH (potassium hydroxide) $Ba(OH)_2$ (barium hydroxide) $Sr(OH)_2$ (strontium hydroxide) $Ca(OH)_2$ (calcium hydroxide)	*Weak bases* $NH_{3(aq)}$ (ammonia water) Most other bases	
Most soluble salts Such as: $NaC_2H_3O_2$ (sodium acetate) NaCl (sodium chloride) K_2SO_4 (potassium sulfate) plus many other salts (See solubility rules in Section 9–9 and inside back cover of text	*Slightly soluble salts* $Pb(C_2H_3O_2)_2$ [lead(II) or plumbous acetate] $HgCl_2$ [mercury(II) or mercuric chloride]	

rizes the strong and weak electrolytes and nonelectrolytes. You must know
examples of strong and weak electrolytes and nonelectrolytes given in this table
to write ionic equations.

Acetic acid is an example of a weak electrolyte. Pure acetic acid (glacial
acetic acid) does not conduct an electric current and the standard bulb does not
glow. Dilution of the acetic acid with water (a nonelectrolyte) gives a solution that
slightly conducts an electric current and the standard bulb has a dull glow. This
conduction of electric current on dilution with water is a result of the partial
ionization of the acetic acid. Pure acetic acid acts as a nonelectrolyte, like water,
with very few ions being present. However, on dilution with water, more ions
are formed as a result of the interaction of the acetic acid with the water. A
water molecule pulls an H^{1+} off of an acetic acid molecule to form the $H_3O^{1+}_{(aq)}$
ion (the hydronium ion, see Section 14–4, Chapter 14) to a small extent. These
new ions, $H_3O^{1+}_{(aq)}$ and $C_2H_3O_2^{1-}_{(aq)}$, are responsible for the observed current
when acetic acid is diluted in water as shown by the following equations:

$$HC_2H_3O_{2(\ell)} \rightleftharpoons H^{1+}_{(\ell)} + C_2H_3O_2^{1-}_{(\ell)} \tag{10–1}$$

does not glow (not enough ions)

$$HC_2H_3O_{2(\ell)} + H_2O_{(\ell)} \rightleftharpoons H_3O^{1+}_{(aq)} + C_2H_3O_2^{1+}_{(aq)} \tag{10–2}$$

glows weakly

10-2   Guidelines for Writing Ionic Equations

ionic equation A method of expressing a chemical reaction involving compounds that exist mostly in ionic form in aqueous solution.

In our discussion of chemical equations (See Section 9–1), we stated that equations may be written in two general ways: as *molecular equations* or as *ionic equations*. We considered molecular equations in detail in Chapter 9, and we will now examine ionic equations. An **ionic equation** is a method of expressing a chemical reaction involving compounds that exist mostly in ionic form in aqueous solution. For *ionic compounds*, the reacting particles are actually *ions*; hence, in ionic equations the ions are written as they *actually* exist in the solution. Therefore, ionic equations give a better representation of a chemical change in aqueous solution than do molecular equations.

In the discussion on balancing molecular equations (see Section 9–3), we suggested a few guidelines to help you balance equations by inspection. We suggest the following guidelines for writing ionic equations.

1. Complete and balance an equation in the form in which it is given to you. If it is given in the molecular form, then complete the equation in the molecular form, balance it, and *then* change it to the ionic form. If it is given in ionic form, complete it and balance it in the ionic form. To complete chemical equations, see Chapter 9.

2. The formulas for compounds written in *molecular form* are
   (a) *Nonelectrolytes*, such as those listed in Table 10–1.
   (b) *Weak electrolytes*, such as those listed in Table 10–1.
   (c) *Solids and precipitates*[4] from aqueous solutions [see solubility rules (Section 9–9) or inside back cover], such as $CaCO_{3(s)}$ and $AgCl_{(s)}$.
   (d) *Gases*. Those such as H_2, N_2, and O_2 are written as the diatomic gas.

3. All *strong electrolytes*, such as those listed in Table 10–1, are written in *ionic* form. As you may have noticed, Table 10–1 is very important in writing ionic equations.

4. When you write compounds in ionic form, use subscripts only to express the composition of polyatomic ions. For example, **1** mol of sulfuric acid (H_2SO_4) in ionic form is written as $2H^{1+} + SO_4^{2-}$ (use subscript 4, since SO_4^{2-} is a polyatomic ion). Write **3** mol of sodium sulfate (**3** Na_2SO_4) as $6 Na^{1+} + 3SO_4^{2-}$.

5. Check ($\sqrt{}$) each ion (monoatomic or polyatomic ion) and each atom to make sure it is balanced on both sides of the equation. *The net charge on each side of the equation must be the same.*

---

[4] A solid and a precipitate most often exist as ions, but both are customarily written in the molecular form. The reason is that the ions in the solid are *not* bound to solvent molecules, and they are *not* separated from ions of the opposite charge in the manner that ions from the dissociation of soluble ionic crystals are separated in aqueous solutions.

net ionic equation An ionic
equation that shows only
those *ions* that have actually
undergone a chemical
change.

6. The **net ionic equation** shows only those *ions* that have undergone a
chemical change. Cross out the ions appearing on both sides of the equation
that have *not* undergone a change; do not include these ions in the net
ionic equation. Include these unaltered ions in the total ionic equation but
not the net ionic equation; they just go along for the ride! Finally, check
($\checkmark$) the net ionic equation for ions, atoms, and charge and to see that the
coefficients are in the lowest possible ratio of whole numbers.

10–3 Examples of Ionic Equations

EXAMPLE 10–1

Complete and balance the following equations, writing them as *total ionic
equations* and as *net ionic equations* [indicate any precipitate by (s) and
any gas by (g)].

(a) $AgNO_{3(aq)} + HCl_{(aq)} \rightarrow$

SOLUTION　Completing and balancing the equation according to
guideline 1 gives

$$AgNO_{3(aq)} + HCl_{(aq)} \longrightarrow AgCl_{(s)} + HNO_3$$

First, consider each of the reactants and products and decide whether
they are to be written in molecular or ionic form:

FORMULA	IDENTIFICATION	CONCLUSION
$AgNO_3$	Salt, strong electrolyte	Ionic form
HCl	Strong acid, strong electrolyte	Ionic form
AgCl	Salt, precipitate (see solubility rules)	Molecular form
HNO_3	Strong acid, strong electrolyte	Ionic form

Write the total ionic equation by applying guidelines 2 and 3. All
compounds here are written in ionic form, except AgCl, which is a
precipitate (see Table 10–1 and solubility rules in Section 9–9). The
total ionic equation is

$$Ag^{1+}_{(aq)} + NO_3^{1-}_{(aq)} + H^{1+}_{(aq)} + Cl^{1-}_{(aq)} \longrightarrow$$
$$AgCl_{(s)} + H^{1+} + NO_3^{1-}$$

Check each ion, atom, and charge according to guideline 4. The total
ionic equation is

$$\overset{\checkmark}{Ag}^{1+}_{(aq)} + \overset{\checkmark}{NO_3}^{1-}_{(aq)} + \overset{\checkmark}{H}^{1+}_{(aq)} + \overset{\checkmark}{Cl}^{1-}_{(aq)} \longrightarrow$$

Charges: $1^+ + 1^- + 1^+ + 1^- = 0$

$$\overset{\checkmark}{Ag}\overset{\checkmark}{Cl}_{(s)} + \overset{\checkmark}{H}^{1+} + \overset{\checkmark}{NO_3}^{1-} \qquad Answer$$

$= 0 + 1^+ + 1^- = 0$

Write the net ionic equation by crossing out ions that appear on both sides of the equation, according to guideline 5. Check the final net ionic equation for ions, atoms, charge, and lowest possible ratio of coefficients:

$$Ag^{1+}_{(aq)} + \cancel{NO_3^{1-}}_{(aq)} + \cancel{H^{1+}}_{(aq)} + Cl^{1-}_{(aq)} \longrightarrow$$

$$AgCl_{(s)} + \cancel{H^{1+}} + \cancel{NO_3^{1-}}$$

The *net ionic equation* is

$$Ag^{1+}_{(aq)} + Cl^{1-}_{(aq)} \longrightarrow AgCl_{(s)} \quad Answer$$

Charges: 1^+ + $1^- = 0$ = 0

Study hint

Another example is the reaction between silver acetate and potassium chloride. See if you get the *same* net ionic equation.

From the net ionic equation, you should note that this reaction is the reaction of any soluble ionic silver salt with a soluble strongly ionic chloride compound.

(b) $NaOH_{(aq)} + H_2SO_{4(aq)} \rightarrow$

SOLUTION   Completing and balancing the equation according to guideline 1 gives

$$2NaOH_{(aq)} + H_2SO_{4(aq)} \longrightarrow Na_2SO_4 + 2H_2O$$

First, consider each of the reactants and products and decide whether they are to be written in molecular or ionic form:

FORMULA	IDENTIFICATION	CONCLUSION
NaOH	Strong base, strong electrolyte	Ionic form
H_2SO_4	Strong acid, strong electrolyte	Ionic form
Na_2SO_4	Salt, soluble (see solubility rules), strong electrolyte	Ionic form
H_2O	Nonelectrolyte	Molecular form

Write the total ionic equation by applying guidelines 2 and 3. All compounds here are written in ionic form, except H_2O (a nonelectrolyte; see Table 10–1).

The *total ionic equation* is

$$2Na^{1+}_{(aq)} + 2OH^{1-}_{(aq)} + 2H^{1+}_{(aq)} + SO_4^{2-}_{(aq)} \longrightarrow$$

$$2Na^{1+} + SO_4^{2-} + 2H_2O$$

Check each ion, atom, and charge according to guideline 4.

$$2Na^{1+}_{(aq)} + 2OH^{1-}_{(aq)} + 2H^{1+}_{(aq)} + SO_4^{2-}_{(aq)} \longrightarrow$$

Charges: $2(1^+)$ + $2(1^-)$ + $2(1^+)$ + $2^- = 0$

$$2Na^{1+} + SO_4^{2-} + 2H_2O \quad Answer$$

$= 2(1^+)$ + 2^- + $0 = 0$

Crossing out the ions that appear on both sides of the equation according to guideline 5 gives the net ionic equation. Check the net ionic equation for ions, atoms, charge, and lowest possible ratio of coefficients:

$$2\cancel{Na^{1+}}_{(aq)} + 2OH^{1-}_{(aq)} + 2H^{1+}_{(aq)} + \cancel{SO_4^{2-}}_{(aq)} \longrightarrow$$
$$2\cancel{Na^{1+}} + \cancel{SO_4^{2-}} + 2H_2O$$

$$2OH^{1-}_{(aq)} + 2H^{1+}_{(aq)} \longrightarrow 2H_2O$$

Dividing both sides of the equation by 2, we find that the *net ionic equation* is

$$\overset{\checkmark\checkmark}{OH^{1-}}_{(aq)} + \overset{\checkmark}{H^{1+}}_{(aq)} \longrightarrow \overset{\checkmark\checkmark}{H_2O} \qquad Answer$$

Charges: 1^- + $1^+ = 0$ = 0

This reaction is a neutralization reaction (see Section 9–10), and as a net ionic equation it is simply the reaction of a hydroxide ion with a hydrogen ion to form water. This, then, is the reaction of any strong acid with a strong base.

Study hint

Another example is the reaction between hydrochloric acid and barium hydroxide. See if you get the *same* net ionic equation.

(c) $Al_{(s)} + H_2SO_{4(aq)} \rightarrow$

SOLUTION   This is a single-replacement reaction involving the electromotive or activity series (see Section 9–8). According to guideline 1, completing and balancing the equation gives

$$2Al_{(s)} + 3H_2SO_{4(aq)} \longrightarrow Al_2(SO_4)_3 + 3H_{2(g)}$$

Again, consider each of the reactants and products and decide whether they are to be written in molecular or ionic form:

FORMULA	IDENTIFICATION	CONCLUSION
Al	Metal, solid	Molecular form
H_2SO_4	Strong acid, strong electrolyte	Ionic form
$Al_2(SO_4)_3$	Salt, soluble (see solubility rules), strong electrolyte	Ionic form
H_2	Gas	Molecular form

Write the total ionic equation by applying guidelines 2 and 3. All substances here are written in ionic form, except Al, a solid and a free metal (*not* an ion), and H_2, a gas. Checking as in guideline 4 gives the following *total ionic equation*:

$$2\overset{\checkmark}{Al}_{(s)} + 6\overset{\checkmark}{H^{1+}}_{(aq)} + 3\overset{\checkmark}{SO_4^{2-}}_{(aq)} \longrightarrow 2\overset{\checkmark}{Al^{3+}} + 3\overset{\checkmark}{SO_4^{2-}} + 3\overset{\checkmark}{H_{2(g)}}$$

Charges: 0 + $6(1^+)$ + $3(2^-) = 0$ = $2(3^+)$ + $3(2^-)$ + $0 = 0$

Answer

Crossing out the ions that appear on both sides of the equation, and checking again as in guideline 5, we have the following *ionic equation*:

$$2Al_{(s)} + 6H^{1+}_{(aq)} + 3\cancel{SO_4^{2-}}_{(aq)} \longrightarrow 2Al^{3+} + 3\cancel{SO_4^{2-}} + 3H_{2(g)}$$

The *net ionic equation* is

$$2\overset{\checkmark}{Al}_{(s)} + 6\overset{\checkmark}{H}^{1+}_{(aq)} \longrightarrow 2\overset{\checkmark}{Al}^{3+} + 3\overset{\checkmark}{H}_{2(g)} \qquad Answer$$

Charges: $0 \;+\; 6(1^+) = 6^+ \;=\; 2(3^+) \;+\; 0 = 6^+$

(d) $NH_{3(aq)} + H_2O_{(\ell)} + Al_2(SO_4)_{3(aq)} \rightarrow$

SOLUTION Completing and balancing the equation according to guideline 1, with $NH_3 + H_2O$ acting as $NH_4^{1+} + OH^{1-}$ gives

$$6NH_{3(aq)} + 6H_2O_{(\ell)} + Al_2(SO_4)_{3(aq)} \longrightarrow$$

$$3(NH_4)_2SO_{4(aq)} + 2Al(OH)_{3(s)}$$

Once again, consider each of the reactants and products and decide whether they are to be written in molecular or ionic form:

FORMULA	IDENTIFICATION	CONCLUSION
NH_3	Weak base, weak electrolyte	Molecular form
H_2O	Nonelectrolyte	Molecular form
$Al_2(SO_4)_3$	Salt, soluble (see solubility rules), strong electrolyte	Ionic form
$(NH_4)_2SO_4$	Salt, soluble (see solubility rules), strong electrolyte	Ionic form
$Al(OH)_3$	Weak base, weak electrolyte and precipitate (see solubility rules)	Molecular form

Write the total ionic equation by applying guidelines 2 and 3. All compounds here are written in ionic form, except $NH_{3(aq)}$, a weak electrolyte; H_2O, a nonelectrolyte; and $Al(OH)_3$, a precipitate (see Table 10–1 and solubility rules in Section 9–9). Checking as in guideline 4 gives the following *total ionic equation*:

$$6\overset{\checkmark\checkmark}{NH}_{3(aq)} + 6\overset{\checkmark\ \checkmark}{H_2O}_{(\ell)} + 2\overset{\checkmark}{Al}^{3+}_{(aq)} + 3\overset{\checkmark}{SO_4^{2-}} \longrightarrow$$

Charges: $0 \;+\; 0 \;+\; 2(3^+) \;+\; 3(2^-) = 0 \;=$

$$6\overset{\checkmark\checkmark}{NH_4}^{1+} + 3\overset{\checkmark}{SO_4}^{2-} + 2\overset{\checkmark\ \checkmark\checkmark}{Al(OH)}_{3(s)} \qquad Answer$$

$6(1^+) \;+\; 3(2^-) \;+\; 0 = 0$

Crossing out the ions that appear on both sides of the equation, and checking again as in guideline 5, we have the following *net ionic equation*:

$$6NH_{3(aq)} + 6H_2O_{(\ell)} + 2Al^{3+}_{(aq)} + 3\cancel{SO_4^{2-}} \longrightarrow$$

$$6NH_4^{1+} + 3\cancel{SO_4^{2-}} + 2Al(OH)_{3(s)}$$

Dividing both sides of the equation by 2, we obtain

$$3\overset{\checkmark\checkmark}{NH}_{3(aq)} + 3\overset{\checkmark\checkmark}{H_2O}_{(\ell)} + \overset{\checkmark}{Al}^{3+}{}_{(aq)} \longrightarrow$$

Charges: 0   +   0   + $3^+ = 3^+$ =

$$3\overset{\checkmark\checkmark}{NH_4}^{1+} + \overset{\checkmark}{Al(\overset{\checkmark}{O}H)}_{3(s)} \qquad Answer$$

$3(1^+)$ + $0 = 3^+$

(e) $Ag^{1+}{}_{(aq)} + H_2S_{(aq)} \rightarrow$

SOLUTION   This is a *double*-replacement reaction (see Section 9–9), not a single-replacement reaction. The silver salt is written in ionic form without showing the anion. Completing and balancing the equation according to guideline 1 gives the following *ionic equation*:

$$2Ag^{1+}{}_{(aq)} + H_2S_{(aq)} \longrightarrow Ag_2S_{(s)} + 2H^{1+}$$

Again, consider each of the reactants and products and decide whether they are to be written in molecular or ionic form:

FORMULA	IDENTIFICATION	CONCLUSION
Ag^{1+}	Ion	Ionic form
H_2S	Weak acid, weak electrolyte	Molecular form
Ag_2S	Salt, precipitate (see solubility rules), weak electrolyte	Molecular form
H^{1+}	Ion	Ionic form

Write the precipitate, Ag_2S, and the weak electrolyte, H_2S, in molecular form, according to guideline 2. Check the ionic equation for ions, atoms, and charge, according to guideline 4:

$$2\overset{\checkmark}{Ag}^{1+}{}_{(aq)} + \overset{\checkmark\checkmark}{H_2S}_{(aq)} \longrightarrow \overset{\checkmark\checkmark}{Ag_2S}_{(s)} + 2\overset{\checkmark}{H}^{1+} \qquad Answer$$

Charges: $2(1^+)$ + $0 = 2^+$ = 0 + $2(1^+) = 2^+$

The *net ionic equation* is the same as this ionic equation, since the same ions do not appear on *both* sides of the equation.

Exercise 10–1

Complete, if necessary, and balance the following equations, writing them as *total ionic equations* and as *net ionic equations* [indicate any precipitate by (s) and any gas by (g)].

(a) $BaCl_{2(aq)} + (NH_4)_2CO_{3(aq)} \rightarrow BaCO_3 + NH^4Cl$
(b) $Fe(NO_3)_{3(aq)} + NH_{3(aq)} + H_2O_{(\ell)} \rightarrow Fe(OH)_3 + NH_4NO_3$
(c) $SrCl_{2(aq)} + K_2CO_{3(aq)} \rightarrow$
(d) $Na_2CO_{3(aq)} + HCl_{(aq)} \rightarrow$
(e) $KCl_{(aq)} + AgNO_{3(aq)} \rightarrow$

Carbon dioxide: what goes around comes around

Solid carbon dioxide is unusual in that it passes directly from the solid phase into the gas phase under normal conditions of temperature and pressure. This is called sublimation. *In this picture, a piece of solid carbon dioxide at the bottom of the graduated cylinder is subliming to form gaseous carbon dioxide (white vapor).*

Name: Carbon dioxide (CO_2) is occasionally called carbonic acid gas or carbonic anhydride. Solid CO_2 is frequently called *dry ice.*

Appearance: Carbon dioxide is a colorless, odorless gas under normal conditions of temperature and pressure. In the solid form (see photograph), it is a cold ($-78°C$) white solid that passes directly from the solid phase into the gas phase (sublimes).

Occurrence: Carbon dioxide can be found all over the earth. It comprises about 0.03 percent of the atmosphere by volume. Carbon dioxide also dissolves in water (oceans, streams, and lakes) to produce a dilute solution of carbonic acid (H_2CO_3):

$$CO_{2(g)} + H_2O_{(l)} \rightleftharpoons H_2CO_{3(aq)}$$

Carbon dioxide also occurs in the earth as carbonate salts, typically limestone (calcite, $CaCO_3$) or dolomite [$CaMg(CO_3)_2$]. The carbon dioxide tied up in such rocks is easily released by heating (Section 9–7) or adding acid (Section 9–9):

$$CaCO_3 \xrightarrow{\Delta} CaO + CO_{2(g)}$$

$$CaCO_3 + 2HCl \longrightarrow CaCl_2 + H_2O + CO_{2(g)}$$

Source: Most commercially available carbon dioxide is produced by one of four methods: (1) the reaction between methane (natural gas, CH_4) and water to give CO_2 and H_2, hydrogen gas; (2) the burning of coal in air to give CO_2; (3) heating limestone ($CaCO_3$); and (4) the fermentation of sugar by yeast to give ethyl alcohol (C_2H_6O) and CO_2.

(1) $CH_4 + 2H_2O \longrightarrow CO_2 + 4H_2$

(2) $C + O_2 \xrightarrow{\Delta} CO_2$

(3) $CaCO_3 \xrightarrow{\Delta} CaO + CO_2$

(4) $C_{12}H_{22}O_{11} + H_2O \xrightarrow{yeast} 4CO_2 + 4C_2H_6O$

Its Role in Our World: Carbon dioxide has many uses in our society. Large amounts of CO_2 are consumed during the industrial production of hydrogen gas (H_2), methanol (CH_3OH), urea (CH_4N_2O, a fertilizer and used in preparation of certain plastics), and ammonia (NH_3). On a more personal level, CO_2 is used in the production of carbonated beverages, and dry ice helps ice cream vendors keep their products from melting.

But carbon dioxide plays a far more important role in our world than this. It is part of the carbon cycle and life on our planet. Green plants convert carbon dioxide, water, and sunlight into sugar (glucose, $C_6H_{12}O_6$) by photosynthesis:

$$6CO_2 + 6H_2O + \text{light energy} \xrightarrow{\text{chlorophyll}}$$
$$C_6H_{12}O_6 + 6O_2$$

The plants are then eaten by plant-eating animals, and the plant-eating animals are in turn eaten by meat-eating animals. Both plants and animals regenerate some carbon dioxide as they live and grow, a process called *respiration* (the reverse of photosynthesis). The death of these living organisms then returns the carbon to the atmosphere as carbon dioxide. Some of the dead matter is converted into coal, oil, natural gas, or rocks. Humans release carbon dioxide to the atmosphere by burning massive amounts of fuels such as oil, gas, coal, and wood.

There is speculation that human activity is in the process of disturbing the balance of processes in the carbon cycle. By releasing large amounts of CO_2 into the atmosphere, the concentration of carbon dioxide in the atmosphere may be slowly increasing. Increased concentrations of CO_2 could trap extra heat in our atmosphere (the Greenhouse Effect) and lead to a gradual (but significant) warming of the earth. The consequences of such a warming are not yet clear, but

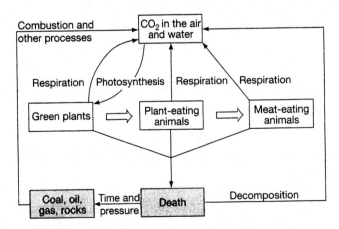

Carbon cycles through our world as life goes on.

some scientists believe that the results could be catastrophic for life on our planet. Currently, world governments are wrestling with these issues, but not enough is yet known about the levels of CO_2 in the atmosphere and the Greenhouse Effect for leaders to agree on what, if anything, should be done.

Unusual Facts: Carbon dioxide actually has a slightly acid taste. The taste is due to the carbonic acid (H_2CO_3) that forms in your mouth as the CO_2 dissolves in your saliva. You can "taste" CO_2 by drinking carbonated mineral water. Humans cannot breathe air containing more than 5 to 10 percent CO_2 without losing consciousness, and prolonged exposure can result in death.

Problems

1. Complete and balance the following equations, writing them as *total ionic equations* and as *net ionic equations* [indicate any precipitate by (s) and any gas by (g)].

   (a) $HgCl_{2(aq)} + H_2S_{(aq)} \rightarrow$
   (b) $MgSO_{4(aq)} + NaOH_{(aq)} \rightarrow$
   (c) $CaO_{(s)} + HCl_{(aq)} \rightarrow$
   (d) $Al(OH)_{3(s)} + HCl_{(aq)} \rightarrow$
   (e) $FeSO_{4(aq)} + (NH_4)_2S_{(aq)} \rightarrow$

2. Complete and balance the following equations, writing them as *total ionic equations* and as *net ionic equations* [indicate any precipitate by (s) and any gas by (g)].

   (a) $MgCl_{2(aq)} + Na_2CO_{3(aq)} \rightarrow$
   (b) $HCl_{(aq)} + Ba(OH)_{2(aq)} \rightarrow$
   (c) $Pb(NO_3)_{2(aq)} + KOH_{(aq)} \rightarrow$
   (d) $Ca_{(s)} + HBr_{(aq)} \rightarrow$
   (e) $AgNO_{3(aq)} + Na_2CO_{3(aq)} \rightarrow$

General Problem

3. The Acropolis in Athens, Greece, is slowly deteriorating. It is composed of marble ($CaCO_3$), which reacts slowly with sulfuric acid from air pollution to form a *salt*, which is *washed* away, destroying this famous historical site. The sulfuric acid is formed from the air pollutant sulfur trioxide and water. Complete and balance the two equations described above, writing them as molecular equations and as total ionic and net ionic equations. (Polymer coating of the marble statues and structures has been suggested as a possible means of protecting these works of art.)

Answers to Exercise

10–1. (a) $Ba^{2+}_{(aq)} + 2Cl^{1-}_{(aq)} + 2NH_4^{1+}_{(aq)} + CO_3^{2-}_{(aq)} \rightarrow$
$$BaCO_{3(s)} + 2NH_4^{1+} + 2Cl^{1-}$$

Net: $Ba^{2+}_{(aq)} + CO_3^{2-}_{(aq)} \rightarrow BaCO_{3(s)}$

(b) $Fe^{3+}_{(aq)} + 3NO_3^{1-}_{(aq)} + 3NH_{3(aq)} + 3H_2O_{(\ell)} \rightarrow$
$$Fe(OH)_{3(s)} + 3NH_4^{1+} + 3NO_3^{1-}$$

Net: $Fe^{3+}_{(aq)} + 3NH_{3(aq)} + 3H_2O_{(\ell)} \rightarrow Fe(OH)_{3(s)} + 3NH_4^{1+}$

(c) $Sr^{2+}_{(aq)} + 2Cl^{1-}_{(aq)} + 2K^{1+}_{(aq)} + CO_3^{2-}_{(aq)} \rightarrow$
$$2K^{1+} + 2Cl^{1-} + SrCO_{3(s)}$$

Net: $Sr^{2+}_{(aq)} + CO_3^{2-}_{(aq)} \rightarrow SrCO_{3(s)}$

(d) $2Na^{1+}_{(aq)} + CO_3^{2-}_{(aq)} + 2H^{1+}_{(aq)} + 2Cl^{1-}_{(aq)} \rightarrow$
$$2Na^{1+} + 2Cl^{1-} + H_2O + CO_{2(g)}$$

Net: $CO_3^{2-}_{(aq)} + 2H^{1+}_{(aq)} \rightarrow H_2O + CO_{2(g)}$
[H_2CO_3 is not stable and forms H_2O and $CO_{2(g)}$.]

(e) $K^{1+}_{(aq)} + Cl^{1-}_{(aq)} + Ag^{1+}_{(aq)} + NO_3^{1-}_{(aq)} \rightarrow AgCl_{(s)} + K^{1+} + NO_3^{1-}$
Net: $Cl^{1-}_{(aq)} + Ag^{1+}_{(aq)} \rightarrow AgCl_{(s)}$

QUIZ   You may use the Periodic Table, the Electromotive or Activity Series, and the Rules for the Solubility of Inorganic Substances in Water.

1. Complete and balance the following equations, writing them as *total ionic equations* and as *net ionic equations* [indicate any precipitate by (s) and any gas by (g)].

(a) $CdCl_{2(aq)} + H_2S_{(aq)} \rightarrow$
(b) $Fe(NO_3)_{3(aq)} + NaOH_{(aq)} \rightarrow$
(c) $Mg_{(s)} + HCl_{(aq)} \rightarrow$

2. Hydrogen sulfide gas has the odor of rotten eggs and is highly poisonous. It can be prepared by the reaction between sodium sulfide and a strong acid—for example, sulfuric acid. Write the balanced net ionic equation for this reaction.

11

STOICHIOMETRY

COUNTDOWN

ELEMENT	ATOMIC MASS UNITS (AMU)
C	12.0
O	16.0
H	1.0
Ca	40.1
N	14.0

5 Calculate the number of moles in each of the following quantities of compounds (Sections 8–2 and 8–3):

(a) 26.4 g CO_2 (0.600 mol)

(b) 4.25 L (STP) CO_2 (0.190 mol)

4 Calculate the number of grams in each of the following quantities (Section 8–2):

(a) 0.240 mol $C_{12}H_{22}O_{11}$ (sugar, sucrose)   (82.1 g)

(b) 0.0150 mol $Ca(OH)_2$ (1.11 g)

3 Calculate the number of liters at STP that the following gases would occupy (Sections 8–2 and 8–3).

(a) 0.165 mol of nitrogen gas (N_2)   (3.70 L)

(b) 4.25 g of nitrogen gas (N_2)   (3.40 L)

2 Balance each of the following equations by inspection (Sections 9–3 and 9–4).

(a) $C_2H_6O_{(\ell)} + O_{2(g)} \xrightarrow{\Delta} CO_{2(g)} + H_2O_{(g)}$
(ethyl alcohol)

$$(1 + 3 \longrightarrow 2 + 3)$$

(b) $Sb_2S_{3(s)} + Fe_{(s)} \xrightarrow{\Delta} Sb_{(s)} + FeS_{(s)}$

$$(1 + 3 \longrightarrow 2 + 3)$$

1 Complete and balance the following equations. Indicate any precipitate by (s) and any gas by (g) (Sections 9–7 and 9–9).

(a) $KClO_{3(s)} \xrightarrow[MnO_2]{\Delta}$

$$[2KClO_{3(s)} \xrightarrow[MnO_2]{\Delta} 2KCl + 3O_{2(g)}]$$

(b) $Pb(C_2H_3O_2)_{2(aq)} + Na_2SO_{4(aq)} \longrightarrow$

$$[Pb(C_2H_3O_2)_{2(aq)} + Na_2SO_{4(aq)} \longrightarrow PbSO_{4(s)} + 2NaC_2H_3O_2]$$

OBJECTIVES

1 Give the distinguishing characteristics of each of the following terms:
(a) stoichiometry (introduction to Chapter 11)
(b) law of conservation of mass (Section 11–1)
(c) limiting reagent (Section 11–4)
(d) excess reagent (Section 11–4)
(e) theoretical yield (Section 11–4)
(f) actual yield (Section 11–4)
(g) percent yield (Section 11–4)
(h) Gay–Lussac's law of combining volumes (Section 11–6)
(i) heat of reaction (Section 11–7)
(j) joule (Section 11–7)
(k) exothermic reaction (Section 11–7)
(l) endothermic reaction (Section 11–7)

2 Given the mass of a reactant or product, a balanced equation, and the Table of Approximate Atomic Masses, calculate the mass of another reactant required or the mass that could be produced of another product (*mass-mass* stoichiometry problems, *direct* examples, Examples 11–1 and 11–2, Exercise 11–1, Problems 1, 2, and 3).

3 Given the mass in grams or moles of a reactant or a product, a balanced equation, and the Table of Approximate Atomic Masses, calculate the number of moles or grams of another reactant required or of moles or grams that could be produced of another product (*mass-mass* stoichiometry problems, *indirect* examples, Examples 11–3 and 11–4, Exercise 11–2, Problems 4, 5, and 6).

4 (a) Given the mass of two or more reactants in a given reaction, a balanced equation, and the Table of Approximate Atomic Masses, determine the limiting reagent.
(b) Given the mass of two or more reactants in a given reaction, a balanced equation, and the Table of Approximate Atomic Masses, calculate the mass of a single product that could be produced.
(*Limiting reagent mass-mass* stoichiometry problems, Example 11–5, Exercise 11–3, and Problems 7 and 8.)

5 Given the actual yield and the theoretical yield, calculate the percent yield (Example 11–6, Exercise 11–4, Problem 9).

6 Given the volume (at STP) of a gaseous reactant or product, a balanced equation, and the Table of Approximate Atomic Masses, calculate the mass in grams or moles of another reactant required or another product produced, and *vice versa* (*mass-volume* stoichiometry problems, Examples 11–7 and 11–8, Exercise 11–5, Problems 10, 11, and 12).

7 Given the volume of a gaseous reactant or product and a balanced equation, calculate the volume of another gaseous reactant required or another gaseous product produced. All gases are measured at the same temperature and pressure (*volume-volume* stoichiometry problems, Example 11–9, Exercise 11–6, Problems 13, 14, and 15).

8 (a) Given a completed chemical equation with the heat of reaction, determine whether the reaction is exothermic or endothermic.

(b) Given the amount of a reactant or product, the Table of Approximate Atomic Masses, and a completed chemical equation with the heat of reaction, calculate the amount of heat energy produced or needed, and *vice versa*.

[Heat of reaction problems, for (a) and (b), Examples 11–10 and 11–11, Exercise 11–7, Problems 16, 17, and 18.]

In our economy we sell goods and services. We attempt to sell these goods or services at a profit. For example, you are going to college now, so you can "sell" your "services" at a later date and support yourself and/or family on a salary ("profit"). The chemical industry operates on the same principle in that it buys chemicals inexpensively and makes new chemicals ("goods") and then sells these new chemicals at a profit. Two of the many questions needing answers in the chemical industry are (1) starting with X and Y amounts of starting chemicals, how much new chemical (product), can *theoretically* be produced, and (2) how much can *actually* be produced by the company's plant operation (see Fig. 11–1)? Chemists working in an applied research laboratory and chemical engineers working in a small-scale chemical manufacturing plant termed the *pilot plant* can answer the second question. We can answer the first question, the theoretical amount, by performing suitable chemical calculations. This type of chemical calculation is called *stoichiometry*.

stoichiometry Measurement of the relative quantities of chemical reactants and products in a reaction.

   Stoichiometry (pronounced stoi'ke̅·om'i·tre̅) is measurement of the relative quantities of chemical reactants and products in a reaction. In Chapter 8, we calculated formula or molecular masses (Section 8–1), moles (Section 8–2), and molar volumes of gases (Section 8–3). In Chapter 9 we completed and balanced chemical equations. In this chapter we use molar information from balanced equations to calculate the amounts of chemicals or energy produced or required in these equations.

Figure 11–1 *Ms. Jones ponders the question, "How much of Z can we get?"*

11–1 Information Obtained from a Balanced Equation

A completed and *balanced* equation affords more information than simply which substances are reactants and which are products. It also gives the relative quantities involved and it is very useful in carrying out certain calculations. Let us consider the oxidation of ethane gas to produce carbon dioxide and water:

$$2C_2H_{6(g)} + 7O_{2(g)} \xrightarrow{\Delta} 4CO_{2(g)} + 6H_2O_{(g)}$$
$$\text{ethane}$$

This balanced equation gives the following information.

1. **Reactants and products:** C_2H_6 (ethane) reacts with O_2 (oxygen) when sufficient heat (Δ) is applied to produce CO_2 (carbon dioxide) and H_2O (gaseous water).

2. **Molecules of reactants and products:** 2 molecules of C_2H_6 need 7 molecules of O_2 to react and produce 4 molecules of CO_2 and 6 molecules of H_2O.

3. **Moles of reactants and products:** 2 mol of C_2H_6 molecules need 7 mol of O_2 molecules to react and produce 4 mol of CO_2 molecules and 6 mol of H_2O molecules.

4. **Volumes of gases:** 2 volumes of C_2H_6 need 7 volumes of O_2 to react and produce 4 volumes of CO_2 and 6 volumes of H_2O, *if all volumes are measured as gases at the same temperature and pressure.*

5. **Relative masses of reactants and products:** 60.0 g of C_2H_6:

$$\left(2 \text{ mol } C_2H_6 \times \frac{30.0 \text{ g } C_2H_6}{1 \text{ mol } C_2H_6} = 60.0 \text{ g } C_2H_6 \right)$$

Study hint

The mole relations (coefficients) are regarded as *exact* values and are *not* considered in determining significant digits.

need 224 g of O_2

$$\left(7 \ \cancel{mol \ O_2} \times \frac{32.0 \ g \ O_2}{1 \ \cancel{mol \ O_2}} = 224 \ g \ O_2\right)$$

to react to produce 176 g of CO_2

$$\left(4 \ \cancel{mol \ CO_2} \times \frac{44.0 \ g \ CO_2}{1 \ \cancel{mol \ CO_2}} = 176 \ g \ CO_2\right)$$

and 108 g of H_2O

$$\left(6 \ \cancel{mol \ H_2O} \times \frac{18.0 \ g \ H_2O}{1 \ \cancel{mol \ H_2O}} = 108 \ g \ H_2O\right)$$

law of conservation of mass Mass can be neither created nor destroyed, and the total mass of the substances involved in a physical or chemical change remains constant.

Notice that the sum of the masses of the reactants (60.0 g + 224 g = 284 g) is equal to the sum of the masses of products (176 g + 108 g = 284 g), obeying the **law of conservation of mass**. This law states that mass is neither created nor destroyed and the total mass of the substances involved in a physical or chemical change remains constant.

Study hint

Before you read on, turn to Examples 8–2, 8–4, 8–8, and 8–9, and study these examples. You must know: (1) how to calculate moles, given grams or liters of a gas at STP, and (2) how to calculate grams or liters of a gas at STP, given moles.

11–2 The Mole Method of Solving Stoichiometry Problems. The Three Basic Steps

There are a number of methods available for solving stoichiometry problems. The method that we consider the best is the *mole method*, which is an application of our general method of problem solving—the *factor-unit* method. Three basic steps are involved in working problems by the *mole method*.

STEP I: Calculate moles of atoms, formula units, molecules, or ions of the element, compound, or ion from the mass or volume (if gases) of the known substance or substances in the problem.

STEP II: Using the coefficients of the substances in the *balanced* equation, calculate moles of the unknown quantities in the problem.

STEP III: From moles of the unknown quantities calculated, determine the mass or volume (for gases) of these unknowns in the units requested by the problem.

As you can see, the key to this method is the *mole*. Think *MOLES!*

The application of these three basic steps is shown in diagram form in Fig. 11–2.

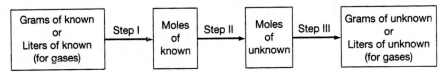

Figure 11-2 *The three basic steps in solving stoichiometry problems. (Prior to Step I and after Step III an additional calculation may be required to convert to or from some mass measurement other than grams.)*

11-3 Types of Stoichiometry Problems

There are three types of stoichemtry problems:

1. Mass–mass ("weight-weight")

2. Mass–volume or volume-mass ("weight"-volume or volume-"weight")

3. Volume–volume

We now wish to apply the three basic steps to the three types of stoichiometry problems.

11-4 Mass–Mass Stoichiometry Problems

In mass–mass stoichiometry problems the quantities of both the known and unknown are given or asked for in mass units. Whether the known quantity is expressed in grams or moles affects which steps we need to apply.

Direct Examples

We shall first consider some *direct examples*, in which the known is expressed in mass units as grams, and the unknown is asked for in mass units as grams; these *direct examples* involve all three basic steps. We emphasize again that the equation must be **balanced** before the calculation is begun.

EXAMPLE 11-1 —————————————————————

Calculate the number of grams of carbon dioxide gas (CO_2) produced by burning 72.0 g of ethane (C_2H_6) gas in oxygen gas (O_2) to produce CO_2 gas and H_2O gas. The balanced equation for the reaction is

$$2C_2H_{6(g)} + 7O_{2(g)} \xrightarrow{\Delta} 4CO_{2(g)} + 6H_2O_{(g)}$$

SOLUTION   Since the equation is balanced, we can proceed to calculate the molecular masses of the substances involved in the calculation, which are CO_2 and C_2H_6.

$$CO_2 = 44.0 \text{ amu}$$

$$C_2H_6 = 30.0 \text{ amu}$$

Organize the data:

Known: 72.0 g of C_2H_6

Unknown: g of CO_2 produced

The calculation will involve converting from grams of known to grams of unknown, so all three basic steps will be involved.

STEP I: Calculate the moles of C_2H_6 molecules given. Since 1 mol has a mass of 30.0 g of C_2H_6,

$$72.0 \text{ g } C_2H_6 \times \frac{1 \text{ mol } C_2H_6}{30.0 \text{ g } C_2H_6} = \frac{72.0}{30.0} \text{ mol } C_2H_6 \text{ given}$$

STEP II: Calculate the moles of carbon dioxide molecules produced. From the balanced equation, the relation of C_2H_6 to CO_2 is given as 2 mol of C_2H_6 to 4 mol of CO_2. Therefore,

$$\frac{72.0}{30.0} \text{ mol } C_2H_6 \times \frac{4 \text{ mol } CO_2}{2 \text{ mol } C_2H_6} = \frac{72.0}{30.0} \times \frac{4}{2} \text{ mol } CO_2 \text{ produced}$$

The most common student error in these problems is to omit step II. *Step II is a most important step.*

STEP III: Calculate the grams of carbon dioxide produced. Since 1 mol of CO_2 has a mass of 44.0 g of CO_2.

$$\frac{72.0}{30.0} \times \frac{4}{2} \text{ mol } CO_2 \times \frac{44 \text{ g } CO_2}{1 \text{ mol } CO_2} = \frac{72.0}{30.0} \times \frac{4}{2} \times 44.0 \text{ g } CO_2$$

$$= 211.2 \text{ g } CO_2 = 211 \text{ g } CO_2 \quad Answer$$

Since our given quantity (72.0 g of C_2H_6) was expressed to three significant digits, our answer is also expressed to three significant digits (211 g of CO_2). [The mole relations (coefficients) are regarded as exact numbers and are not considered in computing significant digits.] The complete solution may be written as follows:

$$72.0 \text{ g } C_2H_6 \times \frac{1 \text{ mol } C_2H_6}{30.0 \text{ g } C_2H_6} \times \frac{4 \text{ mol } CO_2}{2 \text{ mol } C_2H_6} \times \frac{44.0 \text{ g } CO_2}{1 \text{ mol } CO_2} = 211 \text{ g } CO_2 \leftarrow$$

Step I   Step II   Step III

Answer

EXAMPLE 11–2

Calculate the number of grams of chlorine produced by the reaction of 25.0 g of manganese(IV) oxide with excess hydrochloric acid.

$$MnO_2 + 4HCl \longrightarrow MnCl_2 + Cl_2 + 2H_2O$$

SOLUTION   The equation is balanced, so we can proceed to calculate the formula mass of MnO_2 and the molecular mass of Cl_2. The formula mass of MnO_2 is 87.0 amu and the molecular mass of Cl_2 is 71.0 amu. The known quantity is 25.0 g of MnO_2 and the unknown quantity is the maximum number of grams of Cl_2 that may be obtained.

$$25.0 \text{ g } \cancel{MnO_2} \times \underbrace{\frac{1 \text{ mol } \cancel{MnO_2}}{87.0 \text{ g } \cancel{MnO_2}}}_{\text{Step I}} \times \underbrace{\frac{1 \text{ mol } \cancel{Cl_2}}{1 \text{ mol } \cancel{MnO_2}}}_{\text{Step II}} \times \underbrace{\frac{71.0 \text{ g } Cl_2}{1 \text{ mol } \cancel{Cl_2}}}_{\text{Step III}}$$

$$= 20.4 \text{ g } Cl_2 \longleftarrow$$

Answer

Exercise 11-1

Calculate the number of grams of potassium chlorate ($KClO_3$) that could be produced from the reaction of 20.8 g of chlorine gas (Cl_2) with potassium hydroxide (KOH), according to the following balanced equation:

$$3Cl_2 + 6KOH \longrightarrow 5KCl + KClO_3 + 3H_2O$$

Indirect Examples

Sometimes the information supplied is already in moles and we do not need Step I, or sometimes the information to be determined must be expressed in moles and we do not need Step III. Occasionally, *both* Steps I and III may be eliminated.

EXAMPLE 11-3

Calculate the number of moles of oxygen gas (O_2) produced by heating 1.226 g of potassium chlorate ($KClO_3$), according to the following balanced equation:

$$2KClO_3 \xrightarrow{\Delta} 2KCl + 3O_2$$

SOLUTION   The known quantity, 1.226 g of $KClO_3$, is given in grams. Therefore, Step I is needed to calculate moles of $KClO_3$. Step II converts the moles of $KClO_3$ to moles O_2, but Step III is *not* needed since we were asked to calculate moles of O_2. The formula mass of $KClO_3$ is 122.6 amu, as calculated from the atomic masses.

$$\underbrace{1.226 \text{ g } \cancel{KClO_3} \times \frac{1 \text{ mol } \cancel{KClO_3}}{122.6 \text{ g } \cancel{KClO_3}}}_{\text{Step I}} \times \underbrace{\frac{3 \text{ mol } O_2}{2 \text{ mol } \cancel{KClO_3}}}_{\text{Step II}} = 0.01500 \text{ mol } O_2$$

Answer

[Since our given quantity (1.226 g of $KClO_3$) was expressed to four significant digits, our answer is also expressed to four significant digits (0.01500 mol O_2)].

EXAMPLE 11–4

Calculate the number of moles of potassium nitrate (KNO_3) needed to produce 0.0520 mol of oxygen gas (O_2), according to the following balanced equation:

$$2KNO_3 \xrightarrow{\Delta} 2KNO_2 + O_2$$

SOLUTION   The known quantity, 0.0520 mol of O_2, is given in moles. Therefore, Step I is *not* needed and we can proceed to Step II and convert the moles of O_2 to moles of KNO_3. The problem asks for moles of KNO_3, so step III is also *not* needed. Step II is the only step required.

$$\underbrace{0.0520 \;\cancel{\text{mol } O_2} \times \frac{2 \text{ mol } KNO_3}{1 \cancel{\text{mol } O_2}}}_{\text{Step II}} = 0.104 \text{ mol } KNO_3 \qquad Answer$$

The elimination of these same steps can also be applied to mass-volume stoichiometry problems (see Section 11–5).

Exercise 11–2

Calculate the number of grams of potassium chlorate ($KClO_3$) that could be produced from the reaction of 0.200 mol of chlorine gas (Cl_2) with potassium hydroxide, according to the following balanced equation:

$$3Cl_2 + 6KOH \longrightarrow 5KCl + KClO_3 + 3H_2O$$

Limiting Reagent Examples

limiting reagent Reactant in a chemical reaction that is completely used up in the reaction, so called because the amount of this reactant limits the amount of new compounds that can be formed.

excess reagent Reactant in a chemical reaction that is *not* completely used up in the reaction, so called because, when the last amount of the new compound is formed, some of this reactant will be left over.

In carrying out chemical reactions, the quantities of reactants are not usually in exact stoichiometric amounts. One reactant may be used in excess of that theoretically needed for a complete reaction to take place according to the balanced equation. In such cases, the amount of product obtained depends upon the reactant completely used up, which is termed the **limiting reagent**. Many times the more expensive reactant is the limiting ragent and the cheaper reactant is in excess. This cheaper reactant is called the **excess reagent** ("leftover" reactant). The principle of limiting reagent is analogous to a party attended by both men and women. If there are seven (7) men, but only six (6) women, the maximum number of couples we could have would be six (6). The number of women *limits* the number of couples we could obtain, so in this case the women are the limiting reagent and the men are in excess (see Fig. 11–3).

These limiting-reagent types of problems can be easily worked by using the techniques we have been examining. You need only recognize that one of the starting materials will *limit* the amount of product that can be formed. You must take *all* starting materials through Steps I and II to see which reactant will produce the **least** amount of product. This reactant is the limiting reagent.

Figure 11-3 *Limiting Reagent. The principle of limiting reagent is analogous to a party attended by both men and women. If there are seven (7) men, but only six (6) women, the maximum number of couples we could have would be six (6). The women are the "limiting reagent" and the men are the "excess reagent"—one man in excess.*

EXAMPLE 11-5

A 50.0 g sample of calcium carbonate ($CaCO_3$) is allowed to react with 13.0 g of hydrochloric acid (HCl). Calculate the number of grams of calcium chloride that can be produced.

$$CaCO_3 + 2HCl \longrightarrow CaCl_2 + CO_2 + H_2O$$

SOLUTION The equation is balanced and the formula masses of $CaCO_3$, HCl, and $CaCl_2$ are 100.1 amu, 36.5 amu, and 111.1 amu, respectively.

1. First, calculate the moles of each reactant according to Step I:

$$50.0 \text{ g } CaCO_3 \times \frac{1 \text{ mol } CaCO_3}{100.1 \text{ g } CaCO_3} = 0.500 \text{ mol } CaCO_3 \left.\right\} \text{Step I, } CaCO_3$$

$$13.0 \text{ g HCl} \times \frac{1 \text{ mol HCl}}{36.5 \text{ g HCl}} = 0.356 \text{ mol HCl} \left.\right\} \text{ Step I, HCl}$$

2. Second, calculate the moles of $CaCl_2$ that could be produced from each reactant as in Step II.

$$0.500 \text{ mol CaCO}_3 \times \frac{1 \text{ mol CaCl}_2}{1 \text{ mol CaCO}_3} = 0.500 \text{ mol CaCl}_2 \left.\right\} \text{ Step II, CaCO}_3$$

$$0.356 \text{ mol HCl} \times \frac{1 \text{ mol CaCl}_2}{2 \text{ mol HCl}} = 0.178 \text{ mol CaCl}_2 \left.\right\} \text{ Step II, HCl}$$

3. *The reactant that gives the least number of moles of the product is the limiting reagent.* Notice that you carry *both* reactants through Steps I and II. You continue to Step III with the reactant that gives the least number of moles of $CaCl_2$. Hence, HCl is the *limiting reagent* and $CaCO_3$ was used in excess.

$$0.178 \text{ mol CaCl}_2 \times \frac{111.1 \text{ g CaCl}_2}{1 \text{ mol CaCl}_2} = 19.8 \text{ g CaCl}_2 \qquad \textit{Answer}$$

Step III

Exercise 11–3

The industrial production of ethylene glycol, used as an automobile antifreeze and in the preparation of a polyester fiber (Dacron), involves the following reaction:

$$\begin{array}{ccc} CH_2-CH_2 + H_2O & \longrightarrow & CH_2-CH_2 \\ \diagdown O \diagup & & | \quad\quad | \\ & & OH \quad OH \end{array}$$

$$C_2H_4O \qquad\qquad\qquad C_2H_6O_2$$

ethylene oxide   ethylene glycol

Calculate the amount of ethylene glycol that could be obtained from $22\bar{0}$ g of ethylene oxide and $12\bar{0}$ g of water.

theoretical yield Amount of product obtained when we assume that all of the limiting reagent (reactant completely used up) forms products and that none of the products are lost during isolation and purification.

   The amount of the product that we calculated in Examples 11–1 to 11–5 is called the *theoretical yield*. The **theoretical yield** is the amount of product obtained when we assume that all of the limiting reagent forms products, with none of it left over, and that none of the product is lost in its isolation and purification. But such is not generally the case. In organic reactions particularly, side reactions occur giving minor products in addition to the major one. Also, some of the product is lost in the process of its isolation and purification and in transferring it from one container to another. In the chemical industry, this loss in isolation and purification is often minimized by a continuous process in which the materials used in isolation and purification are *recycled*. The amount that is actually obtained is called the **actual yield**. The **percent yield** is the percent of the theoretical yield that is actually obtained, calculated as follows:

$$\text{Percent yield} = \frac{\text{actual yield}}{\text{theoretical yield}} \times 100$$

actual yield Amount of product that is actually obtained in a given reaction.

percent yield Percent of the theoretical yield that is actually obtained in a chemical reaction.

Percent yield

$$= \frac{\text{actual yield}}{\text{theoretical yield}} \times 100$$

Study hint

The difference between a theoretical yield and an actual yield is similar to the difference between the amount you earn from a job and the amount you have left after you pay taxes.

EXAMPLE 11–6

If 205 g of carbon dioxide (CO_2) is actually obtained in Example 11–1, what is the percent yield?

SOLUTION

$$\frac{205 \text{ g } CO_2 = \text{actual yield}}{211 \text{ g } CO_2 = \text{theoretical yield}} \times 100 = 97.2\% \qquad Answer$$

Exercise 11–4

The theoretical yield of the sulfa drug sulfadiazine in a given reaction is 12.8 kg. The amount actually obtained was 12.2 kg; what is the percent yield?

11–5 Mass–Volume Stoichiometry Problems

Next, let us consider mass–volume stoichiometry problems.[1] Like mass–mass stoichiometry problems, mass–volume problems do not always require that you use steps I and/or III of the mole method for solving stoichiometry problems. In these types of problems, *either* the known *or* the unknown is a *gas*. The known may be given in mass units and you will be asked to calculate the unknown in volume units (if a gas), or the known will be given in volume units (if a gas) and you will be asked to calculate the unknown in mass units. In either case, you need to apply the molar volume—that is, *22.4L/1 mol of any gas at STP (0°C and 760 torr),* discussed in Chapter 8 (see Section 8–3).

EXAMPLE 11–7

Calculate the volume in liters of oxygen gas (O_2) measured at 0°C and 760 torr that could be obtained by heating 28.0 g of potassium nitrate

---

[1] The volume of a gas can be calculated at non-STP conditions. We do this using the gas laws, which are discussed in Chapter 12.

(KNO$_3$), according to the following balanced equation:

$$2KNO_3 \xrightarrow{\Delta} 2KNO_2 + O_2$$

SOLUTION   We calculate the formula mass of KNO$_3$ as 101.1 amu from the atomic masses. The conditions 0°C and 76$\overline{0}$ torr are STP conditions; hence, in Step III, the relation *1 mol* O$_2$ molecules at STP occupies *22.4 L* must be used.

$$\underbrace{28.0 \text{ g } \cancel{KNO_3} \times \frac{1 \text{ mol } \cancel{KNO_3}}{101.1 \text{ g } \cancel{KNO_3}}}_{\text{Step I}} \times \underbrace{\frac{1 \text{ mol } \cancel{O_2}}{2 \text{ mol } \cancel{KNO_3}}}_{\text{Step II}} \times \underbrace{\frac{22.4 \text{ L } O_2 \text{ at STP}}{1 \text{ mol } \cancel{O_2}}}_{\text{Step III}}$$

$$= 3.10 \text{ L } O_2 \text{ at STP} \qquad Answer$$

EXAMPLE 11–8

Calculate the number of moles of sodium required to produce 2.45 liters of hydrogen gas (H$_2$) measured at 0°C and 76$\overline{0}$ torr, according to the following balanced equation:

$$2Na + 2H_2O \longrightarrow 2NaOH + H_2$$

SOLUTION   We must first convert the 2.45 liters of H$_2$ gas to moles of H$_2$. Since 0°C and 76$\overline{0}$ torr are STP conditions, we can use 22.4 L of H$_2$ at STP = 1 mol of H$_2$ in Step I. Step II converts the moles of H$_2$ to moles of Na. Since the problem requested moles of Na, Step III is *not* needed.

$$\underbrace{2.45 \text{ L } \cancel{H_2} \text{ at STP} \times \frac{1 \text{ mol } \cancel{H_2}}{22.4 \text{ L } \cancel{H_2} \text{ STP}}}_{\text{Step I}} \times \underbrace{\frac{2 \text{ mol Na}}{1 \text{ mol } \cancel{H_2}}}_{\text{Step II}}$$

$$= 0.219 \text{ mol Na} \qquad Answer$$

Exercise 11–5

Calculate the number of liters of oxygen gas (O$_2$) at STP produced by heating 0.480 mol of potassium chlorate (KClO$_3$). (*Hint:* Write the balanced equation first.)

11–6   Volume–Volume Stoichiometry Problems

Gay-Lussac's law of combining volumes Principle that, at the same temperature and pressure, whenever gases react or gases are formed, they do so in the ratio of small whole numbers by volume.

Volume–volume stoichiometry problems are based on experiments performed by the French chemist and physicist Joseph Louis Gay-Lussac (ga′lu·sak′) (1778–1850). His experimental results are stated in **Gay-Lussac's law of combining volumes,** which states that at the *same temperature and pressure* whenever *gases* react or are formed they do so in the ratio of *small whole numbers by volume*. This ratio of *small whole numbers by volume* is directly proportional to the values of their *coefficients*[2] in the balanced equation.

For example, consider the following reaction, where all components are in the gaseous state and at the same temperature and pressure:

$$CH_{4(g)} + 2O_{2(g)} \xrightarrow{\Delta} CO_{2(g)} + 2H_2O_{(g)}$$

One (1) volume of CH_4 gas (methane) reacts with two (2) volumes of O_2 gas to form one (1) volume of CO_2 gas and two (2) volumes of H_2O vapor. If we had measured these volumes all at STP and assumed that they *all* remain gases at STP, we could have stated that 1 mol (22.4 L) of CH_4 gas reacts with 2 mol (44.8 L) of O_2 to form 1 mol (22.4 L) of CO_2 gas and 2 mol (44.8 L) of H_2O vapor. Note that in all cases, the ratio of the *volumes* is still the same— that is, $1:2:1:2$ for CH_4, O_2, CO_2, and H_2O, respectively. In solving volume–volume stoichiometry problems where all gases are measured at the same temperature and pressure, Steps I and III are not necessary; only Step II is required.

EXAMPLE 11–9

Calculate the volume in liters of oxygen gas (O_2) required and the volume of carbon dioxide (CO_2) and water vapor (H_2O) formed from the complete combustion of 1.50 liters of ethane (C_2H_6), all volumes being measured at 400°C and 760 torr pressure.

$$2C_2H_{6(g)} + 7O_{2(g)} \xrightarrow{\Delta} 4CO_{2(g)} + 6H_2O_{(g)}$$

SOLUTION All the substances are gases measured at the same temperature and pressure, so each volume is related to the corresponding coefficient in the balanced equation:

$$1.50 \text{ L } C_2H_6 \times \frac{7 \text{ L } O_2}{2 \text{ L } C_2H_6} = 5.25 \text{ L } O_2 \qquad \textit{Answer}$$

Step II

$$1.50 \text{ L } C_2H_6 \times \frac{4 \text{ L } CO_2}{2 \text{ L } C_2H_6} = 3.00 \text{ L } CO_2 \qquad \textit{Answer}$$

Step II

$$1.50 \text{ L } C_2H_6 \times \frac{6 \text{ L } H_2O_{(g)}}{2 \text{ L } C_2H_6} = 4.50 \text{ L } H_2O_{(g)} \qquad \textit{Answer}$$

Step II

---

[2] This is the same principle we applied in mass–mass problems, except that here we use volumes instead of moles and *all substances are gases* and are measured at the **same temperature and pressure**.

Exercise 11–6

Calculate the volume in liters of ammonia gas (NH_3) produced from 6.25 liters of nitrogen gas (N_2), all volumes being measured at the same temperature and pressure.

$$N_{2(g)} + 3\ H_{2(g)} \xrightarrow{\Delta} 2\ NH_{3(g)}$$

11–7   Heat in Chemical Reactions

heat of reaction The number of calories or joules of heat energy evolved or absorbed in a given chemical reaction per given amount of reactants and/or products.

joule (J) The standard unit for the measurement of heat energy in the International System of units (SI units).

exothermic reactions Reactions in which heat energy is evolved.

endothermic reactions Reactions in which heat energy is absorbed.

Besides the mass–mass, mass–volume, and volume–volume relations just outlined, energy relationships are also important in chemical reactions. The energy involved is usually observed as heat and is expressed as the heat of reaction. The **heat of reaction** is the number of calories (see Section 6–4) or joules[3] of heat energy evolved or absorbed in a given chemical reaction per given amount of reactants and/or products. In **exothermic reactions** heat energy is *evolved*, whereas in **endothermic reactions** heat energy is *absorbed*.

An example of an exothermic reaction is the combination of 2 mol of hydrogen gas with 1 mol oxygen gas to form 2 mol of water (liquid), accompanied by the evolution of 1.37×10^5 calories or 5.73×10^5 J of heat energy at 25°C. Thus, for this exothermic reaction, the heat of reaction is 1.37×10^5 cal or 5.73×10^5 J for the formation of 2 mol of water (liquid) or 6.85×10^4 cal or 2.87×10^5 J for 1 mol of water (liquid).

$$2H_{2(g)} + O_{2(g)} \longrightarrow 2H_2O_{(\ell)} + 1.37 \times 10^5 \text{ cal or } 5.73 \times 10^5 \text{ J (at 25°C)}$$

Two common examples of exothermic reactions are the preparation of dilute solutions of acids or bases by adding concentrated sulfuric acid to water in the former case and by adding sodium hydroxide pellets to water in the latter. In both cases the flask gets warm. Also, when these two substances (sulfuric acid and sodium hydroxide) react in a neutralization reaction (see Section 9–10) an exothermic reaction occurs. The flask gets warm from the heat that is evolved.

An example of an endothermic reaction is the combination of 1 mol of hydrogen gas with 1 mol of iodine gas, forming 2 mol of gaseous hydrogen iodide, with the *absorption* of 1.24×10^4 calories or 5.19×10^4 J of heat energy at 25°C. The absorption of heat energy is noted by the minus sign in the following balanced equation.

$$H_{2(g)} + I_{2(g)} \rightleftarrows 2HI_{(g)} - 1.24 \times 10^4 \text{ cal or } 5.19 \times 10^4 \text{ J (at 25°C)}$$

[The $\rightleftarrows$ that separates the reactants and products indicates that the reaction is reversible; that is, it goes both ways and is said to be in equilibrium]. Placing the endothermic heat energy on the reactants side gives the following balanced

---

[3] A **joule** (J) is a standard unit for the measurement of heat energy in the International System of Units (SI units). 4.184 joules (J) = 1 calorie (cal)

equation:

$$+1.24 \times 10^4 \text{ cal or } 5.19 \times 10^4 \text{ J} + H_{2(g)} + I_{2(g)} \rightleftharpoons 2HI_{(g)} \text{ (at } 25°C)$$

Thus, for the endothermic reaction, the heat of reaction is 1.24×10^4 cal or 5.19×10^4 J absorbed for the formation of 2 mol of gaseous hydrogen iodide, or 6.20×10^3 cal or 2.60×10^4 J absorbed for 1 mol of HI.

A common example of the endothermic formation of a solution is illustrated by dissolving potassium iodide in water and noticing how cold the flask becomes to the touch.

We can use this heat energy[4] in stoichiometric calculations. The quantity of heat energy, either exothermic or endothermic, is related to the moles of reactants or products in the balanced equation. We therefore use the heat of reaction as we did moles in Step II of our three basic steps (see Section 11–2).

EXAMPLE 11–10

Natural gas (CH_4) burns in air to produce carbon dioxide, water vapor, and heat energy, according to the following balanced equation:

$$CH_{4(g)} + 2O_{2(g)} \longrightarrow CO_{2(g)} + 2H_2O_{(g)} + 213 \text{ kcal (at } 25°C)$$

(a) Is the reaction exothermic or endothermic?
(b) Calculate the number of kilocalories (kcal) of heat energy produced by the burning of 25.0 g of natural gas.

SOLUTION   Heat is evolved, so the reaction is exothermic. The molecular mass of CH_4 is 16.0 amu. The relationship between methane and the heat of reaction is 1 mol of CH_4 to 213 kcal. We therefore solve this problem by using Steps I and II.

$$25.0 \text{ g } CH_4 \times \underbrace{\frac{1 \text{ mol } CH_4}{16.0 \text{ g } CH_4}}_{\text{Step I}} \times \underbrace{\frac{213 \text{ kcal}}{1 \text{ mol } CH_4}}_{\text{Step II}} = 333 \text{ kcal} \qquad Answer$$

EXAMPLE 11–11

Ethylene (C_2H_4) burns in air to produce carbon dioxide, water vapor, and heat energy, according to the following balanced equation:

$$C_2H_{4(g)} + 3O_{2(g)} \xrightarrow{\Delta} 2CO_{2(g)} + 2H_2O_{(g)} + 1.32 \times 10^6 \text{ J (at } 25°C)$$

(a) Is the reaction exothermic or endothermic?

---

[4] The heat energy of reactions is referred to as the *enthalpy of a reaction* using the symbol ΔH. The change in *enthalpy* from reactants to products is ΔH (read "delta H," Δ meaning "change"). For exothermic reactions ΔH is **negative**, such as in the formation of water, $\Delta H = -1.37 \times 10^5$ cal or -5.73×10^5 J at 25°C. The reason for the negative sign is that in going from the reactants (H_2 and O_2) to the product(s) (H_2O), energy is lost (negative) *by* the chemicals, with this energy being given off to the outside (exothermic). For endothermic reactions ΔH is **positive** such as in the formation of hydrogen iodide, $\Delta H = +1.24 \times 10^4$ cal or $+5.19 \times 10^4$ J at 25°C. The reason for the positive signs is that in going from the reactants (H_2 and I_2) to the products(s) (HI), energy must be gained (positive) *from* the outside (endothermic).

(b) Calculate the number of grams of ethylene (C_2H_4) needed to produce 5.23×10^5 J of heat energy.

SOLUTION Heat is evolved, so the reaction is exothermic. The relationship between ethylene and the heat of reaction is 1.32×10^6 J to 1 mol of C_2H_4. The molecular mass of C_2H_4 is 28.0 amu. We therefore solve this problem by using Steps II and III.

$$5.23 \times 10^5 \text{ J} \times \underbrace{\frac{1 \text{ mol } C_2H_4}{1.32 \times 10^6 \text{ J}}}_{\text{Step II}} \times \underbrace{\frac{28.0 \text{ g } C_2H_4}{1 \text{ mol } C_2H_4}}_{\text{Step III}}$$

$$= 11.1 \text{ g } C_2H_4 \qquad \textit{Answer}$$

Exercise 11–7

Given the following balanced equation:

$$2H_{2(g)} + O_{2(g)} \longrightarrow 2H_2O_{(\ell)} + 1.37 \times 10^5 \text{ cal (at 25°C)}$$

(a) Is the reaction exothermic or endothermic?

(b) Calculate the number of grams of hydrogen needed to produce 7.50×10^5 cal of heat energy.

Copper: chemistry and the electronics industry

Name: The name copper derives from the Latin *aes cyprium*, meaning Cyprian metal, because large amounts of copper were produced in ancient times on the island of Cyprus in the Mediterranean Sea. The symbol (Cu) and the modern name evolved from the popular Latin word *cuprum*.

Appearance: A lustrous, beautiful, red-brown metal that is an excellent conductor of electricity and heat.

Occurrence: Copper is one of the few metals that can be found in nature uncombined with other elements. This is one reason that it was one of the first metals used by humans. The main source of copper is from the minerals chalcopyrite ($CuFeS_2$) and chalcocite (Cu_2S). Along with iron and nickel, copper is one of the most important industrial metals.

Source: Metallic copper can be produced from copper(I) sulfide or copper(I) oxide by the following reactions:

$$3Cu_2S + 3O_2 \xrightarrow{\text{heat}} 6Cu + 3SO_2$$

$$2Cu_2O + C \xrightarrow{\text{heat}} 4Cu + CO_2$$

$$Cu_2O + CO \xrightarrow{\text{heat}} 2Cu + CO_2$$

$$Cu_2O + H_2 \xrightarrow{\text{heat}} 2Cu + H_2O$$

Its Role in Our World: Copper is an excellent metal for preparing alloys (mixtures of metals), and more than 1000 different copper alloys have been prepared. Metal alloys can have more strength and improved properties over pure metals. Common copper alloys include brasses (mixtures of copper and zinc with added traces of other metals), bronzes (mixtures of copper and tin with added traces of other metals), copper/nickel alloys, and copper/silver alloys.

The largest use of copper (50 percent of copper use) occurs in the electronics industry, where the electrical conductivity of copper is important. Common uses include electrical wires and contacts.

Other uses of copper and its alloys include silverplating (Cu/Ag/Ni alloy), jewelry, and silverware.

The attractive color and luster of pure copper have made it an important ornamental metal throughout history.

The new superconducting materials (materials that conduct electricity with very little resistance) discovered in the middle 1980s contain copper as a component. While these materials have yet to make a big impact on industry, they may lead to very important products in the future. The material $YBa_2Cu_3O_x$, where $x \approx 7$ and y is yttrium, loses virtually all resistance to the passage of electrical current when cooled to 90 K ($-183°C$). There is speculation that these new materials may lead to high-speed trains of the future.

Unusual Facts: The large number of useful copper alloys derives in some part from the alchemists. Although the alchemists never succeeded in changing other metals into gold, their experiments on mixing different metals led to the preparation of alloys and the first principles of metallurgy.

Problems

Hint: Check each equation to make sure that it is balanced and, if not, balance it. For those questions in which an equation is not given, be sure to write a properly balanced equation. See Sections 9–6 through 9–10 and Section 10–3 for review.

1. Calculate the number of grams of hydrochloric acid (HCl) that is needed to react with 30.0 g of zinc.

$$Zn_{(s)} + 2HCl_{(aq)} \longrightarrow ZnCl_{2(aq)} + H_{2(g)}$$

2. Calculate the number of grams of hydrogen gas (H_2) that could be produced from 8.40 g of aluminum according to the following equation.

$$2Al_{(s)} + 6NaOH_{(aq)} \longrightarrow 2Na_3AlO_{3(aq)} + 3H_{2(g)}$$

3. Calculate the number of grams of oxygen gas (O_2) that could be produced by heating 7.90 g of potassium chlorate ($KClO_3$). (*Hint:* Write the balanced equation first.)

4. Calculate the number of moles of calcium chloride ($CaCl_2$) that would be necessary to prepare 94.0 g of calcium phosphate [$Ca_3(PO_4)_2$].

$$3CaCl_2 + 2Na_3PO_4 \longrightarrow Ca_3(PO_4)_{2(s)} + 6NaCl$$

5. Sodium chloride (0.500 mol) is allowed to react with an excess of sulfuric acid (H_2SO_4). Calculate the number of moles of hydrogen chloride gas (HCl) that could be formed.

$$NaCl + H_2SO_4 \longrightarrow Na_2SO_4 + HCl$$

6. Calculate the number of grams of lead(II) chloride ($PbCl_2$) produced by the reaction of 0.200 mol of chloride ions with excess lead(II) ions (Pb^{2+}). (*Hint:* Write the balanced ionic equation first.)

7. Calculate the number of grams of Ag_2CrO_4 that can be obtained by mixing 1.235 g of silver nitrate and 4.860 g of sodium chromate.

$$2AgNO_3 + Na_2CrO_4 \longrightarrow Ag_2CrO_{4(s)} + 2NaNO_3$$

8. Calculate the number of grams of ammonia (NH_3) that could be produced from 1.65 g of nitrogen gas (N_2) and 1.40 g of hydrogen gas (H_2), according to the following balanced equation:

$$N_{2(g)} + 3H_{2(g)} \xrightarrow{\Delta} 2NH_{3(g)}$$

9. The theoretical yield of the female sex hormone estrone in a given reaction is 18.7 g. The amount actually obtained was 13.7 g; what is the percent yield?

10. Calculate the number of liters of hydrogen sulfide (H_2S) measured at STP that could be prepared from 42.0 g of iron(II) sulfide (FeS).

$$FeS_{(s)} + 2HCl_{(aq)} \longrightarrow FeCl_{2(aq)} + H_2S_{(g)}$$

11. Calculate the number of liters of oxygen gas (O_2) measured at STP that could be obtained by heating 71.0 g of potassium chlorate ($KClO_3$). (*Hint:* Write the balanced equation first.)

12. How many liters of oxygen gas (O_2) at STP can be formed from the decomposition of 0.710 mol of potassium nitrate (KNO_3)? *Hint:* Write the balanced equation first.)

13. Calculate the number of liters of nitrogen gas (N_2) that would be consumed during the production of 4.00 liters of gaseous ammonia, according to the following balanced equation. Both gases are measured at the same temperature and pressure.

$$N_{2(g)} + 3H_{2(g)} \xrightarrow{\Delta} 2NH_{3(g)}$$

14. Calculate the number of liters of ammonia gas (NH_3) measured at STP that could be formed from 49.0 liters of hydrogen gas (H_2), measured at STP. (See Problem 12 for the balanced equation.)

15. Calculate the number of liters of nitrogen dioxide gas (NO_2) measured at STP that could be prepared from 76.0 liters of oxygen gas (O_2) measured at STP.

$$NO_{(g)} + O_{2(g)} \longrightarrow NO_{2(g)} \qquad \text{(unbalanced)}$$

16. Given the following balanced equation:

$$H_{2(g)} + F_{2(g)} \longrightarrow H_2F_{2(g)} + 1.284 \times 10^5 \text{ cal}$$

   (a) Is the reaction exothermic or endothermic?
   (b) Calculate the number of kilocalories of heat energy produced in the reaction of 37.0 g of fluorine gas with sufficient hydrogen gas.

17. Given the following balanced equation:

$$O_{2(g)} + 2F_{2(g)} \longrightarrow 2OF_{2(g)} - 11.0 \text{ kcal}$$

   (a) Is the reaction exothermic or endothermic?
   (b) Calculate the number of grams of fluorine gas needed for the reaction with 2.09 kcal of heat energy and sufficient oxygen gas.

18. Given the following balanced equation:

$$SO_{3(g)} + H_2O_{(\ell)} \longrightarrow H_2SO_{4(\ell)} + 1.30 \times 10^5 \text{ J}$$

   (a) Is the reaction exothermic or endothermic?
   (b) Calculate the number of joules of heat energy produced in the reaction of 62.0 g of sulfur trioxide with sufficient water.

General Problems

19. Methane gas (CH_4) burns in oxygen to produce carbon dioxide gas and water vapor.

   (a) Write the balanced equation of this reaction.
   (b) Calculate the number of moles of hydrogen atoms in 8.00 g of methane.
   (c) Calculate the number of moles of oxygen needed to completely burn 6.25 mol of methane.
   (d) Calculate the number of grams of oxygen needed to completely burn 8.00 g of methane.
   (e) Calculate the number of liters of carbon dioxide gas as STP that could be produced from 12.0 g of methane.
   (f) Calculate the number of liters of oxygen gas required to produce 5.60 L of carbon dioxide gas, both gases being measured at the same temperature and pressure.

(g) Calculate the number of grams of carbon dioxide that could be produced from 17.6 g of methane.

(h) Calculate the percent yield if 40.7 g of carbon dioxide are actually obtained in part (g).

20. A 30.0-g sample of iron is dissolved in concentrated hydrochloric acid (specific gravity 1.18 and 35.0% by mass HCl). How many milliliters of the hydrochloric acid would be necessary to dissolve the iron?

$$Fe_{(s)} + 2HCl_{(aq)} \longrightarrow FeCl_{2(aq)} + H_{2(g)}$$

(*Hint:* A 35.0% by mass HCl means 35.0 g of pure HCl in $10\overline{0}$ g of concentrated hydrochloric acid.)

21. Iron(II) hydroxide (0.320 mol) is treated with 0.250 mol of phosphoric acid.

(a) How many grams of iron(II) phosphate could be produced?

(b) If 34.0 g of iron(II) phosphate is actually obtained, what is the percent yield?

Answers to Exercises

11–1. 12.0 g

11–2. 8.17 g

11–3. $31\overline{0}$ g

11–4. 95.3%

11–5. 16.1 L (Step I is not needed)

11–6. 12.5 L

11–7. (a) exothermic; (b) 21.9 g

QUIZ

ELEMENT	ATOMIC MASS UNITS (amu)
K	39.1
Cl	35.5
O	16.0

1. Potassium is prepared by the action of electrical current on molten (melted) potassium chloride according to the reaction given below. Calculate the number of grams of potassium that could be obtained from $15\overline{0}$ g of potassium chloride.

$$KCl_{(\ell)} \xrightarrow{\text{electric current}} K_{(\ell)} + Cl_{2(g)} \quad \text{(unbalanced)}$$

2. Calculate the number of grams of oxygen gas produced by heating 0.0240 mol of potassium chlorate ($KClO_3$).

3. Calculate the number of liters of oxygen gas at STP produced by heating 1.50 g of potassium chlorate ($KClO_3$).

4. Ammonia is a very important material in the chemical industry. It is used in the production of fertilizers, plastics, and rocket fuels as well as other industrial products. Calculate the maximum numer of liters of ammonia that can be obtained from 5.00 L of hydrogen gas. All gases are measured at STP.

$$N_{2(g)} + H_{2(g)} \xrightarrow{\Delta} NH_{3(g)} \quad \text{(unbalanced)}$$

5. Consider the reaction between 1.2 L of gaseous chlorine (measured at STP) and iodine to give iodine monochloride. Calculate the number of calories of heat energy absorbed in this process if all the chlorine reacts.

$$I_{2(s)} + Cl_{2(g)} \longrightarrow 2ICl_{(g)} - 8.4 \text{ kcal}$$

12

GASES

COUNTDOWN

5 Solve the following linear equations for the unknown (x) (Section 1–5).
  (a) $4x + 7 = 19$ (3);   (b) $ax = 7$ (7/a)

4 Convert the following temperatures to K (Section 2–6).
  (a) $3\overline{0}°C$ (303 K);   (b) $-25°C$ (248 K)

3 Calculate the molecular mass of a gas if 2.85 L measured at STP has a mass of 4.34 g (Section 8–3).

(34.1 amu)

2 Calculate the molecular formula of a compound from the following experimental data: 92.3% carbon, 7.7% hydrogen, and molecular mass of 26.0 amu (Section 8–5).

(C_2H_2)

1 Calcium (0.210 g) reacts with water to produce hydrogen gas.
  (a) Write a balance equation for the reaction (Section 9–8).
  (b) Calculate the volume of hydrogen gas in millimeters produced at STP (atomic mass units: Ca = 40.1 amu, H = 1.0 amu, O = 16.0 amu) [Section 11–5].

[(a) $Ca + 2H_2O \rightarrow Ca(OH)_2 + H_2$
(b) 117 mL]

TASKS

1 Memorize the conditions for standard temperature and pressure (STP), that is, *0°C* and *76̄0* torr or *1.00 atm*.

2 Memorize the ideal-gas equation (*PV = nRT*) and the value of *R* (*0.0821 atm · L/mol · K*).

OBJECTIVES

1 Give the distinguishing characteristics of each of the following terms:
  (a) pressure (Section 12–2)
  (b) Boyle's law (Section 12–3)

(c) Charles' law (Section 12–4)
(d) Gay-Lussac's law (Section 12–5)
(e) Dalton's law of partial pressures (Section 12–7)
(f) ideal-gas equation (Section 12–8)

2 Given a fixed mass of a gas at constant temperature and at a stated volume and pressure, calculate the volume if the pressure is changed. Calculate the pressure if the volume is changed (*Boyle's law*, Example 12–1, Exercise 12–1, Problems 1 and 2).

3 Given a fixed mass of a gas at constant pressure and at a stated volume and temperature, calculate the volume if the temperature is changed. Calculate the temperature if the volume is changed (*Charles' law*, Example 12–2, Exercise 12–2, Problems 3 and 4).

4 Given a fixed mass of a gas at constant volume and at a stated temperature and pressure, calculate the pressure if the temperature is changed. Calculate the temperature if the pressure is changed (*Gay-Lussac's law*, example 12–3, Exercise 12–3, Problems 5 and 6).

5 Given a fixed mass of a gas at a stated volume, temperature, and pressure, calculate the volume if the temperature and pressure are changed. Calculate the temperature if the volume and pressure are changed. Calculate the pressure if the volume and temperature are changed (*combined gas laws*, Example 12–4, Exercise 12–4, Problems 7 and 8).

6 Given a fixed mass of a gas collected over water at a stated volume, temperature, and pressure and the vapor pressure of the water at the stated temperature, calculate the volume of the *dry gas* if the temperature and pressure are changed (*Dalton's law of partial pressures*, Example 12–5, Exercise 12–5, Problems 9 and 10).

7 Given three of the four variables (pressure, volume, mass in moles, and temperature) in the ideal-gas equation, calculate the fourth variable (*ideal-gas equation*, Examples 12–6 and 12–7, Exercise 12–6, Problems 11 and 12).

Our very survival on this planet depends on gases. We breathe a mixture of gases—oxygen, which we must have, also nitrogen, carbon dioxide, argon, and traces of carbon monoxide, helium, methane (CH_4), and water vapor. The oxygen is produced from green plants by the process of photosynthesis. In this process the *carbon dioxide* that we and animals produce is used by plants and converted to *oxygen* and glucose. The production of oxygen is vital to our survival on this planet, and this is one of the reasons the "rain forests" of the world must be preserved.

In Chapter 3, we noted that gases are one of three physical states of matter. We have also learned to write formulas for gases and to balance equations involving gases. Up until now, we have dealt with gases at a single set of conditions where the temperature and pressure have not changed. In this chapter we consider how gases vary with changes in temperature, pressure, and volume.

We begin our discussion of the three physical states of matter with gases for two reasons: (1) gases are the simplest of the three physical state of matter, and (2) more is known about gases than solids or liquids.

Before we consider the properties and changes that occur in the gaseous state, we must first consider the general characteristics of gases. These characteristics are listed below:

1. *Expansion.* Gases expand indefinitely and uniformly to fill all the space in which they are placed.

2. *Indefinite shape or volume.* A given sample of gas has no definite shape or volume but completely fills the vessel in which it is placed.

3. *Compressibility.* Gases can be highly compressed. For example, very large volumes of oxygen can be placed in small pressurized tanks.

4. *Low density.* The densities of gases are much lower than the densities of solids or liquids. Hence, gas densities are measured in grams per liter (g/L) in the metric system rather than grams per milliliter (g/mL) as was done with solids and liquids (see Section 2–7).

5. *Mixability or diffusion.* Two or more nonreacting gases will normally mix completely and uniformly when placed in contact with each other. The gas company takes advantage of this property to facilitate the detection of leaks in natural gas lines. Natural gas is an odorless mixture of gases (mostly methane, CH_4). The gas company adds traces of a foul-smelling gas (C_2H_6S) to natural gas. The C_2H_6S quickly diffuses into the room air and is detected when a leak is present in a line.

12–1 The Kinetic Theory

The kinetic theory has been advanced to explain the characteristics and properties of matter in general. In essence, the theory states that heat and motion are related, that particles of all matter are in **motion** to some degree, and that **heat** is an indication of this motion. We make the following assumptions when we apply the kinetic theory to gases:

1. Gases are composed of very small particles called molecules. The *distance* between these molecules is very *great* compared with the size of the molecules themselves, and the total volume of the molecules is only a small fraction of the entire space occupied by the gas. Therefore, in considering the volume of a gas, we are considering primarily *empty space*. This postulate is the basis of the high compressibility and low density of gases.

2. *No attractive forces* exist between the molecules in a gas. This is what keeps a gas from spontaneously becoming a liquid.

3. These molecules are in a state of constant, *rapid motion*, colliding with each other and with the walls of the container in a completely random manner. For example, oxygen molecules (O_2) at STP are moving about 1000 miles per hour and undergo about 5 billion collisions per second. This postulate is the basis of the complete mixing of two or more different nonreacting gases. The collisions between the gas molecules and the walls of the container account for the pressure exerted by the gas (see Section 12–2).

4. All of these molecular collisions are perfectly *elastic*; consequently, there is no loss of kinetic energy[1] in the system as a whole. Some energy may be transferred from one molecule to the other molecule involved in the collision.

5. The *average kinetic energy* per molecule of the molecules of the gas is proportional to the temperature in kelvins, and the *average kinetic energy per molecule* is *the same at a given temperature for all gases*. The molecules in a gas possess a range of kinetic energies; some molecules have more energy than the average kinetic energy and some molecules have less. Theoretically, at zero kelvin (0 K), molecular motion ceases and the kinetic energy of any particle becomes zero.

Gases that conform to these assumptions are called *ideal gases*, as opposed to *real gases*, such as hydrogen, oxygen, nitrogen, and others. Under moderate conditions of temperature and pressure, real gases behave as ideal gases, but if the *temperature* is very *low* or the *pressure* is very *high*, then real gases deviate considerably from ideal gases. An ideal gas is considered to have the following characteristics: (1) negligible volume of the actual molecules as compared to the volume of the gas itself (assumption 1); (2) no attractive forces between molecules (assumption 2); and (3) perfectly elastic collisions (assumption 4). By avoiding extremely low temperatures and extremely high pressures, we can consider real gases to behave as ideal gases and apply the basic gas laws (see Sections 12–3, 12–4, and 12–5) and the ideal-gas equation (see Section 12–8).

12–2 Pressure of Gases

pressure Force per unit area.

Gases exert pressure, as you have probably observed in inflating your automobile tires or bicycle tires to 32 lb (pounds per square inch, abbreviated *psi*) or 75 lb, respectively. **Pressure** is defined as force per unit area. The pressure of gases is produced by the impact of the gas molecules on the walls of the container.

Gases in the atmosphere (primarily nitrogen, oxygen, and a small amount of argon) also exert a pressure. The atmospheric pressure is measured by a mercury *barometer*, which was first devised in 1643 by Evangelista Torricelli

---

[1] Kinetic energy is the energy derived from the motion of a particle. It is one-half times the mass of the particle times the square of the velocity of the particle ($\frac{1}{2}mv^2$). Any particle or body in motion has a distinct kinetic energy.

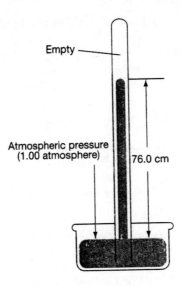

Empty

Atmospheric pressure
(1.00 atmosphere)

76.0 cm

Figure 12–1 *Torricelli's mercury barometer.*

(tŏr·rĕ·cĕl′le), 1608–1647, Italian mathematician and physicist. His barometer consisted of a glass tube at least 76 cm long, sealed at one end, filled with mercury, and then inverted with the open end in a dish of mercury (see Fig. 12–1). At sea level, the mercury level dropped to a height of 76.0 cm in the tube, leaving no air above the mercury level in the tube.

This height of 76.0 cm is called *standard pressure*. Standard pressure is expressed in many units. They are

1. centimeters of mercury: 76.0 cm of Hg

2. millimeters of mercury: 76$\overline{0}$ mm of Hg

3. torr (1 torr = 1 mm of mercury): 76$\overline{0}$ torr

4. inches of mercury: 29.9 in. of Hg

5. atmospheres (atm): 1.00 atm

6. millibars (mbars): 1013 mbars

7. pounds per square inch (psi): 14.7 psi

8. pascals[2] (Pa): 1.013 × 10^5 Pa

Although the *pascal* is the pressure unit recommended by the International System of Units (SI; see Section 2–2), we will generally use either the *torr* (named in honor of Torricelli), the *centimeter* of mercury, the *millimeter* of mercury, or the *atmosphere* as the pressure unit in this book. Conversion to atmospheres from torr, and *vice versa*, is done by knowing that 1 atm = 76$\overline{0}$

---

[2] The pascal is named in honor of Blaise Pascal (1623–1662), a French scientist who formulated principles of physics related primarily to liquids.

torr. Convert $63\overline{0}$ torr to atmospheres as follows:

$$63\overline{0} \text{ torr} \times \frac{1 \text{ atm}}{76\overline{0} \text{ torr}} = 0.829 \text{ atm}$$

In the preceding discussion, we stated that Torricelli's measurement was carried out at sea level. The atmospheric pressure decreases as altitude increases (approximately 25 torr per 1000 ft). At an altitude of 1 mile, the pressure is approximately 630 torr. You may have experienced this decrease in pressure when you go to the mountains or to a higher altitude. As you go to higher altitude and the pressure decreases, your ears "pop" when you yawn. Yawning equalizes the pressure inside and outside of your eardrum by opening a small tube from the middle ear to your mouth.

12–3   Boyle's Law

Boyle's law Principle that, at constant temperature, the volume of a fixed mass of a given gas is *inversely* proportional to the pressure it exerts.

In 1660, British physicist and chemist Robert Boyle carried out experiments on the change in volume of a given amount of gas with the pressure of the gas at constant temperature. From his experiments he formulated the law now referred to as **Boyle's law**: *At constant temperature, the volume of a fixed mass of a given gas is inversely proportional*[3] *to the pressure it exerts.* For example, if the pressure of a given volume of gas is doubled, the volume will be halved; if the pressure is halved, the volume will be doubled, as Fig. 12–2 shows.

Boyle's law may be expressed mathematically as

$$V \propto \frac{1}{P} \quad \text{(temperature constant)}$$

where volume (V) is inversely proportional ($1/P$) to the pressure (P). By introducing a constant of proportionality (k), the value of which depends on the units of P and V as well as the quantity of gas being measured, we can write the equation

$$V = k \times \frac{1}{P}$$

The equation may then be expressed as

$$PV = k \tag{12–1}$$

with the product of the pressure and volume equal to a constant at constant temperature. $P \times V$ is equal to a constant (k) in Equation 12–1, so different conditions of pressure and volume may be expressed for the *same mass* of gas at constant temperature:

$$P_{new} \times V_{new} = k = P_{old} \times V_{old} \tag{12–2}$$

From this equation, the new pressure can be solved as

$$P_{new} = P_{old} \times V_{factor}$$

---

[3] Inversely proportional means that an *increase* in pressure results in a *decrease* in the volume, and a *decrease* in the pressure results in an *increase* in the volume.

2 volumes   1 volume   1/2 volume

Figure 12–2 *A demonstration of Boyle's law. Temperature is constant. (As the volume is decreased the frequency of collisions is increased, resulting in an increase in pressure.)*

$$(12-3)$$

in which the new pressure is equal to the old pressure times a volume factor (V_{factor}), which is an appropriate ratio of the new and old volumes. The new volume can be calculated as

$$V_{new} = V_{old} \times P_{factor} \qquad (12-4)$$

where the new volume is equal to the old volume times a pressure factor (P_{factor}), which is the appropriate ratio of the new and old pressures.

We can evaluate the volume factor and pressure factor by considering the effect that the change in volume or pressure has on the old pressure or volume, and how this change will affect the new pressure or volume. It is not necessary, then, to memorize a formula. In evaluating these factors, it is most important that *both* the numerator and denominator of the factor be expressed in the *same* units.

EXAMPLE 12–1

A sample of gas occupies a volume of 95.2 mL at a pressure of $71\overline{0}$ torr and a temperature of $3\overline{0}°C$. What will be its volume in milliliters at standard pressure and $3\overline{0}°C$?

SOLUTION In working these problems, arrange the data in an orderly form.

$$T = \text{constant}$$

$V_{old} = 95.2 \text{ mL} \qquad P_{old} = 71\overline{0} \text{ torr} \quad \Big|$ pressure increases;

$V_{new} = ? \qquad\qquad P_{new} = 76\overline{0} \text{ torr} \quad \downarrow$ *volume decreases*

From Equation 12–4,

$$V_{new} = V_{old} \times P_{factor}$$

The pressure has increased from $71\overline{0}$ torr to $76\overline{0}$ torr; hence, the new volume will be decreased, and the pressure factor must be written so that the new volume will show a decrease. To reflect this decrease, we must write the pressure factor so that the ratio of pressures is less than 1—hence, $71\overline{0}$ ~~torr~~/$76\overline{0}$ ~~torr~~.

$$V_{new} = 95.2 \text{ mL} \times \frac{71\overline{0} \text{ \sout{torr}}}{76\overline{0} \text{ \sout{torr}}} = 88.9 \text{ ml} \qquad Answer$$

Exercise 12–1

A sample of gas occupies 65.0 mL at a pressure of $72\overline{0}$ torr and a temperature of $3\overline{0}°C$. What will be its pressure in torr if its volume is increased to 75.0 mL and the temperature remains constant?

12–4   Charles' Law

Experiments carried out originally in 1787 by Jacques Charles, a French physicist (1746–1823), and refined in 1802 by Joseph Gay-Lussac (1778–1850) (see Section 11–6), showed that the volume of a gas is increased by $\frac{1}{273}$ of its value at 0°C for every one degree rise in temperature (see Table 12–1).

Although the volume of a gas changes uniformly with changes in temperature, the volume is not directly proportional to the Celsius temperature. If a new temperature scale is devised with a zero point of −273°C (more accurately, −273.15°C) and temperatures are expressed on this scale, then the volume of gas would be directly proportional to the temperature (refer again to Table 12–1). The scale is the *Kelvin temperature scale* and −273°C is assigned the value of *0* K (see Section 2–5). From Table 12–1, the value 0 K is the Kelvin temperature corresponding to *zero* volume of the gas. However, since real gases form liquids and solids on cooling, this zero value is only *theoretical*. Figure 12–3 presents these data as a graph showing the theoretical zero milliliter volume. To

TABLE 12–1   RELATION OF
TEMPERATURE TO VOLUME

t (°C)	*V* (mL)	*T* (K)
273	546	546
100	373	373
10	283	283
1	274	274
0	273	273
−1	272	272
−10	263	263
−100	173	173
−273	0 (theoretically)	0

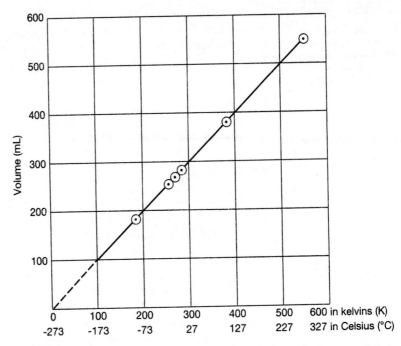

Figure 12–3   *Graph relating temperature to volume of a gas. Zero-milliliter volume is only a theoretical value because real gases form liquids and solids on cooling.*

convert from °C to K, we need only add 273° as we did in Chapter 2[4] (see Section 2–6):

$$K = °C + 273 \qquad\qquad (12\text{--}5)$$

Charles' law states that *at constant pressure, the volume of a fixed mass of a given gas is **directly** proportional[5] to the temperature in kelvins.* For example, if the temperature in kelvins is doubled at constant pressure, the volume is doubled; if the temperature in kelvins is halved, the volume is halved (Fig. 12–4).

Charles' law may be expressed mathematically as

$$V \propto T \text{ (pressure constant)}$$

where volume (V) is directly proportional to the temperature in kelvins (T). An equation can be written by introducing a constant of proportionality (**k**) for a given sample of gas at some constant pressure:

$$V = \mathbf{k}T$$

Charles' law Principle that, at constant pressure, the volume of a fixed mass of a given gas is *directly* proportional to the temperature in kelvins.

---

[4] To simplify the calculations, in this text we shall use 273 to convert °C to K instead of the more accurate 273.15.

[5] Directly proportional means that an *increase* in the temperature results in an *increase* in the volume, and a *decrease* in the temperature results in a *decrease* in the volume.

Figure 12–4 *A demonstration of Charles' law (temperature in kelvins). Pressure is constant. (As the temperature is increased the kinetic energy of the molecules is increased, resulting in an increase in volume.)*

1/2 volume

1 volume   2 volumes

The equation may then be expressed as

$$\frac{V}{T} = \mathbf{k} \tag{12–6}$$

with the volume divided by the temperature in kelvins being equal to a constant at constant pressure. Since V/T is equal to a constant, different conditions of temperature and volume may be expressed for the same mass of a gas at constant pressure:

$$\frac{V_{new}}{T_{new}} = \mathbf{k} = \frac{V_{old}}{T_{old}} \tag{12–7}$$

From this equation, the new temperature in kelvins can be solved as

$$T_{new} = T_{old} \times T_{factor} \tag{12–8}$$

in which the new temperature is equal to the old temperature times a volume factor (V_{factor}), which is an appropriate ratio of the new and old volumes. The new volume can be calculated as

$$V_{new} = V_{old} \times T_{factor} \tag{12–9}$$

in which the new volume is equal to the old volume times a temperature factor in *kelvins*.

EXAMPLE 12–2

A gas occupies a volume of 4.50 liters at 27°C. At what temperatures in °C would the volume be 6.10 liters, the pressure remaining constant?

SOLUTION

P = constant

V_{old} = 4.50 L | volume increases;

t_{old} = 27°C T_{old} = 27 + 273
                                          = $30\overline{0}$ K

V_{new} = 6.10 L ↓ *temperature increases*

t_{new} = ? T_{new} = ?

From Equation 12–8,

$$T_{new} = T_{old} \times V_{factor}$$

The volume increases; therefore, the new temperature will be *greater* and the volume factor must be written so that the new temperature will be greater. The ratio of volumes must be written so that the ratio is *greater than 1*—hence, 6.10 Ł/4.50 Ł.

$$T_{new} = 30\overline{0} \text{ K} \times \frac{6.10 \text{ Ł}}{4.50 \text{ Ł}} = 407 \text{ K}$$

We then convert this kelvin temperature to °C by subtracting the constant, 273;

$$407 \text{ K} = (407 - 273)°\text{C} = 134°\text{C}   \quad Answer$$

Exercise 12–2

A gas occupies a volume of $15\overline{0}$ mL at $3\overline{0}$°C and $72\overline{0}$ torr. What will be its volume in milliliters at 50°C and $72\overline{0}$ torr?

12–5 Gay-Lussac's Law

Gay-Lussac's (gā'lü·sȧk')
law Principle that, at constant volume, the pressure of
a fixed mass of a given gas
is *directly* proportional to
the temperature in *kelvins*.

In 1802, Joseph Gay-Lussac published the results of his experiments, which are now known as Gay-Lussac's law. **Gay-Lussac's law** states that *at constant volume, the pressure of a fixed mass of a given gas is **directly** proportional to the temperature in kelvins*. For example, if the temperature in kelvins is doubled at constant volume, the pressure is doubled; if the temperature in kelvins is halved, the pressure is halved (Fig. 12–5).

This statement may be expressed mathematically as

$$P \propto T \quad \text{(volume constant)}$$

An equation may be written as

$$P = kT$$

and also as

$$\frac{P}{T} = k \qquad\qquad (12\text{–}10)$$

Figure 12–5 *A demonstration of Gay-Lussac's law (temperature in kelvins). Volume is constant. (As the temperature is increased the kinetic energy of the molecules is increased and the frequency of collisions is increased, resulting in an increase in pressure.)*

For different conditions of pressure and temperature, an equation may be written as

$$\frac{P_{new}}{T_{new}} = k = \frac{P_{old}}{T_{old}} \tag{12–11}$$

Hence, solving for P_{new} and T_{new}, the equations are

$$P_{new} = P_{old} \times T_{factor} \tag{12–12}$$

and

$$T_{new} = T_{old} \times P_{factor} \tag{12–13}$$

Table 12–2 summarizes Boyle's, Charles', and Gay-Lussac's Laws.

TABLE 12–2   SUMMARY OF BOYLE'S, CHARLES', AND GAY-LUSSAC'S LAWS

LAW	GIVEN	EFFECT[a]
Boyle's law	$P \uparrow$	$V \downarrow$
(T constant)	$P \downarrow$	$V \uparrow$
	$V \uparrow$	$P \downarrow$
	$V \downarrow$	$P \uparrow$
Charles' law	$T \uparrow$	$V \uparrow$
(P constant)	$T \downarrow$	$V \downarrow$
	$V \uparrow$	$T \uparrow$
	$V \downarrow$	$T \downarrow$
Gay-Lussac's law	$T \uparrow$	$P \uparrow$
(V constant)	$T \downarrow$	$P \downarrow$
	$P \uparrow$	$T \uparrow$
	$P \downarrow$	$T \downarrow$

[a] P = pressure, V = volume, T = temperature, $\uparrow$ = increase, $\downarrow$ = decrease.

EXAMPLE 12–3

The temperature of 1 liter of a gas originally at STP is changed to $20\overline{0}$ °C at constant volume. Calculate the final pressure of the gas in torr.

SOLUTION

V = constant

			temperature
$P_{old} = 76\overline{0}$ torr	$t_{old} = 0°C$	$T_{old} = 0 + 273$	increases;
		$= 273$ K	
$P_{new} = ?$	$t_{new} = 20\overline{0}°C$	$T_{new} = 20\overline{0} + 273$	*pressure*
		$= 473$ K	*increases*

From Equation 12–12,

$$P_{new} = P_{old} \times T_{factor}$$

The temperature increases at constant volume, so the pressure will increase and we must write the temperature factor so that the new pressure will be greater. To reflect this increase, we must write the temperature factor so that the ratio of temperatures is greater than 1—hence, 473 K/273 K.

$$P_{new} = 76\overline{0} \text{ torr} \times \frac{473 \text{ K}}{273 \text{ K}} = 1320 \text{ torr (to three significant digits)}$$

Answer

Exercise 12–3

A gas occupies 225 mL at $3\overline{0}°C$ and $64\overline{0}$ torr. At what temperature in degrees Celsius (°C) would the pressure be 1.00 atm if the volume remains constant?

12–6 The Combined Gas Laws

Boyle's and Charles' laws can be combined into one mathematical expression:

$$\frac{P_{new}V_{new}}{T_{new}} = \frac{P_{old}V_{old}}{T_{old}} \tag{12–14}$$

Solving Equation 12–14 for V_{new}, P_{new}, and T_{new} gives

$$V_{new} = V_{old} \times P_{factor} \times T_{factor} \tag{12–15}$$

$$P_{new} = P_{old} \times V_{factor} \times T_{factor} \tag{12–16}$$

$$T_{new} = T_{old} \times V_{factor} \times P_{factor} \tag{12–17}$$

Study hint

Consider each factor *separately*. It is just like working two simpler problems of the type we have already covered.

We will consider each factor separately in these equations (12–15 through 12–17) and their effect on the new volume, pressure, or temperature. In equation 12–15, the new volume is equal to the old volume multiplied by a pressure factor and a temperature factor. If the pressure increases, the pressure ratio must be less than 1, because increasing the pressure would decrease the old volume. If the pressure decreases, the pressure ratio must be greater than 1, because a decrease in pressure would increase the old volume. If the temperature increases, the ratio of the temperature in kelvins must be greater than 1, because the temperature change would increase the old volume. Conversely, if the temperature decreases, the temperature ratio must be less than 1. By applying similar reasoning to Equations 12–16 and 12–17, we can solve for the new pressure and temperature.

The gas laws apply only when behavior of real gases closely resembles that of an ideal gas (see Section 12–1). Under certain conditions of temperature and/or pressure the properties of a real gas deviate markedly from those of an ideal gas. Other equations have been developed to handle such cases, but for our purposes in this book we shall consider that for most practical purposes real gases generally behave like ideal gases.

EXAMPLE 12–4

A certain gas occupies $5\overline{0}0$ mL at $76\overline{0}$ torr and $0°C$. What volume in milliliters will it occupy at 10.0 atm and $1\overline{0}0°C$?

SOLUTION

$V_{old} = 5\overline{0}0$ mL $P_{old} = 76\overline{0}$ torr $= 1.00$ atm | pressure increases;

$V_{new} = ?$ $P_{new} = 10.0$ atm   ↓ volume decreases

$T_{old} = 0 + 273 = 273$ K | temperature increases;

$T_{new} = 1\overline{0}0 + 273 = 373$ K ↓ volume increases

From Equation 12–15,

$$V_{new} = V_{old} \times P_{factor} \times T_{factor}$$

Because the units of P_{old} must be the same as those of P_{new}, both pressures must be expressed in the same units. The pressure factor should make the new volume less (1.00 atm/10.0 atm), whereas the temperature factor should make the new volume greater (373 K/273 K). The result is a *new volume* that is *less*, due to the magnitude of the pressure factor. Hence, the decrease in volume due to the pressure factor has a greater effect on the new volume than the increase in volume produced by the temperature factor.

$$V_{new} = 5\overline{0}0 \text{ mL} \times \frac{1.00 \text{ atm}}{10.0 \text{ atm}} \times \frac{373 \text{ K}}{273 \text{ K}} = 68.3 \text{ ml} \qquad Answer$$

Notice that the effect of each factor is considered *independently* and the final volume depends on correctly applying *both* factors.

Exercise 12–4

A certain gas occupies 20.0 liters at $5\overline{0}°C$ and $78\overline{0}$ torr. Under what pressure in torr would this gas occupy 75.0 liters at $0°C$?

12–7 Dalton's Law of Partial Pressures

John Dalton, whose atomic theory was discussed in Section 4–2, was also keenly interested in meteorology. This interest led him to study gases, and in 1801 he announced his conclusions, which are now known as **Dalton's law of partial pressures**. This law states that *each gas in a mixture of gases exerts a partial pressure equal to the pressure it would exert if it were the only gas present in the same volume; the total pressure of the mixture is then the **sum** of the partial pressures of all the gases present.* For example, if two gases, such as oxygen and nitrogen, are present in a 1-liter flask, and the pressure of the oxygen is $25\overline{0}$ torr and that of the nitrogen is $30\overline{0}$ torr, then the total pressure is $55\overline{0}$ torr.

Dalton's law of partial pressures may be expressed mathematically as

$$P_{\text{total}} = P_1 + P_2 + P_3 \qquad (12\text{--}18)$$

where P_1, P_2, and P_3 are the partial pressures of the individual gases in the mixture.

An application of Dalton's law of partial pressures is the collection of a gas over water. The gas will contain a certain amount of water vapor. The pressure exerted by the water vapor in the gas will be a constant value *at any given temperature* if sufficient time has been allowed to permit equilibrium conditions to be established. The total pressure at which the volume of the "wet" gas is measured must be equal to the sum of the gas pressure and the water vapor pressure at the temperature at which the gas is collected and measured, or mathematically,

$$P_{\text{total}} = P_{\text{gas}} + P_{\text{water}} \quad \text{(Dalton's law of partial pressures)} \qquad (12\text{--}19)$$

The vapor pressure of water varies with temperature, but it has a constant and predictable value at any given temperature. Therefore, the pressure of the "dry" gas can be calculated by subtracting the known equilibrium vapor pressure of water at the given temperature from the total pressure of the "wet" gas mixture.

$$P_{\text{gas}} = P_{\text{total}} - P_{\text{water}} \qquad (12\text{--}20)$$

The vapor pressure of water at various temperatures is found in Appendix 4.

EXAMPLE 12–5

The volume of a certain gas, collected over water, is $15\overline{0}$ mL at $3\overline{0}°$C and 720.0 torr. Calculate the volume in milliliters of the *dry* gas at STP.

SOLUTION The first step in the calculation is to determine the pressure of the dry gas at the initial volume ($15\overline{0}$ mL) and temperature ($3\overline{0}°$C). The pressure of the wet gas (720.0 torr) is equal to the sum of the pressure of the dry gas and the vapor pressure of water at the initial temperature. From Appendix 4, the vapor pressure of water at $3\overline{0}°$C is 31.8 torr. The pressure of the dry gas is therefore equal to $P_{\text{total}} - P_{\text{water}} = 720.0$ torr $- 31.8$

torr = 688.2 torr. Thus, if the water vapor were removed—that is, if the gas were dry—the pressure of the gas would have measured 688.2 torr in a volume of $15\bar{0}$ mL at $3\bar{0}°C$. With these data, the next step is to work a combined gas law problem to calculate the volume of the dry gas as STP, as follows:

$V_{old} = 15\bar{0}$ mL

$V_{new} = ?$

$P_{old} = 720.0$ torr $- 31.8$ torr $= 688.2$ torr | *pressure increases;*

$P_{new} = 76\bar{0}$ torr ↓ *volume decreases*

$T_{old} = 3\bar{0} + 273 = 303$ K | *temperature decreases:*

$T_{new} = 0 + 273 = 273$ K ↓ *volume decreases*

From Equation 12–15,

$$V_{new} = V_{old} \times P_{factor} \times T_{factor}$$

$$V_{new} = 15\bar{0} \text{ mL} \times \frac{688.2 \text{ torr}}{76\bar{0} \text{ torr}} \times \frac{273 \text{ K}}{303 \text{ K}} = 122 \text{ mL}   \quad \textit{Answer}$$

Exercise 12–5

The volume of a certain gas, collected over water, is 175 mL at 27°C and 635.0 torr. Calculate the volume in milliliters of the *dry* gas at STP (see Appendix 4).

12–8   Ideal-Gas Equation

In the gas laws [Boyle's (Section 12–3), Charles' (Section 12–4), and Gay-Lussac's (Section 12–5)] the mass was fixed and one of the three variables, temperature, pressure, or volume, was also constant. Using a new equation, the *ideal-gas equation*, we can vary not only the temperature, pressure, and volume, but also the mass of the gas. The **ideal-gas equation** is given below:

ideal-gas equation Formula that allows scientists to vary not only the temperature, pressure, and volume of a gas, but also its mass; it is expressed mathematically as $PV = nRT$, where P is pressure, V is volume, n is the amount of gas in moles, T is temperature, and R is the universal gas constant. In using $R = 0.0821$ atm·L/mol·K, the pressure (P) must be expressed in atmospheres, the volume (V) in liters, the amount (n) in moles, and the temperature (T) in kelvins.

$$\mathbf{PV = nRT} \tag{12–21}$$

where $\mathbf{P}$ = pressure, $\mathbf{V}$ = volume, $\mathbf{n}$ = quantity of gas in *moles*, $\mathbf{T}$ = temperature, and $\mathbf{R}$ = the universal gas constant. The numerical value of $\mathbf{R}$ can be obtained by substituting known values of $\mathbf{P}$, $\mathbf{V}$, $\mathbf{n}$, and $\mathbf{T}$ in the expression $\mathbf{R} = \mathbf{PV/nT}$. Because we know that at STP [0°C (273 K) and 1.00 atm] 1 mol ($\mathbf{n} = 1.00$) of an ideal gas occupies 22.4 liters, $\mathbf{R}$ is evaluated as (1.00 atm $\times$ 22.4 L/1.00 mol $\times$ 273 K) = 0.0821 atm·L/mol·K. You must know the ideal gas equation and the numerical value of $\mathbf{R}$ and its *units* to work problems involving the ideal-gas equation with its four variables: moles, temperature, pressure, and volume.

EXAMPLE 12–6

Calculate the volume in liters of 1.75 mol of oxygen gas at 27°C and 1.25 atm.

SOLUTION Use the ideal-gas equation, $PV = nRT$, and solve for V (volume) as follows:

$$PV = nRT \qquad \text{Divide both sides of the equation by } P.$$

$$\frac{PV}{P} = \frac{nRT}{P}$$

$$V = \frac{nRT}{P}$$

Substituting into this linear equation the values for n (1.75 mol), R (0.0821 atm · L/mol · K), T (27°C + 273 = $30\overline{0}$ K), and P (1.25 atm), we obtain the following:

$$V = \frac{1.75 \ \cancel{mol} \times 0.0821 \ \dfrac{atm \cdot L}{\cancel{mol} \cdot \cancel{K}} \times 30\overline{0} \ \cancel{K}}{1.25 \ \cancel{atm}} = 34.5 \text{ L} \qquad Answer$$

(See Section 1–5 for solving and substituting into a linear equation.)

EXAMPLE 12–7

Calculate the temperature in degrees Celsius (°C) of 0.652 mol of oxygen gas at a pressure of 1.20 atm and occupying a 15.0-L cylinder at 30°C.

SOLUTION Use the ideal-gas equation, $PV = nRT$, and solve for T (temperature), as follows:

$$PV = nRT \qquad \text{Place the unknown } T \text{ on the left,}$$

$$nRT = PV \qquad \text{Divide both sides of the equation by } nR.$$

$$\frac{nRT}{nR} = \frac{PV}{nR}$$

$$T = \frac{PV}{nR}$$

Substituting into the linear equation above the values for n (0.652 mol), R (0.0821 atm · L/mol · K), P (1.20 atm), and V (15.0 L). We obtain the following:

$$T = \frac{1.20 \ \cancel{atm} \times 15.0 \ \cancel{L}}{0.652 \ \cancel{mol} \times 0.0821 \ \dfrac{atm \cdot L}{mol \cdot K}} = 336 \text{ K}$$

(Note that the units of R cancel out again, leaving only K. In division of fractions, you invert and multiply; therefore,

$$\frac{1}{\dfrac{1}{K}} = 1 \times \frac{K}{1} = K$$

The temperature requested is to be expressed in degrees Celsius (°C). Therefore, °C = K − 273 from Equation 2–4 in Section 2–6, and the temperature in degrees Celsius (°C) is as follows:

$$336 \text{ K} = (336 - 273)°C = 63°C \qquad \textit{Answer}$$

Exercise 12–6

Calculate the pressure in atmospheres for 0.750 mol of oxygen gas occupying an 11.0-L cylinder at 27°C.

Hydrogen: lighter than air

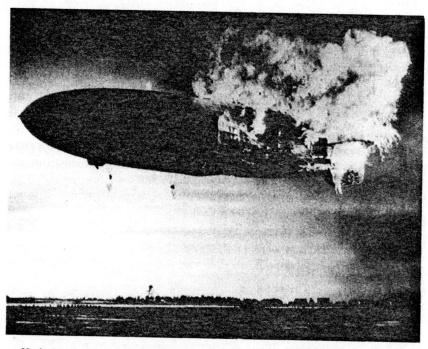

Hydrogen gas is lighter than air, so it was used to fill dirigibles (rigid framework blimps) in the first part of the century. Hydrogen is also very combustible, which was vividly illustrated by the explosion of the Hindenburg, *the largest dirigible ever built, in 1937 at Lakehurst, New Jersey. Understandably, modern blimps (nonrigid or semirigid framework) use helium as the flotation gas.*

Name: Hydrogen derives from the greek *hydro-*, meaning water, and the French *-gene*, meaning to form. The element was named by the French chemist and physicist, Antoine Laurent Lavoisier (1743–1794).

Appearance: Hydrogen occurs as a colorless and

odorless diatomic gas, H_2. Hydrogen is the lightest element and has a density of only 0.0899 g/L.

Occurrence: Hydrogen comprises only 0.1 percent of the earth's crust, but it is 10.8 percent of the oceans. On earth, hydrogen gas (H_2) occurs in the free (uncombined) state only to a very small extent. Hydrogen is most commonly found as a component of water (H_2O). However, hydrogen (as uncombined hydrogen atoms in free space or in stars) is the most common element in the universe, and it accounts for 80 percent of the mass and 94 percent of the molecules in the universe.

Source: Hydrogen gas is obtained from the electrolytic breakdown of water (1), or from the reaction between steam and various carbon-containing compounds (2 and 3):

$$(1) \ 2H_2O_{(l)} \ \xrightarrow{\text{electric current}} \ 2H_{2(g)} + O_{2(g)}$$

$$(2) \ CH_{4(g)} + H_2O_{(g)} \ \xrightarrow[\text{catalyst}]{800°} \ CO_{(g)} + 3H_{2(g)}$$

$$(3) \ CO_{(g)} + H_2O_{(g)} \ \xrightarrow[\text{catalyst}]{400°} \ CO_{2(g)} + H_{2(g)}$$

Its Role in Our World: Hydrogen gas is a very important material in the chemical industry. It is used in the production of a number of chemical compounds, including ammonia (NH_3) and methanol (CH_3OH). It can also be used to remove sulfur-containing pollutants from coal, gas, and oil.

$$N_{2(g)} + 3H_{2(g)} \ \xrightarrow[\text{catalyst}]{400°} \ 2NH_{3(g)}$$

$$CO_{(g)} + 2H_{2(g)} \ \xrightarrow[\text{catalyst}]{400°} \ CH_3OH_{(g)}$$

Hydrogen is also used to prepare a number of metallic elements from their oxides at high temperatures. For example:

$$FeO_{(s)} + H_{2(g)} \ \longrightarrow \ Fe_{(s)} + H_2O_{(g)}$$

$$Cr_2O_{3(s)} + 3H_{2(g)} \ \longrightarrow \ 2Cr_{(s)} + 3H_2O_{(g)}$$

$$NiO_{(s)} + H_{2(g)} \ \longrightarrow \ Ni_{(s)} + H_2O_{(g)}$$

Hydrogen can help make things hot or cool things off. Hydrogen gas is used in high-temperature welding torches ($\approx$2000°C) like the oxy-hydrogen torch. Liquid hydrogen is used as a low-temperature coolant (-252°C or 21 K).

Unusual Facts: Hydrogen gas has been touted as the "fuel of the future," because it burns cleanly and completely in air to give water and *no* pollutants. Scientists have suggested the use of hydrogen in the internal combustion engine. The technological problems that must be solved for hydrogen to serve as a fuel include finding way to use sunlight to turn water into hydrogen and oxygen, and finding a way to *safely* and efficiently store and transport large quantities of hydrogen gas.

Problems

(The vapor pressure of water at various temperatures is given in Appendix 4.)

1. A sample of gas has a volume of $4\overline{0}0$ mL when measured at 25°C and $3\overline{0}0$ torr. What will be its volume in milliliters at 25°C and 195 torr?

2. What final pressure in torr must be applied to a sample of gas having a volume of $2\overline{0}0$ mL at $2\overline{0}$°C and $75\overline{0}$ torr pressure to permit the expansion of the gas to a volume of $6\overline{0}0$ mL at $2\overline{0}$°C?

3. A gas occupies a volume of $1\overline{0}0$ mL at 27°C and $74\overline{0}$ torr. What volume in milliliters will the gas have at 5°C and $74\overline{0}$ torr?

4. A gas occupies a volume of $1\overline{0}0$ mL at 27°C and $63\overline{0}$ torr. At what temperature in degrees Celsius (°C) would the volume be 80.0 mL at $63\overline{0}$ torr?

5. A sample of gas occupies 5.00 liters at $2\overline{0}0$ torr and 27°C. Calculate its pressure in torr if the temperature is changed to 127°C while the volume remains constant.

6. A gas occupies a volume of 50.0 mL at 27°C and $63\overline{0}$ torr. At what temperature in degrees Celsius (°C) would the pressure be $76\overline{0}$ torr if the volume remains constant?

7. A sample of gas occupies a volume of 395 mL at 60.0 torr and 27°C. What volume in milliliters will it occupy at STP?

8. A given sample of a gas has a volume of 5.20 liters at 27°C and $64\overline{0}$ torr. Its volume and temperature are changed to 2.10 liters and $10\overline{0}$°C, respectively. Calculate the pressure in torr at these conditions.

9. The volume of a sample of oxygen gas collected over water is 155 mL at 25°C and 600.0 torr. Calculate the volume in milliliters of *dry* oxygen gas at STP.

10. The volume of a sample of nitrogen gas collected over water is 285 mL at 25°C and 700.0 torr. Calculate the volume in milliliters of *dry* nitrogen gas at STP.

11. Calculate the volume in milliliters for 0.0300 mol of nitrogen gas at $3\overline{0}$°C and 1.20 atm.

12. Calculate the number of moles of nitrogen gas in a 5.00-L cylinder at 27°C and 1.05 atm.

General Problems

13. An organic gas has a mass of 0.443 g with a volume of 0.650 L at 37°C and 438 torr. Its composition is 80.0% carbon and 20.0% hydrogen.

    (a) Calculate the volume of the gas in liters at STP.
    (b) Calculate the molecular mass of the gas (see Section 8–3).
    (c) Calculate the molecular formula of the gas (see Section 8–3).

14. Halothane (Fluothane) is a nonflammable, nonirritating, general anesthetic and in many instances is superior to ethyl ether. At 57°C and $64\overline{0}$ torr, 0.529 g of the gas occupies a volume of 86.4 mL. Its composition is 12.2% carbon, 0.5% hydrogen, 40.5% bromine, 18.0% chlorine, and 28.9% fluorine.

    (a) Calculate the volume of halothane in milliliters at STP.
    (b) Calculate the molecular mass of halothane (see Section 8–3).
    (c) Calculate the molecular formula of halothane (see Section 8–3).

15. Potassium chlorate (10.0 g) produces oxygen gas when heated.

    (a) Write a balanced equation for the reaction (see Section 9–7).
    (b) Calculate the volume of oxygen in liters produced at STP (see Section 11–5).
    (c) Convert the volume of oxygen produced at STP to a volume in liters measured at 35°C and $63\overline{0}$ torr.

16. Magnesium (0.520 g) reacts with excess hydrochloric acid to produce hydro gen gas.

   (a) Write a balanced equation for the reaction (see Section 9–8).
   (b) Calculate the volume of hydrogen in liters produced at STP (see Section 11–5).
   (c) Convert the volume of hydrogen produced at STP to a volume in liters measured at 25°C and $64\bar{0}$ torr.

17. Calculate the number of grams of nitrogen gas (N_2) in a 10.0-L cylinder at $3\bar{0}$°C and 1.10 atm. (*Hint*: See Section 8–2.)

Answers to Exercises

12–1. 624 torr

12–2. $16\bar{0}$ mL

12–3. 87°C

12–4. 176 torr

12–5. 127 mL

12–6. 1.68 atm

QUIZ

1. A sample of a gas occupies a volume of 125 mL at 0°C and $64\bar{0}$ torr. Calculate the volume of the gas in milliliters at 0°C and $78\bar{0}$ torr.

2. A gas sample occupies 229 mL at 27°C and $63\bar{0}$ torr. Calculate the volume of the gas in milliliters at STP.

3. The volume of a sample of oxygen gas collected over water is 258 mL at 27°C and 620.0 torr. Calculate the volume in milliliters of *dry* oxygen gas at STP. (The vapor pressure of water at 27°C is 26.7 torr.)

4. Calculate the pressure in atmospheres exerted by 0.250 mol of hydrogen gas in a 5.00-L cylinder at 35°C.

5. Marble or limestone ($CaCO_3$) reacts with aqueous hydrochloric acid to give carbon dioxide. Calculate the number of liters of carbon dioxide produced at 18°C and 715 torr from 87.0 g of limestone and excess hydrochloric acid. (Atomic mass units: Ca = 40.1 amu, C = 12.0 amu, O = 16.0 amu.)

13

SOLUTIONS

COUNTDOWN

ELEMENT	ATOMIC MASS UNITS (amu)
C	12.0
H	1.0
O	16.0
Ba	137.3

5 Carry out the following conversions (Section 2–3).
(a) 475 mL to L   (0.475 L);   (b) 268 mg to g   (0.268 g)

4 Calculate the formula or molecular mass of the following compounds (Section 8–1).
(a) $C_3H_8O_3$ (glycerol)   (92.0 amu)
(b) $Ba(OH)_2$ (barium hydroxide)   (171.3 amu)

3 Calculate the number of moles in each of the following quantities of compounds (Section 8–2).
(a) 14.8 g glycerol ($C_3H_8O_3$)   (0.161 mol)
(b) 29.4 g barium hydroxide [$Ba(OH)_2$]   (0.172 mol)

2 Given 0.950 mol of glycerol ($C_3H_8O_3$), calculate the following (Section 8–2):
(a) the number of grams of glycerol   (87.4 g)
(b) the number of molecules of glycerol   (5.72×10^{23} molecules)

1 Calculate the percent metal in an oxide if 0.200 g of metal combines with 0.310 g of oxygen to form the oxide (Section 8–4).

(39.2%)

TASK **1** Memorize the four methods for expressing concentrations of solutions given in Table 13–1.

OBJECTIVES

1 Given the distinguishing characteristics of each of the following terms:
(a) solution (Section 13–1)
(b) solute (Section 13–1)
(c) solvent (Section 13–1)
(d) percent by mass (Section 13–3)
(e) molality (Section 13–4)
(f) molarity (Section 13–5)
(g) normality (Section 13–6)

2 Given any two of the following three quantities for a solution: (a) the mass of solute; (b) the mass of solvent or solution; and (c) the percent concentration of solution by mass. Calculate the third quantity (Examples 13–1 and 13–2, Exercises 13–1 and 13–2, Problems 1 and 2).

3 Given the formula of a solute, the Table of Approximate Atomic Masses, and any two of the following three quantities for a solution: (a) the mass of solute; (b) the mass of solvent or solution; and (c) the molal concentration (molality). Calculate the third quantity (Examples 13–3 through 13–5, Exercises 13–3 through 13–5, and Problems 3 through 5).

4 Given the formula of a solute, the Table of Approximate Atomic Masses, and any two of the following three quantities for a solution: (a) the mass of solute; (b) the volume of solution; and (c) the molar concentration (molarity). Calculate the third quantity (Examples 13–6 through 13–8, Exercises 13–6 through 13–8, Problems 6 through 8).

5 Given the formula of a solute, the Table of Approximate Atomic masses, the use of the solute in a reaction, and any two of the following three quantities for a solution: (a) the mass of solute; (b) the volume of solution; and (c) the normal concentration (normality). Calculate the third quantity (Examples 13–9 through 13–11, Exercises 13–9 through 13–11, Problems 9 through 11).

6 Given the formula of solute and the use of the solute in a reaction, interconvert between the normal and molar concentrations of the solutions (Example 13–12, Exercise 13–12, Problem 12).

Every morning you probably encounter a solution. If you have coffee or tea with sugar in the morning, you are preparing a solution. The concentration of the solution depends on your taste. Possibly, you prefer to get your morning "shot" of caffeine with soda pop such as Coke® or Pepsi.® The soda pop is also a solution. Read the ingredients on the side of the can sometime!

13–1 Solutions

solution Homogeneous mixture composed of two or more pure substances; its composition can be varied usually *within certain limits*.

As we noted in Chapter 3, by definition a **solution** is homogeneous throughout. It is composed of two or more pure substances, and its composition can be *varied*, usually *within certain limits*. For example, a solution of sugar dissolved in water could contain 1 g, 2 g, or 3 g of sugar in 100 g of water and still be a "sugar solution." A solution consists of a solute dissolved in a solvent. The term **solute** usually refers to the component present in *lesser* quantity in the solution. The term **solvent** refers to the component present in *greater* quantity in the solution. For example, in a 5.00% sugar solution in water, sugar is the solute and water is the solvent. This solution is called an *aqueous* solution because the solvent is water.

A covalently bonded substance typically disperses in the solvent as individual *molecules*, whereas an ionic substance usually dissolves as individual *ions* among the solvent molecules. Often, the molecules or ions will associate or attach themselves to one or more solvent molecules. As Figure 13–1a shows, in an aqueous sugar solution, sugar molecules will leave the solid mass of sugar at the bottom of the container and disperse themselves into the bulk of the water. As they do so, they become loosely tied to a number of water (solvent) molecules. These dissolved molecules are considered to be *hydrated* (solvated) by water molecules. More molecules continue to dissolve in the water and become hydrated until either all the sugar present dissolves, or the water can hold no more sugar. Thus, there is a limit to the extent to which most solutes dissolve in a solvent.

solute Substance dissolved in a solution; it is usually present in lesser quantity than the solvent.

solvent Dissolving substance in a solution; it is usually present in greater quantity than the solute.

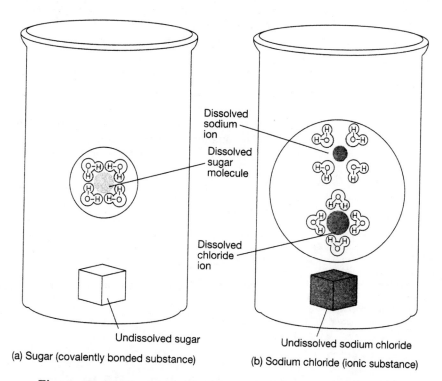

Dissolved sodium ion

Dissolved sugar molecule

Dissolved chloride ion

Undissolved sugar

(a) Sugar (covalently bonded substance)

Undissolved sodium chloride

(b) Sodium chloride (ionic substance)

Figure 13–1   *The process of dissolving: (a) sugar dissolving in water to form hydrated sugar molecules, $C_{12}H_{22}O_{11}(H_2O)_x$: (b) sodium chloride dissolving in water to form hydrated sodium ions, $Na^{1+}(H_2O)_x$ (relatively negative oxygen atom of water attracted to positive sodium ions) and hydrated chloride ions, $Cl^{1-}(H_2O)_y$ (relatively positive hydrogen atom of water attracted to negative chloride ions).*

In an aqueous sodium chloride solution, the individual sodium ions (Na^{1+}) and chloride ions (Cl^{1-}) will leave the solid mass of sodium chloride at the bottom of the container and disperse into the water. Each ion becomes surrounded by water molecules as shown in Figure 14–1b. Note that the sodium ions are surrounded by water molecules that have aimed the negative end of the water molecule (oxygen atom) toward the positive sodium ion, while the chloride ions are surrounded by water molecules that have aimed the positive end of the water molecule (hydrogen atom) toward the negative chloride ion. The energy released in the formation of the hydrated ions overcomes the strong attractive forces between the sodium cations and the chloride anions in solid crystalline NaCl.

13–2 Concentration of Solutions

The terms "concentrated" (conc or concd) and "dilute" (dil) are sometimes used to express concentrations, but these are at best very qualitative. Concentrated hydrochloric acid contains approximately 37 g of hydrogen chloride per $\overline{100}$ g of solution, whereas concentrated nitric acid has approximately 72 g of hydrogen nitrate per $\overline{100}$ of solution. Dilute solutions are less concentrated, but beyond this little more can be said concerning them. A dilute solution of hydrochloric acid, for example, could be 1.00 g, 5.00 g, or 10.0 g of hydrogen chloride per $\overline{100}$ g of solution, depending on the particular purpose for which the acid is intended. Obviously, we must use more quantitative terms for expressing concentrations.

In the next four sections, we discuss the more common quantitative methods used to express the concentration of solutions. The particular method for expressing the concentration of a solution is generally determined by the *eventual use* of the solution. These methods for expressing concentrations are

1. Percent by mass

2. Molality

3. Molarity

4. Normality

percent by mass Measure of the concentration of a solution expressed as parts by mass of solute per $\overline{100}$ parts by mass of *solution*:

Percent by mass =

$$\frac{\text{mass of solute}}{\text{mass of } solution} \times 100$$

Each method has an advantage over the other methods depending on the eventual use. For example, if you want to know the number of pounds of salt in a given mass of ocean water, it is more convenient to express the concentration in percent by mass.

Study hint

When you use percent by mass, be sure that the mass units for the solute and solvent are the *same*.

13–3 Percent by Mass

The **percent by mass** of a solute in a solution is the parts by mass of solute per $\overline{100}$ parts by mass of *solution*:

$$\text{Percent by mass} = \frac{\text{mass of solute}}{\text{mass of } solution} \times 100$$

The mass of *solution* is equal to the mass of the *solute* plus the mass of the *solvent*. For example, a 20.0% solution of sodium sulfate would contain 20.0 g of sodium sulfate in $10\overline{0}$ g of solution (80.0 g of water), as shown in Fig. 13-2. In chemistry, concentrations expressed as percentages are understood to mean *percent by mass* unless specified otherwise.

EXAMPLE 13-1

Calculate the percent of sodium chloride if 15.0 g of sodium chloride (NaCl) is dissolved in enough water to make 165 g of solution.

SOLUTION   Since the total mass of the solution is 165 g, the percent of sodium chloride is readily obtained as

$$\frac{15.0 \text{ g NaCl}}{165 \text{ g solution}} \times 100$$

$$= 9.09 \text{ parts of sodium chloride per } 10\overline{0} \text{ parts of solution}$$

$$= 9.09\% \text{ NaCl} \quad \textit{Answer}$$

Exercise 13-1

Calculate the percent of calcium chloride if 21.4 g of calcium chloride ($CaCl_2$) is dissolved in 94.2 g of water.

EXAMPLE 13-2

Calculate the number of grams of sugar ($C_{12}H_{22}O_{11}$) that must be dissolved in $100\overline{0}$ g of water to prepare a 20.0% sugar solution.

SOLUTION   In this solution, there would be 20.0 g of sugar for every 80.0 g of water (100.0 g solution − 20.0 g sugar = 80.0 g water), and the number of grams of sugar needed for $100\overline{0}$ g of water is calculated

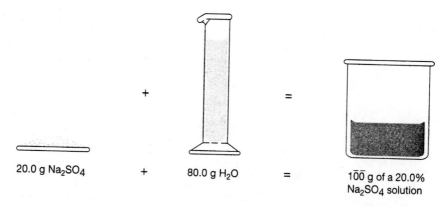

20.0 g Na$_2$SO$_4$ +   80.0 g H$_2$O   = $10\overline{0}$ g of a 20.0%
                                                                          Na$_2$SO$_4$ solution

Figure 13-2   *A 20.0% by mass aqueous solution of sodium sulfate (Na$_2$SO$_4$).*

as

$$1\overline{000} \text{ g H}_2\text{O} \times \frac{20.0 \text{ g sugar}}{80.0 \text{ g H}_2\text{O}} = 25\overline{0} \text{ g sugar needed for } 1\overline{000} \text{ g water}$$

Answer

Exercise 13–2

Calculate the number of grams of sodium chloride (NaCl) that must be dissolved in $12\overline{0}$ g of water to prepare a 5.65% sodium chloride solution.

13–4 Molality

molality (*m*) Measure of the concentration of solution expressed as the number of moles of solute per *kilogram* of *solvent*:

m = molality

$$= \frac{\text{moles of solute}}{\text{kilogram of } \textit{solvent}}$$

Molality (abbreviated as *m*) is the number of moles of solute per *kilogram* of solvent. This method of expressing concentration is based on the mass of solute (expressed as moles) per unit mass (1.00 kg) of *solvent*:

$$m = \text{molality} = \frac{\text{moles of solute}}{\text{kilogram of } \textit{solvent}}$$

In the preparation of a one-molal (*m*) aqueous solution of sodium sulfate, one mole of sodium sulfate (142.1 g) would be dissolved in 1.000 kg ($1\overline{000}$ g) of water, as shown in Fig. 13–3. Note that the total volume of the solution is not known. However, we can find the mass of the solution by adding the *mass of the solute* and the *mass of the solvent*. From a knowledge of the density of the solution we can calculate the total volume. In the expression of concentration in terms of molality, we must know the masses of the solute and the solvent; their volumes are not involved.

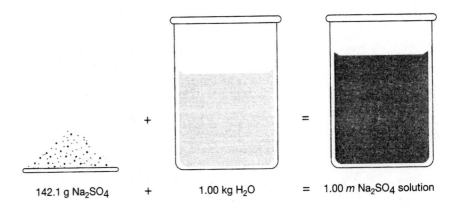

142.1 g Na$_2$SO$_4$ +   1.00 kg H$_2$O   =   1.00 *m* Na$_2$SO$_4$ solution

Figure 13–3 *A one-molal (1.00-m) aqueous solution of sodium sulfate (Na$_2$SO$_4$).*

EXAMPLE 13–3

Calculate the molality of a phosphoric acid solution containing 32.7 g of phosphoric acid (H_3PO_4) in $10\overline{0}$ g of water.

SOLUTION The molality of the solution must express the concentration of H_3PO_4 as moles per kilogram of water. The formula mass of H_3PO_4 is 98.0 amu; calculate the molality as:

$$\frac{32.7 \text{ g } H_3PO_4}{10\overline{0} \text{ g } H_2O} \times \frac{1 \text{ mol } H_3PO_4}{98.0 \text{ g } H_3PO_4} \times \frac{1000 \text{ g } H_2O}{1 \text{ kg } H_2O} = \frac{3.34 \text{ mol } H_3PO_4}{1 \text{ kg } H_2O}$$

$$= 3.34 \text{ } m \qquad Answer$$

Exercise 13–3

Calculate the molality of a sugar solution containing 78.0 g of sugar ($C_{12}H_{22}O_{11}$) in 125 g of water.

EXAMPLE 13–4

Calculate the number of grams of glycerol ($C_3H_8O_3$) necessary to prepare $50\overline{0}$ g of a 2.00 m solution of glycerol in water.

SOLUTION The molecular mass of glycerol ($C_3H_8O_3$) is 92.0 amu. A 2.00-m glycerol solution would contain 2.00 mol (2.00 mol $\times$ 92.0 g/mol = 184 g) of glycerol in 1.000 kg ($10\overline{0}0$ g) of water. The *total mass* of this solution would be 1184 g (184 g glycerol + $10\overline{0}0$ g water). Calculate the mass of glycerol necessary for $50\overline{0}$ g of a 2.00 m solution as

$$50\overline{0} \text{ g solution} \times \frac{184 \text{ g glycerol}}{1184 \text{ g solution}} = 77.7 \text{ g glycerol} \qquad Answer$$

Exercise 13–4

Calculate the number of grams of sugar ($C_{12}H_{22}O_{11}$) necessary to prepare 225 g of a 2.00 m solution of sugar in water.

EXAMPLE 13–5

Calculate the number of grams of water that must be added to 75.0 g of sugar ($C_{12}H_{22}O_{11}$) in the preparation of a 1.25 m solution.

SOLUTION The number of grams of water is required, so we must use the inverse factor for molality. A 1.25-m sugar solution would contain 1.25 mol of sugar ($C_{12}H_{22}O_{11}$) in 1.000 kg of water. The molecular mass of sugar is 342 amu. Calculate the number of grams of water as

$$75.0 \text{ g } C_{12}H_{22}O_{11} \times \frac{1 \text{ mol } C_{12}H_{22}O_{11}}{342 \text{ g } C_{12}H_{22}O_{11}} \times \frac{1 \text{ kg } H_2O}{1.25 \text{ mol } C_{12}H_{22}O_{11}}$$

$$\times \frac{1000 \text{ g } H_2O}{1 \text{ kg } H_2O} = 175 \text{ g } H_2O$$

$$Answer$$

Exercise 13–5

Calculate the number of grams of water that must be added to 47.0 g of glycerol ($C_3H_8O_3$) in the preparation of a 1.65 m solution.

13–5 Molarity

Molarity or *molar concentration* (abbreviated as M) is the number of moles of solute per *liter of solution*[1]:

$$M = \text{molarity} = \frac{\text{moles of solute}}{\text{liter of } solution}$$

This method of expressing concentration is very useful when volumetric equipment (graduated cylinders, burets, etc.) is used to measure a quantity of the solution. From the volume measured, a simple calculation gives the mass of solute used.

To prepare one liter of a one-molar (M) aqueous solution of sodium sulfate, one mole of sodium sulfate (142.1 g) is dissolved in water. *Enough* water is then added to bring the volume of the solution to one *liter*, as shown in Fig. 13–4. An important point to note here is that no information is stated as to the amount of solvent added, only that the solution is made to bring the total volume to one liter. We can calculate the amount of water used if we know the density of the resulting solution.

EXAMPLE 13–6

Calculate the molarity of an aqueous phosphoric acid (H_3PO_4) solution containing 284 g of phosphoric acid is $5\overline{0}0$ mL of solution.

SOLUTION   The formula mass of phosphoric acid is 98.0 amu; therefore, calculate the molarity as

$$\frac{284 \text{ g } H_3PO_4}{5\overline{0}0 \text{ mL solution}} \times \frac{1 \text{ mol } H_3PO_4}{98.0 \text{ g } H_3PO_4} \times \frac{1000 \text{ mL solution}}{1 \text{ L solution}}$$

$$= \frac{5.80 \text{ mol } H_3PO_4}{1 \text{ L solution}} = 5.80 \text{ } M$$

Answer

Exercise 13–6

Calculate the molarity of an aqueous sodium chloride solution containing 128 g of sodium chloride (NaCl) in 775 mL of solution.

---

[1] Another term similar to molarity is *formality* (F). This term is used for solutions in which the solute exists as ions. In this book we shall use the term "molarity" without regard to the type of bonding found in the solute.

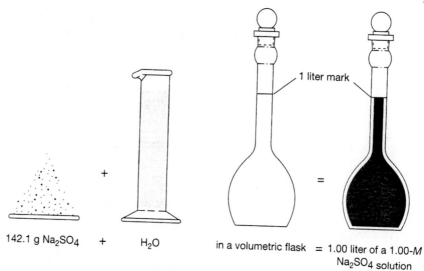

142.1 g Na₂SO₄   +   H₂O   in a volumetric flask = 1.00 liter of a 1.00-*M* Na₂SO₄ solution

Figure 13–4   *A one-molar (1.00-M) aqueous solution of sodium sulfate (Na₂SO₄).*

EXAMPLE 13–7

(a) Calculate the number of grams of sugar ($C_{12}H_{22}O_{11}$) required to prepare 225 mL of a 1.25 *M* aqueous sugar solution. (b) Explain how this solution would be prepared.

SOLUTION   (a) The molecular mass of sugar ($C_{12}H_{22}O_{11}$) is 342 amu. A 1.25 *M* sugar solution would contain 1.25 mol of sugar in 1.00 L of solution. Calculate the number of grams of sugar needed for preparing 225 mL of a 1.25 *M* solution as

$$225 \text{ mL solution} \times \frac{1 \text{ L solution}}{1000 \text{ mL solution}} \times \frac{1.25 \text{ mol } C_{12}H_{22}O_{11}}{1 \text{ L solution}}$$

$$\times \frac{342 \text{ g } C_{12}H_{22}O_{11}}{1 \text{ mol } C_{12}H_{22}O_{11}} = 96.2 \text{ g } C_{12}H_{22}O_{11} \qquad Answer$$

(b) The sugar (96.2 g) is dissolved in sufficient water to make the total volume of the solution equal to 225 mL.

Exercise 13–7

(a) Calculate the number of grams of sodium chloride (NaCl) required to prepare $28\overline{0}$ mL of a 1.65 *M* aqueous sodium chloride solution. (b) Explain how this solution would be prepared.

EXAMPLE 13–8

Calculate the number of liters of 6.00 *M* aqueous sodium hydroxide solution required to provide $30\overline{0}$ g of sodium hydroxide (NaOH).

SOLUTION   The formula mass of NaOH is 40.0 amu. In a 6.00 *M* NaOH solution, there are 6.00 mol of NaOH per 1.00 L of solution. Calculate the number of liters of 6.00 *M* solution required to provide $\overline{300}$ g of NaOH as

$$\overline{300} \text{ g NaOH} \times \frac{1 \text{ mol NaOH}}{40.0 \text{ g NaOH}} \times \frac{1.00 \text{ L solution}}{6.00 \text{ mol NaOH}} = 1.25 \text{ L solution}$$

Answer

Exercise 13–8

Calculate the number of milliliters of a 3.60 *M* aqueous phosphoric acid solution required to provide 40.0 g of phosphoric acid (H_3PO_4).

13–6   Normality

normality (*N*) Measure of the concentration of a solution expressed as the number of equivalents of solute per liter of *solution*:

N = normality

$= \dfrac{\text{equivalents of solute}}{\text{liter of } \textit{solution}}$

Normality (abbreviated as *N*) is the number of equivalents (abbreviated eq) of solute per liter of *solution*:

$$N = \text{normality} = \frac{\text{equivalents of solute}}{\text{liter of } \textit{solution}}$$

The equivalent mass in grams (one equivalent) of the solute is based on the reaction involved and is defined by either the acid–base concept or the oxidation–reduction concept, depending on the ultimate use of the solution. Here, however, we limit our discussion of equivalents and normality to applications using the acid–base concept of equivalence.

One equivalent of any *acid* is equal to the mass in grams of that acid capable of supplying 6.02×10^{23} (Avogadro's number; see Section 8–2) of hydrogen ions (1 mol). *One equivalent* of any *base* is equal to the mass in grams of that base that will combine with 6.02×10^{23} hydrogen ions (1 mol) or supply 6.02×10^{23} hydroxide ions (1 mol). Thus, *one equivalent of any acid will exactly combine with one equivalent of any base. One equivalent* of any *salt* is defined by the reaction the salt undergoes and is equal to the mass in grams of the salt capable of supplying 6.02×10^{23} positive charges or 6.02×10^{23} negative charges.

The equivalent mass in grams (one equivalent) of an acid is determined by dividing the molar mass (See Section 8–2) of the acid by the number of moles of hydrogen ion per mole of acid *used in the reaction*. The equivalent mass in grams (one equivalent) of a base is determined by dividing the molar mass of the base by the number of moles of hydrogen ions combining with one mole of the base *in the reaction*. The equivalent mass in grams (one equivalent) of a salt is determined by dividing the molar mass of the salt by the number of moles of positive or negative charges per mole of the salt *used in the reaction*. In all cases, the reaction must be considered.

Consider the following examples:

One equivalent (eq) of H_2SO_4 if $2H^{1+}$ are replaced $=$

$$\frac{\text{molar mass of } H_2SO_4}{2} = \frac{98.1 \text{ g}}{2} = 49.0 \text{ g} \qquad \text{(equivalent mass; see footnote 2)}$$

One equivalent (eq) of H_2SO_4 if $1H^{1+}$ is replaced $=$

$$\frac{\text{molar mass of } H_2SO_4}{1} = \frac{98.1 \text{ g}}{1} = 98.1 \text{ g} \qquad \text{(equivalent mass; see footnote 2)}$$

One equivalent (eq) of $Ca(OH)_2$ if $2OH^{1-}$ are replaced $=$

$$\frac{\text{molar mass of } Ca(OH)_2}{2} = \frac{74.1 \text{ g}}{2} = 37.0 \text{ g} \qquad \text{(equivalent mass)}$$

One equivalent (eq) of $Ca(OH)_2$ if $1OH^{1-}$ is replaced $=$

$$\frac{\text{molar mass of } Ca(OH)_2}{1} = \frac{74.1 \text{ g}}{1} = 74.1 \text{ g} \qquad \text{(equivalent mass)}$$

One equivalent (eq) of Na_2SO_4 if $2Na^{1+}$ are replaced $=$

$$\frac{\text{molar mass of } Na_2SO_4}{2} = \frac{142.1 \text{ g}}{2} = 71.0 \text{ g} \qquad \text{(equivalent mass)}$$

One equivalent (eq) of Na_2SO_4 if $1Na^{1+}$ are replaced $=$

$$\frac{\text{molar mass of } Na_2SO_4}{1} = \frac{142.1 \text{ g}}{1} = 142.1 \text{ g} \qquad \text{(equivalent mass)}$$

Since we are dividing the molar mass by whole numbers, a one normal (1.00 N) solution of a compound will then bear a certain whole-number ratio to a one molar (1.00 M) solution of the same compound. A one normal sodium chloride (NaCl) solution where the sodium ion is replaced has a concentration of one molar, because there is one equivalent (eq) of sodium chloride in one mole of sodium chloride. However, a one normal sodium sulfate (Na_2SO_4) solution where both sodium ions are replaced has a concentration of 0.500 M, because there are two equivalents of sodium sulfate in *one mole* of sodium sulfate.

$$\frac{1.00 \text{ eq } Na_2SO_4}{1 \text{ L solution}} \times \frac{1 \text{ mol } Na_2SO_4}{2 \text{ eq } Na_2SO_4} = \frac{0.500 \text{ mol } Na_2SO_4}{1 \text{ L solution}} = 0.500 \text{ M}$$

Notice that there will always be ONE mole of a substance that contains a variable whole number of equivalents.

To prepare a one normal aqueous solution of sodium sulfate where both sodium ions are to be replaced, dissolve one equivalent (142.1 g/2 = 71.0 g) in

---

[2] Examples of the first two reactions shown are as follows:

$2NaOH + H_2SO_4 \longrightarrow Na_2SO_4 + 2H_2O \qquad$ (*both* H^{1+} replaced)

$NaOH + H_2SO_4 \longrightarrow NaHSO_4 + H_2O \qquad$ (*one* H^{1+} replaced)

water. Add *enough* water to bring the volume of the solution *to one liter*, as shown in Fig. 13–5. Table 13–1 reviews the four different types of solutions discussed previously.

EXAMPLE 13–9

Calculate the normality of an aqueous phosphoric acid solution containing $28\overline{0}$ g of phosphoric acid (H_3PO_4) in 1.00 L of solution in reactions that replace all three hydrogen ions.

SOLUTION   The molar mass of phosphoric acid is 98.0 g. One equivalent of H_3PO_4 is (98.0 g/3) = 32.7 g, because 3 mol of hydrogen ions are used for each mole of the acid.

$$\frac{28\overline{0} \text{ g } H_3PO_4}{1.00 \text{ L solution}} \times \frac{1 \text{ eq } H_3PO_4}{32.7 \text{ g } H_3PO_4} = \frac{8.56 \text{ eq } H_3PO_4}{1.00 \text{ L solution}} = 8.56 \text{ N}\quad Answer$$

Exercise 13–9

Calculate the normality of an aqueous sulfuric acid solution containing 127 g of sulfuric acid (H_2SO_4) in $40\overline{0}$ mL of solution in reactions that replace both hydrogen ions.

EXAMPLE 13–10

Calculate the number of grams of sulfuric acid (H_2SO_4) necessary to prepare $50\overline{0}$ mL of 0.100 N aqueous sulfuric acid solution in reactions that replace both hydrogen ions.

SOLUTION   The molar mass of sulfuric acid is 98.1 g. One equivalent of H_2SO_4 is (98.1 g/2) = 49.0 g, because 2 mol of hydrogen ions are used for each mole of the acid. In a 0.100 N H_2SO_4 solution, there would be 0.100 equivalent of H_2SO_4 in 1.00 L of solution. Calculate the number of grams needed to prepare $50\overline{0}$ mL of 0.100 N H_2SO_4 solution as

$$50\overline{0} \text{ mL solution} \times \frac{1 \text{ L solution}}{1000 \text{ mL solution}} \times \frac{0.100 \text{ eq } H_2SO_4}{1 \text{ L solution}}$$

$$\times \frac{49.0 \text{ g } H_2SO_4}{1 \text{ eq } H_2SO_4} = 2.45 \text{ g } H_2SO_4\quad Answer$$

The sulfuric acid is dissolved in sufficient water to make the total volume of the *solution* equal to $50\overline{0}$ mL.

Exercise 13–10

Calculate the number of grams of calcium hydroxide [$Ca(OH)_2$] necessary to prepare $25\overline{0}$ mL of 0.125 N aqueous calcium hydroxide solution in reactions that replace both hydroxide ions.

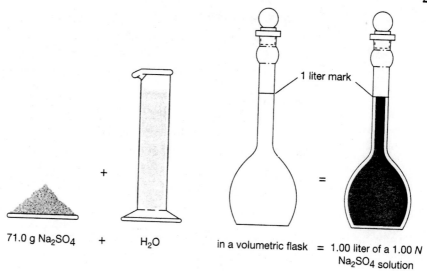

Figure 13–5 *A one-normal (1.00-N) aqueous solution of sodium sulfate (Na₂SO₄) replacing both sodium ions.*

EXAMPLE 13–11

Calculate the number of milliliters of a 2.00 N aqueous sulfuric acid (H_2SO_4) solution required to provide 134 g of sulfuric acid for a reaction that replaces both hydrogen ions in the sulfuric acid.

SOLUTION A 2.00 N sulfuric acid solution contains 2.00 equivalents of H_2SO_4 per liter of solution. One equivalent H_2SO_4 is (98.1 g/2) = 49.0 g, because 2 mol of hydrogen ions are used for each mole of the acid. Calculate the number of milliliters needed as

$$134 \text{ g } \cancel{H_2SO_4} \times \frac{1 \text{ eq } \cancel{H_2SO_4}}{49.0 \text{ g } \cancel{H_2SO_4}} \times \frac{1 \text{ L } \cancel{\text{solution}}}{2.00 \text{ eq } \cancel{H_2SO_4}} \times \frac{1000 \text{ ml solution}}{1 \text{ L } \cancel{\text{solution}}}$$

= 1370 mL solution (to three significant digits)   *Answer*

TABLE 13–1 EXPRESSING CONCENTRATIONS OF SOLUTION

$$\text{Percent by mass} = \frac{\text{mass of solute}}{\text{mass of } \textit{solution}} \times 100$$

$$m = \text{molality} = \frac{\text{moles of solute}}{\text{kilograms of } \textbf{solvent}}$$

$$M = \text{molarity} = \frac{\text{moles of solute}}{\text{liter of } \textit{solution}}$$

$$N = \text{normality} = \frac{\text{equivalents of solute}}{\text{liter of } \textit{solution}}$$

Exercise 13–11

Calculate the number of milliliters of a 1.10 N aqueous calcium hydroxide [Ca(OH)$_2$] solution required to provide 12.5 g of calcium hydroxide for a reaction that replaces both hydroxide ions in the calcium hydroxide.

EXAMPLE 13–12

Calculate the normality of 2.00 M aqueous sulfuric acid (H$_2$SO$_4$) solution in reactions that replace both hydrogen ions.

SOLUTION   A 2.00 M sulfuric acid solution contains 2.00 mol of H$_2$SO$_4$ per liter of solution. There are 2 eq per mol of H$_2$SO$_4$, so the normality is calculated as

$$\frac{2.00 \text{ mol H}_2\text{SO}_4}{1 \text{ L solution}} \times \frac{2 \text{ eq H}_2\text{SO}_4}{1 \text{ mol H}_2\text{SO}_4} = \frac{4.00 \text{ eq H}_2\text{SO}_4}{1 \text{ L solution}} = 4.00 \ N \quad \textit{Answer}$$

Exercise 13–12

Calculate the molarity of a 3.00 N aqueous phosphoric acid (H$_3$PO$_4$) solution in reactions that replace all three hydrogen ions.

Soap: chemistry cleans up

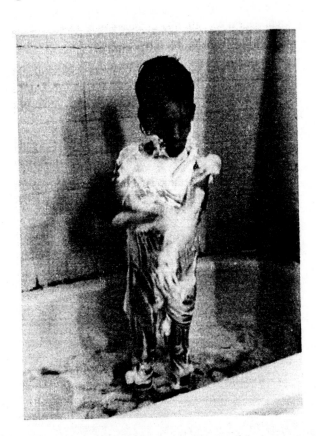

Soap was invented for little children!

Name: The word "soap" probably derives from *saipon*, from the Germanic family of languages in northern Europe. The German *seifen*, the French *savon*, the Italian *sapone*, and the Spanish *jabon* all came from *saipon*.

Appearance: Soap is a soft, waxy solid that dissolves to some extent in water.

Occurrence: Credit is generally given to the Phoenicians for inventing soap about 2600 years ago. However, it is believed that soap was first prepared 5000 years ago from wood ashes and animal fat. Until the early 1700s soap was prepared in small batches in homes or as a cottage industry for home and local use. At that time it began to be prepared on a larger scale as an industry. All this time, soap was made one batch at a time. It wasn't until 1938 that a large-scale continuous process was perfected by the Procter and Gamble Company and soap could be made efficiently on an industrial scale. A number of other continuous processes have been developed since 1938.

Source: Soap was historically made by heating animal fats (triglycerides) with soda ash (Na_2CO_3). A more efficient process involves treating animal fats with sodium or potassium hydroxide solutions to give soap and glycerol (glycerine). A typical example would be

Most soaps involve R groups that contains a total of 11, 13, 15, or 17 carbon atoms. Soft soaps involve potassium hydroxide (KOH) instead of sodium hydroxide (NaOH), and thus the soap contains potassium ions (K^{1+}) instead of sodium ions (Na^{1+}).

Its Role in Our World: Soap can be envisioned as a long skinny molecule ($-CH_2-CH_2-$ ⋯ $-CH_2-CH_3$ portion, a "tail" shown as ∿∿∿∿∿) with a negatively charged head (a charged "head" shown as ⊖). The charged head likes to be in water, since water molecules can hydrate (solvate, Section 13–1) the negative charge. The long, skinny tail does not like to be in water, since the bonds in the tail are not at all polar (Section 6–5) and do not hydrate well. In fact, the tails prefer to be in contact with grease or oil if possible.

charged head—

likes water ⊖∿∿∿∿∿∿ uncharged tail—
                    does not like water

The structure of soap molecules enables soap to suspend dirt and grease in water solution. It does this by forming little aggregates (called *micelles*) of soap around each bit of greasy dirt; the tails of the soap molecules surround the greasy dirt,

triglycerides (fats)   glycerol   soap

$$R = -\overset{\overset{\displaystyle H}{|}}{\underset{\underset{\displaystyle H}{|}}{C}} - \overset{\overset{\displaystyle H}{|}}{\underset{\underset{\displaystyle H}{|}}{C}} - \cdots - \overset{\overset{\displaystyle H}{|}}{\underset{\underset{\displaystyle H}{|}}{C}} - \overset{\overset{\displaystyle H}{|}}{\underset{\underset{\displaystyle H}{|}}{C}} - H$$

variable number of CH_2 units

and the charged heads maintain contact with the water solution that surrounds the micelle. In this way the greasy dirt is "dissolved" in the water solution.

Detergents also form micelles and suspend greasy dirt, the only difference being that the charged head has a different structure in detergents. Detergents are more important in laundry and dishwashing formulations, while soap is most useful for cleaning faces and bodies.

Unusual Facts: Animal fat has gotten a bad reputation in terms of healthy eating habits. However, animal fats (beef and lamb fat, lard) are the most common fats for soap making. Other fats and oils used in soap production include coconut oil, palm oil, and palm kernel oil.

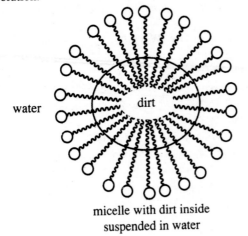

water

dirt

micelle with dirt inside
suspended in water

Problems

1. Calculate the percent of the solute in each of the following aqueous solutions.
   (a) 8.60 g of sodium chloride (NaCl) in 98.0 g of solution
   (b) 25.0 g of potassium carbonate (K_2CO_3) in 100.0 g of water

2. Calculate the grams of solute that must be dissolved in
   (a) $35\overline{0}$ g of water in the preparation of a 15.0% potassium sulfate (K_2SO_4) solution
   (b) 20.0 g of water in the preparation of a 10.0% sodium chloride (NaCl) solution

3. Calculate the molality of each of the following solutions.
   (a) 175 g of ethyl alcohol (C_2H_6O) in $65\overline{0}$ g of water
   (b) 3.50 g of sulfuric acid (H_2SO_4) in 10.0 g of water

4. Calculate the number of grams of solute necessary to prepare the following aqueous solutions.
   (a) $35\overline{0}$ g of a 0.500 m aqueous solution of ethyl alcohol (C_2H_6O)
   (b) $70\overline{0}$ g of a 0.600 m aqueous solution of sulfuric acid (H_2SO_4)

5. Calculate the number of grams of water that must be added to:
   (a) 60.0 g of glucose ($C_6H_{12}O_6$) in the preparation of a 2.00 m solution
   (b) 85.0 g of sugar ($C_{12}H_{22}O_{11}$) in the preparation of a 6.00 m solution

6. Calculate the molarity of each of the following aqueous solutions.
   (a) 75.0 g of ethyl alcohol (C_2H_6O) in $45\overline{0}$ mL of solution
   (b) 2.60 g of sodium chloride (NaCl) in 50.0 mL of solution

7. Calculate the number of grams of solute necessary to prepare the following solutions. Explain how each solution would be prepared.

   (a) $50\overline{0}$ mL of a 0.100 M aqueous sodium hydroxide (NaOH) solution
   (b) $25\overline{0}$ mL of a 0.0200 M aqueous calcium chloride ($CaCl_2$) solution

8. Calculate the number of milliliters of solution required to provide the following.

   (a) 5.00 g of sodium bromide (NaBr) from a 0.100 M aqueous solution.
   (b) 7.65 g of calcium chloride ($CaCl_2$) from a 1.40 M aqueous solution.

9. Calculate the normality of each of the following aqueous solutions.

   (a) 8.75 g of sodium hydroxide (NaOH) in $45\overline{0}$ mL of solution
   (b) 2.00 g of barium hydroxide [$Ba(OH)_2$] in $50\overline{0}$ mL of solution in reactions that replace both hydroxide ions

10. Calculate the number of grams of solute necessary to prepare the following aqueous solutions.

    (a) $25\overline{0}$ mL of a 0.0100 N sulfuric acid (H_2SO_4) solution in reactions that replace both hydrogen ions
    (b) 135 mL of a 0.800 N phosphoric acid (H_3PO_4) solution in reactions that replace all three hydrogen ions

11. Calculate the number of milliliters of solution required to provide the following.

    (a) 60.0 g of sulfuric acid (H_2SO_4) from a 4.00 N aqueous solution in reactions that replace both hydrogen ions
    (b) 75.0 g of calcium chloride ($CaCl_2$) from a 2.00 N aqueous solution in reactions that replace both chloride ions

12. Calculate the requested concentration of the following solutions.

    (a) normality of a 0.00570 M aqueous calcium hydroxide [$Ca(OH)_2$] solution in reactions that replace both hydroxide ions
    (b) molarity of a 3.78 N aqueous calcium chloride ($CaCl_2$) solution in reactions that replace both chloride ions

General Problems

13. Laboratory concentrated sulfuric acid is approximately 98.0% H_2SO_4 and its density is 1.83 g/mL. Calculate the following concentrations of the aqueous sulfuric acid solution.

    (a) molality
    (b) molarity (*Hint*: Use the density of the solution to get the volume.)
    (c) normality (both hydrogen ions replaced)

14. Concentrated hydrochloric acid is approximately 37.0% HCl and its density is 1.18 g/mL. Calculate the following concentrations of the aqueous hydrochloric acid solution:

(a) molality
(b) molarity (*Hint*: Use the density of the solution to get the volume.)
(c) normality

15. Radiator fluid is typically about 25% by mass ethylene glycol ($C_2H_6O_2$) in water. If a particular car has fluid that has 2.50 kg of $C_2H_6O_2$ in 7.89 kg of water, calculate the molality of this ethylene glycol solution.

Answers to Exercises

13–1. 18.5%
(*Hint*: 115.6 g of *solution*)

13–3. 1.82 *m*

13–5. 31$\overline{0}$ g

13–2. 7.19 g

13–4. 91.4 g

13–6. 2.82 *M*

13–7. (a) 27.0 g; (b) Sodium chloride (27.0 g) is dissolved in sufficient water to make the total volume of the solution equal to 28$\overline{0}$ mL.

13–8. 113 mL

13–10. 1.16 g

13–12. 1.00 *M*

13–9. 6.48 *N*

13–11. 307 mL

QUIZ

ELEMENT	ATOMIC MASS UNITS (amu)
H	1.0
C	12.0
O	16.0
Na	23.0
S	32.1

1. When rinsing wounds, physicians use an "isotonic" solution of aqueous sodium chloride (NaCl) to minimize any pain. This solution is composed of 0.933 g of NaCl in 100.00 g of water. Calculate the percent by mass of sodium chloride in the isotonic solution.

2. Calculate the number of grams of glucose ($C_6H_{12}O_6$) required to prepare 35$\overline{0}$ g of a 0.500 *m* aqueous glucose solution.

3. A 10.0 mL sample of blood contians 32.2 mg of sodium ions (Na^{1+}). Calculate the molarity of the sodium ions in the blood.

4. Calculate the normality of a solution containing 92.5 g of sodium sulfate (Na_2SO_4) in 42$\overline{0}$ mL of solution.

5. Concentrated phosphoric acid is 14.6 *M* in phosphoric acid. Calculate the normality of the solution in reactions where only two hydrogen ions (H^{1+}) are replaced.

14

MORE ADVANCED TOPICS

COUNTDOWN

ELEMENT	ATOMIC MASS UNITS (AMU)
Na	23.0
C	12.0
O	16.0
Ca	40.1
H	1.0
Cl	35.5
Pb	207.2

5 Give the *relative charge* on the following subatomic particles (Section 4–3).

(a) proton

(b) neutron

(c) electron

$(+1)$

(0)

(-1)

4 Calculate the oxidation number for the element indicated in each of the following compounds or ions (Section 6–3).

(a) Cl in $HClO_2$

(b) S in H_2SO_3

(c) Cl in ClO_3^{1-}

(d) Cr in $Cr_2O_7^{2-}$

$(+3 \text{ or } 3^+)$

$(+4 \text{ or } 4^+)$

$(+5 \text{ or } 5^+)$

$(+6 \text{ or } 6^+)$

3 Calculate the number of moles in each of the following quantities of compounds. Express your answer in scientific notation (Sections 1–4 and 8–2).

(a) 0.250 g sodium carbonate (Na_2CO_3)

$(2.36 \times 10^{-3} \text{ mol})$

(b) 62.5 mg calcium hydroxide [$Ca(OH)_2$]

$(8.43 \times 10^{-4} \text{ mol})$

2 Calculate the number of grams in each of the following (Section 8–2).

(a) 0.0125 mol C_2H_6O (ethyl alcohol)

(0.575 g)

(b) 0.0240 mol sodium hydroxide (NaOH)

(0.960 g)

1 (a) Complete and balance the following equation, writing it as a *total ionic equation* and as a *net ionic equation* [indicate any precipitate by (s)] (Sections 10–2 and 10–3).

$$Pb(NO_3)_{2(aq)} + HCl_{(aq)} \xrightarrow{\text{cold}}$$

$$[\text{Total: } Pb^{2+}{}_{(aq)} + 2NO_3{}^{1-}{}_{(aq)} + 2H^{1+}{}_{(aq)} + 2Cl^{1-}{}_{(aq)} \xrightarrow{\text{cold}}$$

$$PbCl_{2(s)} + 2H^{1+} + 2NO_3{}^{1-}$$

$$\text{Net: } Pb^{2+}{}_{(aq)} + 2Cl^{1-}{}_{(aq)} \xrightarrow{\text{cold}} PbCl_{2(s)}$$

(b) If 0.245 g of chloride ion reacts with excess lead(II) ion, calculate the number of grams of lead(II) chloride that could be prepared.

(0.960 g)

TASKS

1 Memorize the definitions of oxidation, reduction, oxidizing agent, and reducing agent.

2 Memorize what species to add in order to balance H or O atoms in either acidic or basic media when balancing oxidation–reduction reactions by the ion–electron method.

3 (a) Memorize the definition of pH, pOH, and their relationship to each other.

(b) Memorize the pH range for acidity, for basicity, and the neutral pH point.

OBJECTIVES

1 Give the distinguishing characteristics of each of the following terms:

(a) oxidation number (Section 14–1)

(b) oxidation (Section 14–1)

(c) reduction (Section 14–1)

(d) reducing agent (Section 14–1)

(e) oxidizing agent (Section 14–1)

(f) titration (Section 14–3)

(g) equivalence point (Section 14–3)

(h) pH (Section 14–5)

(i) pOH (Section 14–5)

2 Given the formulas of the reactants and products in oxidation–reduction equations, balance the equations by the ion–electron method (Example 14–1, Exercise 14–1, Problem 1).

3 Given the results of a titration of a specific volume of an unknown concentration of an acid or base solution with the volume of another solution of known molarity concentration, calculate the unknown molarity concentration. Given the results of a titration of a solution of unknown concentration of an acid or base solution with a known *mass* of a base or acid, calculate the unknown molarity concentration (Examples 14–2 and 14–3, Exercises 14–2 and 14–3, Problems 2 and 3).

4 Given the hydrogen ion concentration of a solution, calculate the pH and pOH of the solution (Example 14–4, Exercise 14–4, Problem 4).

In this chapter we cover three more advanced topics: oxidation–reduction reactions, titration, and pH. The last two topics involve acid–base chemistry (Section 9–10).

These advanced topics are related in a sense, for they both involve the transfer of a simple chemical species from one compound to another. Oxidation–reduction reactions involve the transfer of *electrons*, while acid–base reactions (neutralization reactions) involve the transfer of *hydrogen ions* (H^{1+}).

You have probably never heard of oxidation–reduction reactions, but they are a vital part of your life. For example, the battery that starts your car depends on an oxidation–reduction reaction. Oxidation–reduction reactions are also involved in the batteries in your flashlight and calculator, the development of your photographic film, the metabolism of your food, and in photosynthesis, the process that creates sugar from carbon dioxide, water, and light. Without photosynthesis, there would be no life on earth. Oxidation–reduction reactions can be tricky to balance, and we will learn how to balance these types of equations (Sections 14–1 and 14–2).

Acid–base reactions may also seem unfamiliar, but they are involved in the metabolism of your food, the health of oceans, lakes, and rivers, and the ability of the soil to support crops. Your very survival depends on how much acid and base are present in your blood (Section 14–5). Titration is a technique that allows you to determine the concentration of acid in a solution (Section 14–3), and pH is a way of expressing the strength of a solution of acid or base (Section 14–5).

14–1 Definitions of Oxidation and Reduction. Oxidizing and Reducing Agents

oxidation number A positive or negative whole number used to describe the combining capacity of an element in a compound.

Before we consider the definitions of oxidation and reduction, we need to review oxidation numbers as outlined in Section 6–3. In that section we defined **oxidation number** (ox no) as a positive or negative whole number used to describe the combining capacity of an element in a compound.[1] The change in oxidation number from one state to another implies the number of *electrons lost* or *gained*. An example of a *positive change* would be in going from the free state (zero oxidation number) to the combined state with a positive oxidation number. Electrons are lost (remember, electrons are negatively charged). An example of a *negative change* would be in going from the free state to the combined state with a negative oxidation number. Electrons are gained.

Oxidation at one time referred only to the combination of an element with oxygen, but the term has been expanded, and **oxidation** is now defined as a chemical change in which a substance loses electrons, or one or more elements in it *increases in oxidation number*. If electrons (negative) are lost from an element, the resulting element will have an increase in oxidation number.

oxidation A chemical change in which a substance loses electrons, or one or more elements in it increase in oxidation number.

reduction A chemical change in which a substance gains electrons, or one or more elements in it decrease in oxidation number.

Reduction is a chemical change in which a substance gains electrons, or one or more elements in it *decreases in oxidation number*. If an element gains electrons (negative), the resulting element will have a decrease (algebraic) in oxidation number. Figure 14–1 may help you differentiate these two terms. A useful relationship to memorize is that *reduction* involves a *decrease in oxidation number*. The opposite, then, would be true for oxidation, or an increase in oxidation number is involved in oxidation. These can also be translated into gaining or losing electrons, since a gain of negative electrons will result in a decrease in oxidation number (reduction), and a loss of electrons will result in a gain in oxidation number (oxidation).

Study hint

Remember that "LEO the lion goes GER" (grr . . . .). *Loss of Electrons = Oxidation* (*L E O*), and *Gain of Electrons = Reduction* (*G E R*).

In a given reaction, whenever a substance is oxidized, it loses electrons to another substance, which is thereby reduced; hence, *oxidation accompanies reduction and reduction accompanies oxidation*. The equation is therefore called an oxidation–reduction equation (or *redox* equation).

In an oxidation–reduction equation the substance *oxidized* is called the **reducing agent** (reductant), because it induces a reduction in another substance. The substance being *reduced* is called the **oxidizing agent** (oxidant), since it produces an oxidation in another substance.

reducing agent The substance oxidized in an oxidation–reduction reaction.

oxidizing agent The substance reduced in an oxidation–reduction reaction.

A simple example of a combination reaction will illustrate this point:

$$Ca_{(s)} + S_{(s)} \xrightarrow{\Delta} CaS_{(s)} \tag{14–1}$$

Calcium metal (zero oxidation number) combines with sulfur (zero oxidation number) to form calcium sulfide (2$^+$ oxidation number for calcium and 2$^-$ oxidation number for sulfur). The calcium has, therefore, *increased in oxidation number* (or lost electrons) and is *oxidized*.

$$Ca \longrightarrow Ca^{2+} + 2e^- \tag{14–2}$$

The *sulfur* has *decreased in oxidation number* (or gained electrons) and is *reduced*.

$$S + 2e^- \longrightarrow S^{2-} \tag{14–3}$$

Since calcium has been oxidized, it is called the *reducing agent*; since sulfur has

---

[1] Fractional average oxidation numbers of atoms in compounds do exist, as mentioned in Chapter 6 (footnote 1).

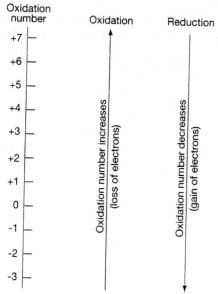

Figure 14–1   *Oxidation and reduction, change in oxidation number, and electron transfer*

been reduced, it is called the *oxidizing agent*. This, then, is an example of an oxidation–reduction equation.

14–2   Balancing Oxidation–Reduction Equations. The Ion Electron Method

The ***ion electron method*** is one of the methods used to balance oxidation–reduction equations. The technique in this method involves *two partial equations* representing *half-reactions*: One equation describes the *oxidation process* and the other equation describes the *reduction process*. The two partial equations are then added to produce a final balanced equation. Although we artificially divide the original reaction into two partial equations, these partial equations do not take place alone; whenever oxidation occurs, so does reduction. These partial equations are useful in electrochemistry, as you will learn in more advanced courses.

As we did in balancing molecular equations (see Section 9–3) and in writing ionic equations (see Section 10–2), we shall suggest a few guidelines to follow when balancing oxidation–reduction equations by the *ion electron method*.

1. Write the equation in *net ionic* form (See Section 10–2) *without* attempting to balance it.

2. Determine by inspection the elements that undergo a change in oxidation number and then write two partial equations: an *oxidation half-reaction* and a *reduction half-reaction*.

3. Balance the atoms on each side of the partial equations. In **acid** solution, you may add H^{1+} ions and H_2O molecules. For each hydrogen atom (H) needed, add a H^{1+} ion. For each oxygen atom (O) needed, add a H_2O molecule, with $2H^{1+}$ ions being shown on the other side of the partial equation. In **basic** solution, you may add OH^{1-} ions and H_2O molecules. For each hydrogen atom (H) needed, add an H_2O molecule with an OH^{1-} ion written on the other side of the partial equation. For *each* oxygen atom (O) needed, add *two* OH^{1-} ions with *one* H_2O molecule written on the other side of the partial equation. The following summarizes these additions in acid and base:

$$
\begin{array}{lcl}
 & \textit{Need} & \textit{Add} \\
\textit{In acid:} & H & H^{1+} \\
 & \boxed{O} & \boxed{H_2O \longrightarrow 2H^{1+}} \\
\\
\textit{In base:} & H & H_2O \longrightarrow OH^{1-} \\
 & \boxed{O} & \boxed{2OH^{1-} \longrightarrow H_2O}
\end{array}
$$

(To remember what to add in a given type of solution, you may find it helpful to make "flash cards" of the preceding summary.)

4. Balance the partial equation electrically by adding electrons to the appropriate side of the equation so that the *charges* on both sides of the partial equations are *equal*. These two partial equations are defined as follows: The *oxidation* half-reaction equation, in which the reactant loses electrons, is written with the electrons on the *products* side; the *reduction* half-reaction equation in which the reactant *gains* electrons is written with the electrons on the reactant side.

$$\textit{Oxidation:} \quad M \longrightarrow M^{1+} + 1e^-$$
$$\textit{Reduction:} \quad X + 1e^- \longrightarrow X^{1-}$$

5. Multiply each *entire* partial equation by an appropriate number so that the **electrons lost in one partial equation** (oxidation half-reaction) **are equal to the electrons gained in the other partial equation** (reduction half-reaction).

6. Add the two partial equations and eliminate those electrons, ions, or water molecules that are common to both sides of the equation.

7. Place a check ($\checkmark$) above each atom on both sides of the equation to ensure that the equation is balanced. Also check the net charges on both sides of the equation to see that they are equal. Check the equation to see that the coefficients are the lowest possible ratios.

Now, let us consider some examples. In this text, the products of the oxidation–reduction reaction will be given (due to the often complicated nature of these products). Thus, you will be asked only to balance the equations. Also, to simplify writing the equation, the physical states will normally not be included in the equation.

EXAMPLE 14–1 ──────

Balance the following oxidation–reduction equations using the ion electron method. Indicate the substances that are oxidized and reduced, and the oxidizing and reducing agents.

(a) $Fe^{2+} + MnO_4^{1-} \longrightarrow Fe^{3+} + Mn^{2+}$ (in acid solution)

SOLUTION Guideline 1 does not apply, because the equation is already in ionic form, so use guideline 2. The following two partial equations for the half reactions result:

(1) $$Fe^{2+} \longrightarrow Fe^{3+}$$

(2) $$MnO_4^{1-} \longrightarrow Mn^{2+}$$

Balance the atoms for each partial equation according to guideline 3. In partial equation (2), add $4H_2O$ to the products to balance the O atoms in MnO_4^{1-} and then add $8H^{1+}$ to the reactants, because the reaction is carried out in acid.

(1) $$Fe^{2+} \longrightarrow Fe^{3+}$$

(2) $$8H^{1+} + MnO_4^{1-} \longrightarrow Mn^{2+} + 4H_2O$$

Then, balance the two partial equations electrically by adding electrons (negatively charged) to the appropriate sides according to guideline 4. In partial equation (1), electrons are lost; hence, this equation represents the *oxidation* half-reaction. In partial equation (2), electrons are gained; this equation represents the *reduction* half-reaction.

(1)   *Oxidation*: $Fe^{2+} \longrightarrow Fe^{3+} + 1e^-$

Charges: $2^+ = 3^+ + 1^- = 2^+$

(2) *Reduction*: $8H^{1+} + MnO_4^{1-} + 5e^- \longrightarrow Mn^{2+} + 4H_2O$

Charges: $8^+ + 1^- + 5^- = 2^+ = 2^+ = 2^+$

Multiply the *entire* partial equation (1) by 5, so that the gain of electrons in partial equation (2) is equal to the loss, according to guideline 5, giving the following partial equations:

(1) *Oxidation*: $5Fe^{2+} \longrightarrow 5Fe^{3+} + 5e^-$

(2) *Reduction*: $8H^{1+} + MnO_4^{1-} + 5e^- \longrightarrow Mn^{2+} + 4H_2O$

(A common error is to forget to multiply the *entire* partial equation by the appropriate factor). Add the two partial equations, and eliminate the electrons on both sides of the equation according to guideline

6. The following equation results:

$$5Fe^{2+} + 8H^{1+} + MnO_4^{1-} + \cancel{5e^-} \longrightarrow$$
$$5Fe^{3+} + \cancel{5e^-} + Mn^{2+} + 4H_2O$$

Check each atom and the charges on both sides of the equation to obtain the final balanced equation according to guideline 7:

$$5\overset{\checkmark}{Fe^{2+}} + 8\overset{\checkmark}{H^{1+}} + \overset{\checkmark\checkmark}{MnO_4^{1-}} \longrightarrow 5\overset{\checkmark}{Fe^{3+}} + \overset{\checkmark}{Mn^{2+}} + 4\overset{\checkmark\checkmark}{H_2O} \; Answer$$

Charges: $5(2^+) + 8^+ \quad + 1^- = 17^+ = 5(3^+) + 2^+ = 17^+$

In partial equation (1), the Fe^{2+} is oxidized, so it is the *reducing agent*; in partial equation (2), MnO_4^{1-} is reduced, so it is the *oxidizing agent*.

(b) $Zn + HgO \longrightarrow ZnO_2^{2-} + Hg$ (in basic solution)

SOLUTION Guideline 1 does not apply, because the equation is already in ionic form, so use guideline 2. We can write the following two partial equations for the half reactions:

(1) $Zn \longrightarrow ZnO_2^{2-}$

(2) $HgO \longrightarrow Hg$

Balance the atoms for each partial equation according to guideline 3. In partial equation (1), two oxygen atoms are required in the reactants. Because the solution is basic, add $4OH^{1-}$ ions to the reactants and $2H_2O$ molecules to the products to balance the atoms. In partial equation (2), one oxygen atom is required in the products, so add $2OH^-$ ions to the products and one H_2O molecule to the reactants to balance the atoms.

(1) $Zn + 4OH^{1-} \longrightarrow ZnO_2^{2-} + 2H_2O$

(2) $HgO + H_2O \longrightarrow Hg + 2OH^{1-}$

According to guideline 4, balance the two partial equations electrically by adding electrons to the appropriate side. (Remember that a free metal has zero charge; Zn and Hg are neutral, zero oxidation number.) In partial equation (1), the electrons are lost; hence, this equation represents the *oxidation* half-reaction. In partial equation (2), electrons are gained; hence, this equation represents the *reduction* half-reaction.

(1) *Oxidation*: $Zn + 4OH^{1-} \longrightarrow ZnO_2^{2-} + 2H_2O + \mathbf{2e^-}$

     Charges: $4^- = \qquad 2^- \quad + \qquad 2^- = 4^-$

(2) *Reduction*: $HgO + H_2O + \mathbf{2e^-} \longrightarrow Hg + 2OH^{1-}$

     Charges: $2^- = \qquad\qquad 2^-$

In the two partial equations, the number of electrons lost is equal to the number of electrons gained according to guideline 5. Therefore, add the two partial equations and eliminate the electrons and ions on both sides of the equation according to guideline 6, to obtain the following equation:

$$Zn + 4OH^{1-} + \cancel{2e^{-}} + HgO + H_2O \longrightarrow$$
$$ZnO_2^{2-} + 2H_2O + \cancel{2e^{-}} + Hg + 2OH^{1-}$$

The resulting OH^{1-} ions present on the left side are $2OH^{1-}$ ($4OH^{1-}$ on the left minus $2OH^{1-}$ on the right), and the resulting H_2O molecules on the right side are $1H_2O$ ($2H_2O$ on the right minus $1H_2O$ on the left). Hence, the following equation results:

$$Zn + 2OH^{1-} + HgO \longrightarrow ZnO_2^{2-} + H_2O + Hg$$

Check each atom and the charge on both sides of the equation to obtain the final balanced equation according to guideline 7:

$$\overset{\checkmark}{Zn} + 2\overset{\checkmark\checkmark}{OH}^{1-} + \overset{\checkmark\checkmark}{HgO} \longrightarrow \overset{\checkmark\checkmark}{ZnO_2^{2-}} + \overset{\checkmark\checkmark}{H_2O} + \overset{\checkmark}{Hg} \ \textit{Answer}$$

Charges: 2^- $= 2^-$

In partial equation (1), the *Zn* is oxidized, so it is the *reducing agent*; in partial equation (2), *HgO* is reduced, so it is the *oxidizing agent*.

(c) $NaI + Fe_2(SO_4)_3 \longrightarrow I_2 + FeSO_4 + Na_2SO_4$

(in aqueous solution)

SOLUTION   Apply guideline 1 by writing the equation in *net ionic form without* attempting to balance it. Refer to the solubility rules of inorganic substances inside the back cover of the book. The following net ionic equation results:

$$\cancel{Na^+} + I^{1-} + Fe^{3+} + \cancel{SO_4^{2-}} \longrightarrow$$
$$I_2 + Fe^{2+} + \cancel{SO_4^{2-}} + \cancel{Na^+} + \cancel{SO_4^{2-}}$$

Net ionic: $I^{1-} + Fe^{3+} \longrightarrow I_2 + Fe^{2+}$

(Note that we make no attempt to balance the ions.) Write two partial equations according to guideline 2:

(1) $I^{1-} \longrightarrow I_2$

(2) $Fe^{3+} \longrightarrow Fe^{2+}$

Balance the atoms for each partial equation according to guideline 3:

(1) $2I^{1-} \longrightarrow I_2$

(2) $Fe^{3+} \longrightarrow Fe^{2+}$

Then balance these two equations electrically, according to guideline

4. In partial equation (1), electrons are lost; hence, this equation represents the *oxidation* half-reaction. In partial equation (2), electrons are gained; hence, this equation represents the *reduction* half-reaction.

(1)   *Oxidation:* $2I^{1-} \longrightarrow I_2 + 2e^-$

        Charges: 2^- $=$ 2^-

(2)   *Reduction:* $Fe^{3+} + 1e^- \longrightarrow Fe^{2+}$

        Charges: 3^+ $+ 1^-$ $=$ 2^+

In the two partial equations, the number of electrons lost must be equal to the number of electrons gained according to guideline 5; therefore, multiply partial equation (2) by 2.

(1)   *Oxidation:* $2I^{1-} \longrightarrow I_2 + \mathbf{2e^-}$

(2)   *Reduction:* $2Fe^{3+} + \mathbf{2e^-} \longrightarrow 2Fe^{2+}$

Add the two partial equations and eliminate the electrons on opposite sides of the equation according to guideline 6, to obtain the following equation:

$$2I^{1-} + 2Fe^{3+} + \cancel{2e^-} \longrightarrow I_2 + \cancel{2e^-} + 2Fe^{2+}$$

$$2I^{1-} + 2Fe^{3+} \longrightarrow I_2 + 2Fe^{2+}$$

Check each atom and the charge on both sides of the equation to obtain the final balanced equation according to guideline 7:

$$2\overset{\checkmark}{I}^{1-} + 2\overset{\checkmark}{Fe}^{3+} \longrightarrow \overset{\checkmark}{I_2} + 2\overset{\checkmark}{Fe}^{2+} \qquad Answer$$

Charges: $2^- + 2(3^+) = 4^+ = 2(2^+) = 4^+$

In partial equation (1), the I^{1-} is oxidized, so it is the *reducing agent*; in partial equation (2), Fe^{3+} is reduced, so it is the *oxidizing agent*.

(d) $CrO_2 + ClO^{1-} \longrightarrow CrO_4^{2-} + Cl^{1-}$ (in basic solution)

SOLUTION Guideline 1 does not apply, because the equation is already in ionic form, so use guideline 2. We can write the following two partial equations for the half reactions:

(1) $CrO_2 \longrightarrow CrO_4^{2-}$

(2) $ClO^{1-} \longrightarrow Cl^{1-}$

Balance the atoms for each partial equation according to guideline 3:

(1) $CrO_2 + 4OH^{1-} \longrightarrow CrO_4^{2-} + 2H_2O$

(2) $ClO^{1-} + H_2O \longrightarrow Cl^{1-} + 2OH^{1-}$

Balance these two equations electrically according to guideline 4. In partial equation (1), electrons are lost, so this equation represents the *oxidation* half-reaction. In partial equation (2), electrons are gained; therefore this equation represents the *reduction* half-reaction.

(1) *Oxidation*: $CrO_2 + 4OH^{1-} \longrightarrow CrO_4^{2-} + 2H_2O + 2e^-$

   Charges: 4^- $=$ 2^- $+$ $2^- = 4^-$

(2) *Reduction*: $ClO^{1-} + H_2O + 2e^- \longrightarrow Cl^{1-} + 2OH^{1-}$

   Charges: 1^- $+$ $2^- = 3^- = 1^-$ $+ 2^- = 3^-$

According to guideline 5, the number of electrons lost must equal the number of electrons gained. Now, add the two partial equations and eliminate the electrons and ions on both sides of the equation according to guideline 6, to obtain the following equation:

$CrO_2 + 4OH^{1-} + ClO^{1-} + H_2O + 2\cancel{e^-} \longrightarrow$

$$CrO_4^{2-} + 2H_2O + 2\cancel{e^-} + Cl^{1-} + 2OH^{1-}$$

The resulting OH^{1-} ions present on the left side are $2OH^{1-}$ ($4OH^{1-}$ on the left minus $2OH^{1-}$ on the right), and the resulting H_2O molecules on the right side are $1H_2O$ ($2H_2O$ on the right minus $1H_2O$ on the left). The following equation results:

$$CrO_2 + 2OH^{1-} + ClO^{1-} \longrightarrow CrO_4^{2-} + H_2O + Cl^{1-}$$

Check each atom and the charge on both sides of the equation to obtain the final balanced equation according to guideline 7:

$$\overset{\checkmark\checkmark}{CrO_2} + 2\overset{\checkmark\checkmark}{OH}^{1-} + \overset{\checkmark\checkmark}{ClO}^{1-} \longrightarrow \overset{\checkmark\checkmark}{CrO_4}^{2-} + \overset{\checkmark\checkmark}{H_2O} + \overset{\checkmark}{Cl}^{1-} \text{ Answer}$$

Charges: 2^- $+ 1^-$ $= 3^- = 2^- +$ $1^- = 3^-$

In partial equation (1), the CrO_2 is oxidized, so it is the *reducing agent*; in partial equation (2), the ClO^{1-} is reduced, so it is the *oxidizing agent*.

Exercise 14–1

Balance the following oxidation–reduction equations using the ion electron method. Indicate the substances that are oxidized and reduced, and the oxidizing and reducing agents.

(a) $Cr_2O_7^{2-} + Fe^{2+} \longrightarrow Cr^{3+} + Fe^{3+} + H_2O$ (in acid solution)

(b) $MnO_4^{1-} + ClO_2^{1-} \longrightarrow MnO_2 + ClO_4^{1-}$ (in basic solution)

(c) $S^{2-} + NO_3^{1-} \longrightarrow S + NO$ (in acid solution)

(d) $Sn^{2+} + IO_3^{1-} \longrightarrow Sn^{4+} + I^{1-}$ (in acid solution)

14–3 Titration

In neutralization reactions, an acid or an acid oxide reacts with a base or a bas
oxide. In most of these reactions, water is one of the products. The formatic
of water acts as the driving force behind the neutralization. This neutralizatic
can be represented by a general equation,

$$HX + MOH \longrightarrow MX + HOH \qquad (14\text{-}$$

where HX is an acid and MOH is a base. Water is one of the products. No
that the reaction involves the transfer of a hydrogen ion (H^{1+}) from the ac
(HX) to the **OH**.

Neutralization reactions are often carried out by a procedure called *titratio.*
Titration is a procedure for determining the *concentration* of an acid or ba:
by adding a base or an acid of *known concentration.* For example, if we ac
just enough of a base of known concentration to *exactly* neutralize a give
quantity of the original acid, we can use stoichiometry (Chapter 11) to determir
the concentration of the original acid. The point at which we have exact
neutralized the original substance is called the **equivalence point** (or *end point*).
is determined by a change in color of an appropriately selected indicator.[2] Indic
tors usually used are methyl orange (red in acid and yellow in base), methyl re
(red in acid and yellow in base), and phenolphthalein (pronounced fe′ nol·tha
en, colorless in acid and red in base). See Table 14–1. An instrument called
pH meter may also be used, which we will discuss in Section 14–5. Regardle
of how the neutralization point is determined, *at the equivalence point, the numb
of moles of added base or acid is exactly enough to neutralize the original ac
or base.*

In a titration, a measured amount of the acid or base of unknown concentr
tion is placed in an Erlenmeyer flask and a drop or two of an appropriate indicat
is added. To this solution, a solution of an acid or base (whichever is appropriat
of known concentration is slowly added from a buret[3] until the color of tl
indicator just changes. This is called the *equivalence point (end point).*[4] F
example, we may want to determine the concentration of an aqueous solution
sodium hydroxide. We place a measured volume or mass of the solution in a
Erlenmeyer flask and add a drop or two of the indicator phenolphthalein. Tl
solution turns red because phenolphthalein is red in basic solution. Then, v

titration A procedure for determining the *concentra-tion* of an acid or base by adding a base or an acid of *known concentration.*

equivalence point The point at which an acid or base is exactly neutralized in the titration process; it is also called the *end point.*

---

[2] An *indicator* is a compound that changes color as the concentration of H^{1+} in the soluti
changes.

[3] A *buret* is a piece of glassware that permits the chemist to add small measured amounts
a solution to another solution. The quantities added may vary in size according to the chemis
need. The quantities may be as large as 50.00 mL or as small as 0.01 mL (See Fig. 14–2).

[4] The equivalence point is not always the end point. As we have described above, when t
color of the indicator changes, this is the *end point.* It should be the point where all the acid or ba
has been neutralized—the *equivalence point.* If the indicator changes color *and* all the acid or bʁ
has been neutralized, the end point and the equivalence point are the *same.* In this book we w
consider them the same.

TABLE 14-1   COLOR CHANGES OF INDICATORS

INDICATOR	APPROXIMATE pH AT WHICH COLOR CHANGES*	COLOR	
		IN ACID	IN BASE
Methyl orange	4	Red	Yellow
Methyl red	5	Red	Yellow
Phenolphthalein	9	Colorless	Red

ᵃ The choice of an indicator depends on the pH (see Section 14-5) of the aqueous solution of the salt formed when the acid or base is neutralized.

slowly add a hydrochloric acid solution of known concentration from a buret until the red color of the indicator *just* fades. A colorless solution results at the equivalence point (end point), as shown in Fig. 14-2. By measuring the volume of the hydrochloric acid solution used, we can determine the amount of sodium hydroxide in the unknown sample.

EXAMPLE 14-2

In the titration of 30.0 mL of sodium hydroxide solution of unknown concentration, 45.2 mL of 0.100 M hydrochloric acid was required to neutralize the sodium hydroxide solution to a phenolphthalein equivalence point. Calculate the molarity of the sodium hydroxide solution.

SOLUTION   As with stoichiometry problems, the first thing we must know is the balanced equation:

$$NaOH + HCl \longrightarrow NaCl + H_2O$$

Next, apply the stoichiometry procedure (Chapter 11) and the units of molarity to calculate the moles of sodium hydroxide neutralized with 45.2 mL of 0.100 M hydrochloric acid solution:

$$45.2 \text{ mL solution} \times \frac{1 L}{1000 \text{ mL}} \times \frac{0.100 \text{ mol HCl}}{1 L \text{ solution}} \times \frac{1 \text{ mol NaOH}}{1 \text{ mol HCl}}$$

> determined from the balanced equation

$$= 0.00452 \text{ mol NaOH}$$

Finally, calculate the concentration of the sodium hydroxide solution in moles per liter from the 30.0 mL of the sodium hydroxide solution that was used:

$$\frac{0.00452 \text{ mol NaOH}}{30.0 \text{ mL solution}} \times \frac{1000 \text{ mL}}{1 L} = \frac{0.151 \text{ mol NaOH}}{1 L \text{ solution}}$$

$$= 0.151 \ M \qquad Answer$$

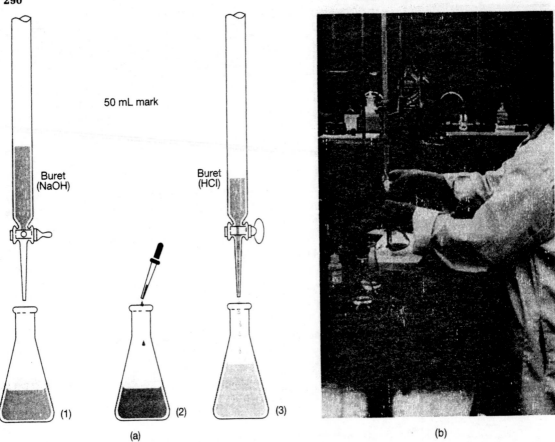

Figure 14–2 *Titration of a sodium hydroxide solution of unknown concentration. (a) (1) Measure an exact amount of the sodium hydroxide solution of unknown concentration into an Erlenmeyer flask. (2) Add to this solution one or two drops of phenolphthalein (an indicator) to give a red solution. (3) From a buret add hydrochloric acid solution of known concentration until the red color just fades and a nearly colorless solution appears at the* equivalence point (end point). *(b) The actual titration. (Photograph courtesy of David S. Seese)*

EXAMPLE 14–3

Pure sodium carbonate is used as a standard in determining the molarity of an acid. If 0.875 g of pure sodium carbonate was dissolved in water and the solution was titrated with 35.6 mL of hydrochloric acid to a methyl orange equivalence point, calculate the molarity of the hydrochloric acid solution.

SOLUTION

Equation:

$$Na_2CO_3 + 2HCl \longrightarrow 2NaCl + CO_2 + H_2O$$

Moles of HCl (the formula mass of Na_2CO_3 is 106.0 amu):

$$0.875 \text{ g Na}_2\text{CO}_3 \times \frac{1 \text{ mol Na}_2\text{CO}_3}{106.0 \text{ g Na}_2\text{CO}_3} \times \frac{2 \text{ mol HCl}}{1 \text{ mol Na}_2\text{CO}_3} = 0.0165 \text{ mol HCl}$$

determined from the balanced equation

Molarity of HCl:

$$\frac{0.0165 \text{ mol HCl}}{35.6 \text{ mL solution}} \times \frac{100 \text{ mL}}{1 \text{L}} = \frac{0.463 \text{ mol HCl}}{1 \text{L solution}} = 0.463 \ M \qquad Answer$$

Exercise 14–2

Household ammonia is a solution of ammonia in water. In the titration of 2.00 mL of household ammonia, 34.9 mL of 0.110 M hydrochloric acid solution was required to neutralize this solution to a methyl red equivalence point. Calculate the molarity of the ammonia solution.

Exercise 14–3

If 1.51 g of potassium carbonate (100.0% pure) is titrated with 15.4 mL of hydrochloric acid to a methyl orange equivalence point, calculate the molarity of the hydrochloric acid solution.

14–4   Ionization of Water

In neutralization reactions (see Sections 9–10 and 14–3) we mentioned that water in most cases was one of the products and was the driving force behind the reaction. The reason for this is that water is only *slightly* ionized, as shown by its very slight conduction of an electric current using sensitive instruments (see Section 10–1). The following equations illustrate this slight ionization:

$$HOH + H_2\ddot{O}: \ \rightleftharpoons \ H_3O^{1+} + OH^{1-} \qquad (14\text{–}5)$$

hydronium   hydroxide
ion   ion

or, simply,

$$HOH \ \rightleftharpoons \ H^{1+}_{(aq)} + OH^{1-}_{(aq)} \qquad (14\text{–}6)$$

hydrogen   hydroxide
ion   ion

Equations 14–5 and 14–6 represent the equilibrium of water with its respective ions, H_3O^{1+} or $H^{1+}_{(aq)}$ and $OH^{1-}_{(aq)}$.

In 1 L of pure water at 25°C, there are 1.0×10^{-7} mol/L of hydrogen ions (hydronium ions) and 1.0×10^{-7} mol/L of hydroxide ions. In any aqueous solution at 25°C, the product of the concentration of the hydrogen ion in moles per liter and the concentration of the hydroxide ion in moles per liter is *equal* to a *constant*, K_w. This constant K_w is called the *ion product constant* for water and it is equal to 1.0×10^{-14}.[5] Hence, for any aqueous solution at 25°C,

$$[\text{H}^{1+}][\text{OH}^{1-}] = K_w = 1.0 \times 10^{-14} \ (\text{mol}^2/\text{L}^2) \qquad (14\text{--}7)$$

where the brackets represent the concentration in moles per liter of the substance whose formula is enclosed in the brackets. Changes in the hydrogen ion concentration will result in a corresponding change in the hydroxide ion concentration, such that the product of these concentrations in any aqueous solution at 25°C will always be equal to the constant K_w, $1.0 \times 10^{-14} \ (\text{mol}^2/\text{L}^2)$.

If a solution has a *hydrogen ion* concentration *larger* than 1.0×10^{-7} mol/L (that is, a smaller negative exponent), such as 1.0×10^{-5} mol/L, the solution is termed "acidic." If the *hydroxide ion* concentration is *larger* than 1.0×10^{-7} mol/L, such as 1.0×10^{-5} mol/L, or the hydrogen ion concentration is less than 1.0×10^{-7} mol/L, such as 1.0×10^{-9} mol/L, the solution is called "basic." If the product of the hydrogen ion concentration in moles per liter $[\text{H}^{1+}]$ and the hydroxide ion concentration in moles per liter $[\text{OH}^{1-}]$ is equal to 1.0×10^{-14}, a neutral solution can occur only when the hydrogen ion concentration is *equal* to the hydroxide ion concentration—that is, when *each* is equal to 1.0×10^{-7} mol/L. Hence, using the reasoning above, we have

$$[\text{H}^{1+}] > 1.0 \times 10^{-7} \text{ mol/L} = \textbf{\textit{acidic}} \qquad (> = \text{greater than})$$

$$\left. \begin{array}{c} [\text{OH}^{1-}] > 1.0 \times 10^{-7} \text{ mol/7} \\ \\ \text{or} \\ \\ [\text{H}^{1+}] < 1.0 \times 10^{-7} \text{ mol/L} \end{array} \right\} = \textbf{\textit{basic}} \qquad (< = \text{less than})$$

$$[\text{H}^{1+}] = [\text{OH}^{1-}] = 1.0 \times 10^{-7} \text{ mol/L} = \textbf{\textit{neutral}}$$

Carbonated soft drinks have a hydrogen ion concentration of approximately 1×10^{-4} mol/L and are acidic.

14–5 pH and pOH

pH Quantitative way of expressing the acidic or basic nature of solutions using the negative logarithmic value of their hydrogen ion (H^{1+}) concentration,

$$\textbf{pH} = -\log[\text{H}^{1+}]$$

A solution with pH < 7 is acidic; one with pH > 7 is basic; one with pH = 7 is neutral.

For convenience, hydrogen ion concentration is expressed in terms of pH, which is defined as the negative logarithm of the hydrogen ion concentration:

$$\textbf{pH} = -\log[\text{H}^{1+}] \qquad (14\text{--}8)$$

---

[5] As you may discover in more advanced chemistry courses, equilibrium and ionization constants like K_w *do not have units*. It is useful in elementary treatments, however, to ascribe a set of units to such equilibrium constants to help you work the problems. We will place these "convenience units" in parentheses following the equilibrium constant.

The hydroxide ion concentration may be expressed in terms of pOH, which is defined as the negative logarithm of the hydroxide ion concentration:

$$pOH = -\log[OH^{1-}] \qquad (14\text{–}9)$$

pOH Quantitative way of expressing the acidic or basic nature of solutions using the negative logarithmic value of their hydroxide ion (OH^{1-}) concentration,

$$pOH = -\log[OH^{1-}]$$

A solution with pOH < 7 is basic; one with pOH > 7 is acidic; one with pOH = 7 is neutral.

A *logarithm* is the exponent (see Section 1–3) portion of a number expressed as a power of 10. That is, the logarithm of 10^8 is 8, and the logarithm of 10^{-6} is −6. (Appendix 2 will help you use your calculator to calculate logarithms of numbers.)

The general range for the hydrogen ion concentration falls between 1.0 M and 1.0×10^{-14} M, which converts into a pH range of 0 to 14 for most common solutions. Solutions with a pH below 7 are acidic, whereas solutions with a pH above 7 are basic. You should note that the *lower* the pH number is, the *higher* the hydrogen ion concentration. A solution is *neutral* when the hydrogen ion concentration and hydroxide ion concentration are equal, which corresponds to a pH of 7:

$$0 \longleftrightarrow 7.00 \longleftrightarrow 14$$
$$\uparrow$$

Acidic   NEUTRAL   Basic

Another relationship that is useful is one that relates pH and pOH to each other. We can derive this expression as follows:

K_w expression: $\qquad\qquad K_w = [H^{1+}][OH^{1-}] = 1.0 \times 10^{-14}$

Take the logarithms
of both sides: $\qquad\qquad \log\{[H^{1+}][OH^{1-}]\} = \log(1.0 \times 10^{-14})$

Simplify: $\qquad\qquad \log[H^{1+}] + \log[OH^{1-}] = -14.00$

Multiply by −1: $\qquad\qquad -\log[H^{1+}] - \log[OH^{1-}] = 14.00$

Finally: $\qquad\qquad$ **pH + pOH = 14.00**

Thus, if you know the pH, you can calculate the pOH very easily and *vice versa*.

Table 14–2 summarizes the relationships among $[H^{1+}]$, $[OH^{1-}]$, pH, and pOH in aqueous solutions and represents a number of common substances and their pH. You should notice that the pH of your blood is slightly *basic*. Moreover, it has a very narrow pH range—a mere 0.2 pH unit (7.3 to 7.5). If the pH of the blood goes much below 7.3, *acidosis* occurs; if it falls below 7.0, death may occur. If the pH goes above 7.5, *alkalosis* occurs; if it goes above 7.8, death may result. The pH is maintained within a narrow range by *buffers*, solutions of substances that prevent a large change in the pH. The three types of buffers in the blood are

1. carbonic acid (H_2CO_3) and sodium hydrogen carbonate ($NaHCO_3$)

2. sodium dihydrogen phosphate (NaH_2PO_4) and disodium hydrogen phosphate (Na_2HPO_4)

3. certain proteins

TABLE 14-2   RELATIONSHIP AMONG [H^{1+}], [OH^{1-}], pH, AND pOH; AND
pH OF SOME COMMON EXAMPLES

[H^{1+}] (mol/L)	[OH^{1-}] (mol/L)	pH[a]	pOH	ACID OR BASE STRENGTH	COMMON EXAMPLES (APPROXIMATE pH RANGE)[a]
10^0(1)	10^{-14}	0	14	Strongly acid	1 M HCl (0)
10^{-1}	10^{-13}	1	13		Gastric juice (1–3)
10^{-2}	10^{-12}	2	12		Limes (1.8–2.0)
					Soft drinks (2.0–4.0)
					Lemons (2.2–2.4)
10^{-3}	10^{-11}	3	11	Weakly acidic	Dill pickles (3.2–3.6)
10^{-4}	10^{-10}	4	10		Acid rain (below 5.6)
10^{-5}	10^{-9}	5	9		Urine (4.5–8.0)
10^{-6}	10^{-8}	6	8		Sour milk (6.0–6.2)
					Milk (6.5–6.7)
					Saliva (6.5–7.5)
10^{-7}	**10^{-7}**	**7**	**7**	**Neutral**	Blood (7.3–7.5)
10^{-8}	10^{-6}	8	6	Weakly basic	
10^{-9}	10^{-5}	9	5		
10^{-10}	10^{-4}	10	4		Milk of magnesia (9.9–10.1)
10^{-11}	10^{-3}	11	3		Household ammonia (11.5–12.0)
10^{-12}	10^{-2}	12	2	Strongly basic	
10^{-13}	10^{-1}	13	1		
10^{-14}	10^0(1)	14	0		1 M NaOH (14)

(ACIDIC applies to the upper rows; BASIC applies to the lower rows.)

[a] The *normal* pH range is from 0 to 14, although solutions with negative pH (to −2) exist, as do solutions having a pH greater than 14 (to 16).

Now, with some knowledge of pH and its applications, let us consider the calculation of pH and pOH, given the hydrogen ion concentration in moles per liter.

EXAMPLE 14-4

Calculate the pH and pOH of the following solutions.

(a) The hydrogen ion concentration is

$$4.6 \times 10^{-6} \text{ mol/L}.$$

SOLUTION

$$pH = -\log[H^{1+}] = -\log[4.6 \times 10^{-6}]$$

Using Appendix 2 (Your Calculator), find the log of 4.6×10^{-6} as follows: Enter the decimal (4.6); then press the EXP, EE, or EEX key, then the positive value of the exponent (6), the change sign key $+/-$ or CHS to make the exponent negative, and finally the *log* key and the change sign key $+/-$ or CHS.

$$pH = -[-5.3372] = 5.3372, 5.34 \quad \textit{Answer}^6$$

$$pOH = 14.00 - 5.34 = 8.66 \quad \textit{Answer}$$

(b) The hydrogen ion concentration in Gatorade®, a popular antithirst drink, is 8.0×10^{-4} mol/L.

SOLUTION

$$pH = -\log[H^{1+}] = -\log[8.0 \times 10^{-4}]$$

Use Appendix 2 (Your Calculator) to find the log of 8.0×10^{-4} as follows: Enter the decimal (8.0); then press the EXP, EE, or EEX key, then the positive value of the exponent (4), the change sign key $+/-$ or CHS to make the exponent negative, and finally the *log* key and the change sign key $+/-$ or CHS.

$$pH = -[-3.0969] = 3.0969, 3.10 \quad \textit{Answer}$$

$$pOH = 14.00 - 3.10 = 10.90 \quad \textit{Answer}$$

(c) The hydrogen ion concentration in commercial tomato juice is 25×10^{-6} mol/L.

SOLUTION

$$pH = -\log[H^{1+}] = -\log[25 \times 10^{-6}]$$

Use Appendix 2 (Your Calculator) to find the log of 25×10^{-6} as follows: Enter 25; then press EXP, EE, or EEX key, then the positive value of the exponent (6), the change sign key $+/-$ or CHS to make the exponent negative, and finally the *log* key and the change sign key $+/-$ or CHS.

$$pH = -[-4.6021] = 4.6021, 4.60 \quad \textit{Answer}$$

$$pOH = 14.00 - 4.60 = 9.40 \quad \textit{Answer}$$

The pH of a solution can be determined directly with a pH meter,[7] as shown in Fig. 14–3, where the pH is usually read by *approximating* to the *hundredths'* place.

---

[6] The logarithm of a number with *two* significant digits will have *two* significant digits to the *right* of the *decimal* point.

[7] Before a pH meter can be used to collect data, it must be calibrated with standard solutions of known pH. This ensures that the data obtained are reliable.

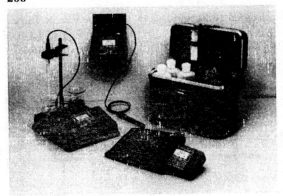

(a)

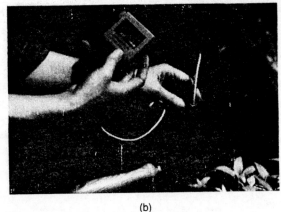

(b)

Figure 14–3 *Modern pH meters make pH measurements with a high degree of accuracy both (a) in the laboratory and (b) in the field. [Courtesy of Beckmann Instruments, Inc. (a) and Dr. E. R. Degginger (b)].*

Exercise 14–4

Calculate the pH and pOH of the following solutions.

(a) The hydrogen ion concentration is 6.4×10^{-7} mol/L.
(b) The hydrogen ion concentration is 9.6×10^{-10} mol/L.
(c) The hydrogen ion concentration is 4.0×10^{-6} mol/L.

In this book we have covered a few of the topics basic to an understanding of chemistry. Many topics have been omitted on purpose. A more advanced chemistry course would include considerably more material. In an attempt to help you improve your mathematics background, we have introduced the *factor-unit* method of problem solving. We hope that you will try to use it not only in other science courses but also in various mathematical operations you may encounter.

Nitrogen oxides: chemistry and smog

Name: Nitrogen monoxide (NO) and nitrogen dioxide (NO_2) are often collectively abbreviated NO_x, hence the nickname NOX.

Appearance: Nitrogen monoxide (nitric oxide) is a colorless gas, whereas nitrogen dioxide is a reddish-brown gas. The red-brown tinge of photochemical smog results from the presence of NO_2 gas in the atmosphere.

Occurrence: Nitrogen oxides are normally a very small fraction of the atmosphere. The activities of human society, however, have served to increase the amounts of NO and NO_2 to higher levels than normal. Gases such as NO and NO_2

that are the result of human activities are called *pollutants*.

Source: Nitrogen monoxide (NO) and nitrogen dioxide (NO_2) are major components of automobile exhaust. When air is used as a source of oxygen in gasoline engines, the nitrogen (N_2) and oxygen (O_2) in air react with each other as they pass through the hot chambers of the engine to produce a mixture of nitrogen monoxide and nitrogen dioxide.

$$N_{2(g)} + O_{2(g)} \longrightarrow 2NO_{(g)}$$

$$2NO_{(g)} + O_{2(g)} \longrightarrow 2NO_{2(g)}$$

The nitrogen oxides NO and NO₂ play key roles in the production of photochemical smog.

Its Role in Our World: The smog found in modern cities is more properly termed *photochemical smog*. Smog in Los Angeles is typically a haze and occurs in the hot months. Photochemical smog can be very irritating to lung tissues. Three key elements combine to produce photochemical smog: (1) increased concentrations of pollutants, (2) weather conditions that trap the pollutants at ground level, and (3) sunlight to promote the necessary chemical reactions.

The most important pollutants in smog are nitrogen monoxide, nitrogen dioxide, unburned hydrocarbons (compounds such as gasoline that contain carbon and hydrogen), and hydrocarbon derivatives (contain carbon, hydrogen, and oxygen). These last two pollutants are released into the atmosphere when humans burn coal or hydrocarbon fuels to produce energy. One of the most important single sources of such pollution is the automobile engine. In addition, hydrocarbons are released into the atmosphere during the refining and transportation of petroleum products and even during the filling of fuel tanks at gas stations. In large cities with heavy commuter traffic populations, these emissions can build up quickly in the atmosphere each morning and provide the "fuel" for smog production.

Weather and geographical considerations can conspire to trap these pollutants over the city. Large cities that lie in a basin (such as Los Angeles) or are surrounded by mountains (such as Las Vegas, Nevada) provide geographical traps for the polluted air. In addition, when air near the ground is cooler than air at the higher altitude (the reverse of normal), the less dense warm air does not mix well with the denser, cooler air below, where the pollutants are generated. This condition is called a *temperature inversion*. Los Angeles suffers temperature inversions regularly because the cooler sea air moves in during the day and pushes warmer air up to higher altitudes.

The last key element is sunlight. Sunlight promotes the reactions that transform nitrogen oxides (NO and NO_2), oxygen, and hydrocarbons into ozone (O_3) and the hydrocarbon derivatives that constitute photochemical smog. The basic features of smog formation are (a) nitrogen monoxide is converted into nitrogen dioxide; (b) sunlight acts upon the resulting nitrogen dioxide to give oxygen atoms; (c) the oxygen atoms react with oxygen gas to give ozone.

(a) $NO_{(g)} + O_{2(g)} +$ hydrocarbons $\longrightarrow$

$\qquad NO_{2(g)} +$ hydrocarbon derivatives

(b) $NO_{2(g)} +$ light $\longrightarrow NO_{(g)} + O_{(g)}$

(c) $\qquad O_{(g)} + O_{2(g)} \longrightarrow O_{3(g)}$

A variety of hydrocarbon derivatives are produced in the course of a day. The most offensive are aldehydes, of which formaldehyde (H-CHO) and acetaldehyde (CH_3-CHO) are the principal components. The ozone and aldehydes cause most of the irritation to human lungs on smoggy days. **Unusual Facts:** Monitoring the concentrations of nitrogen monoxide, nitrogen dioxide, ozone, and aldehydes reveals an interesting pattern. The concentration of nitrogen monoxide peaks early (7 a.m.) during the morning commute. The nitrogen dioxide concentration grows more slowly and peaks at about 10 a.m. Meanwhile, the levels of ozone and aldehydes rise later in the day (12 noon to 2 p.m.).

Problems

1. Balance the following oxidation–reduction equations by the ion electron method. Indicate the substances that are oxidized and reduced, and the oxidizing and reducing agents.

   (a) $I^{1-} + SO_4^{2-} \rightarrow H_2S + I_2 + H_2O$ (in acid solution)
   (b) $Cu + NO_3^{1-} \rightarrow Cu^{2+} + NO + H_2O$ (in acid solution)
   (c) $Cu + SO_4^{2-} \rightarrow Cu^{2+} + SO_2 + H_2O$ (in acid solution)
   (d) $Cl_2 \rightarrow ClO_3^{1-} + Cl^{1-} + H_2O$ (in basic solution)

2. If 32.0 mL of a dilute solution of lime water (calcium hydroxide) required 12.4 mL of 0.100 M hydrochloric acid solution for neutralization to a methyl red equivalence point, calculate the molarity of the lime water.

3. If 0.625 g of pure sodium carbonate was dissolved in water and the solution titrated with 30.8 mL of hydrochloric acid to a methyl orange equivalence point, calculate the molarity of the hydrochloric acid solution.

4. Calculate the pH and pOH of the following solutions.

   (a) The hydrogen ion concentration is 1.0×10^{-8} mol/L.
   (b) The hydrogen ion concentration in household ammonia is 2.0×10^{-12} mol/L.
   (c) The hydrogen ion concentration in commercial milk is 2.0×10^{-7} mol/L.
   (d) The hydrogen ion concentration in sour milk is 6.3×10^{-7} mol/L.
   (e) The hydrogen ion concentration in seawater is 5.3×10^{-9} mol/L.

General Problems

5. In one of the tests to determine the alcohol content in the breath, the police officer uses a chemical test. The equation is

$$C_2H_6O + Cr_2O_7{}^{2-} \longrightarrow C_2H_4O_2 + Cr^{3+} + H_2O \qquad \text{(in acidic solution)}$$

ethyl
alcohol   acetic
                                                    acid

The basis for the test is the color change from the red–orange dichromate ion $(Cr_2O_7{}^{2-})$ to the blue–green chromium(III) ion $[Cr^{3+}]$.

(a) Balance the equation by the ion–electron method.
(b) If 1.11 mg of chromium(III) ion is produced, calculate the number of milligrams of alcohol present in the breath.

6. If 0.200 g of pure sodium hydroxide is titrated with 5.70 mL of sulfuric acid solution to a phenolphthalein equivalence point, calculate

(a) the molarity of the sulfuric acid solution
(b) the normality of the sulfuric acid solution (both hydrogen ions replaced)

7. Iodine is generally made from sodium iodate ($NaIO_3$), which is mined in naturally occurring salt deposits found in Chile.

(a) Balance the following equation in *acid* solution by the ion electron method.

$$IO_3{}^{1-} + HSO_3{}^{1-} \longrightarrow I_2 + SO_4{}^{2-}$$

(b) Calculate the number of grams of iodine produced from 0.900 kg of sodium iodate.

Answers to Exercises

14–1. (a) $1 + 6 + 14H^{1+} \to 2 + 6 + 7H_2O$
$Cr_2O_7{}^{2-}$: reduced, oxidizing agent
Fe^{2+}: oxidized, reducing agent
(b) $4 + 3 + 2H_2O \to 4 + 3 + 4OH^{1-}$
$MnO_4{}^{1-}$: reduced, oxidizing agent
$ClO_2{}^{1-}$: oxidized, reducing agent
(c) $3 + 2 + 8H^{1+} \to 3 + 2 + 4H_2O$
$NO_3{}^{1-}$: reduced, oxidizing agent
S^{2-}: oxidized, reducing agent
(d) $3 + 1 + 6H^{1+} \to 3 + 1 + 3H_2O$
$IO_3{}^{1-}$: reduced, oxidizing agent
Sn^{2+}: oxidized, reducing agent

14–2. 1.92 *M*   **14–3.** 1.42 *M*

14–4. (a) 6.19, 7.81; (b) 9.02, 4.98; (c) 5.40, 8.60

QUIZ   **1.** Balance the following oxidation–reduction equations by the ion electron method. Indicate the substances that are oxidized and reduced, and the oxidizing and reducing agents. This reaction is used as a quantitative test for arsenic compounds.

$$Zn + AsO_4^{3-} \longrightarrow Zn^{2+} + AsH_{3(g)} + H_2O \qquad \text{(in acid solution)}$$

2. "Muriatic acid," a solution of hydrochloric acid, is sold for use in controlling the acidity in swimming pools. A 2.00 mL sample of commercial muriatic acid required 51.8 mL of a 0.287 M sodium hydroxide solution to reach a phenolphthalein equivalence point. Calculate the molarity of the muriatic acid.

3. The hydrogen ion concentration in a solution in a chemical processing plant must be maintained between 0.012 and 0.020 M. Calculate this hydrogen ion range in pH units.

$\mathbf{A}$ppendix 1

SI Units and Some Conversion Factors

We have introduced the International System of Units (SI) in this text; the scientific community uses this system to a limited extent now, and in a few years it may be the only system in use. The SI is derived from seven *base units*, as Table A1–1 shows.

In addition to these base units, there are two *supplementary units*, the radian (rad) and steradian (sr), used to define angular measurement.

Multiple and submultiple *prefixes* are used to indicate orders of magnitude. These prefixes define either a fractional or multiple value of the base unit; thus, a kilometer is 1000 meters, and a millimeter is 0.001 (or 10^{-3}) meter. Table A1–2 shows these prefixes and their orders of magnitude.

A series of *derived units*, which define various physical quantities used in scientific measurements, are derived from the seven base and two supplementary units. Table A1–3 shows some of these units.

Units currently being used in chemical measurements (and used in this text) that are not *exactly* defined in terms of SI units are the *atmosphere*, *torr*, and *calorie*. The SI committee recommends that these units be abandoned; however, many chemists will probably use these units for some time.

Table A1–4 lists conversion factors that may be used to convert from non-SI units to the recognized SI unit.

TABLE A1–1 BASE UNITS

QUANTITY MEASURED	UNIT NAME	SI SYMBOL FOR UNIT
Length	meter	m
Mass	kilogram	kg
Time	second	s
Electric current	ampere	A
Thermodynamic temperature	kelvin	K
Amount of substance	mole	mol
Luminous intensity	candela	cd

TABLE A1–2   SI PREFIXES

FACTOR	PREFIX	SI SYMBOL
10^{12}	tera	T
10^9	giga	G
10^6	mega	M
10^3	kilo	k
10^2	hecto[a]	h
10^1	deka[a]	da
10^{-1}	deci[a]	d
10^{-2}	centi[a]	c
10^{-3}	milli	m
10^{-6}	micro	μ
10^{-9}	nano	n
10^{-12}	pico	p
10^{-15}	femto	f
10^{-18}	atto	a

[a] Avoid these prefixes when possible.

TABLE A1–3   DERIVED UNITS

PHYSICAL QUANTITY	UNIT	SI SYMBOL	DEFINITION
Acceleration	meter/second^2	m/s^2	m/s^2
Area	meter^2	m^2	m^2
Density	kilogram/meter^3	kg/m^3	kg/m^3
Electric capacitance	farad	F	$A \cdot s/V$
Electric potential difference	volt[a]	V	$J/A \cdot s = W/A$
Electric resistance	ohm	Ω	$V/A = kg \cdot m^2/s^3 \cdot A^2$
Energy	joule	J	$N \cdot m = kg \cdot m^2/s^2$
Force	newton	N	$kg \cdot m/s^2$
Power	watt	W	$J/s = kg \cdot m^2/s^3$
Quantity of electricity	coulomb	C	$A \cdot s$
Pressure	pascal	Pa	$N/m^2 = kg/m \cdot s^2$
Quantity of heat	joule	J	$N \cdot m = kg \cdot m^2/s^2$
Specific heat	joule/kilogram-kelvin	$J/kg \cdot K$	$J/kg \cdot K$
Velocity	meter/second	m/s	m/s
Volume[b]	cubic meter	m^3	m^3

[a] Also the unit for expressing electromotive force.
[b] In 1964, the liter was adopted as a special name for the cubic decimeter (dm^3), but its use for measurement of extremely precise volumes is discouraged.

TABLE A1–4   SELECTED CONVERSION FACTORS

TO CONVERT FROM	TO	MULTIPLY BY
calorie (cal)	joule (J)	4.184
atmosphere (atm)	pascal (Pa)	1.013×10^5
torr (torr)	pascal (Pa)	1.333×10^2
inch (in.)	meter (m)	2.54×10^{-2}
pound-mass (lbm)	kilogram (kg)	4.536×10^{-1}

Appendix **2**

Your Calculator

Modern technology has made it possible for you to purchase highly reliable and convenient calculators at a modest cost. As a result, longhand calculations, slide rules, and logarithm tables have become a thing of the past. Calculators equipped with the functions $+$, $-$, $\times$, $\div$, and $\sqrt{}$ are sufficient for almost every calculation in this text, except Chapter 14, where you will need the function log x. In college chemistry you will need the function log x, and also the function 10^x.

Although a number of calculator designs are available, we will describe the use of the two most common types. The first is most easily recognized by the presence of an "=" key. The second most common type has an "ENTER" key.

Two notes of caution. Calculators are fast and convenient, and they do not make mistakes. However, you can make a mistake in using the calculator. Always check to see if your answer is reasonable. For example, let's say you multiplied 8.2 $\times$ 2.3 and got 188.6. Is this reasonable? Well, you know that 8 $\times$ 2 = 16, so that 8.2 $\times$ 2.3 should be a bit more than 16. Your answer, 188.6, *makes no sense*—you must have made an error in using the calculator. Repeat your calculation if you have any doubts. The correct answer is 18.86. In the original calculation, you almost certainly misplaced a decimal point and entered 82 $\times$ 2.3 or 8.2 $\times$ 23!

Second, calculators will give you far more significant digits than you can justify. Always round off your answers to the correct number of significant digits when you are done. *Do not record the answer shown on the calculator without considering what the correct number of significant digits should be.* In the following examples, the tables show the numbers as displayed on the calculator. Round off any displayed number before presenting it as an answer.

Entering Numbers. Exponential Notation

Entering numbers into a calculator is very simple. You just type them in as they appear on the paper, being sure to enter any decimals points in the correct position. Practice a few times by entering 12.41, 12000.1, 273, and 0.000572.

You might wish to enter the second and fourth numbers in the last paragraph in exponential or scientific notation. To do this, you will need to identify two keys on your particular brand of calculator:

■ Exponent key: On the "=" type of calculator, two common labels for this key are "EE" (**E**nter **E**xponent) or "EXP" (**EXP**onent). On the "ENTER" type of calculator, the key is often labeled "EEX" (**E**nter **EX**ponent). Consult the manual for your calculator if you have any doubts.

■ Change sign key: On the "=" type of calculator, the most common label for this key is "+/−." On the "ENTER" type of calculator, the key is often labeled "CHS" (**CH**ange **S**ign).

In either case, you key in the number and then press the exponent key. Two zeros should appear to the right of the entered number. To complete the operation, just key in the desired exponent (or power) to which 10 is to be taken, and, if the exponent to which 10 is taken is a *negative* number, press the change sign key after the exponent has been keyed in. Examples are shown in the following table:

"=" OR "ENTER" TYPE

DESIRED NUMBER	PRESS	ON DISPLAY	
1.765×10^4	1.765	*1.765*	
	EXP, EE, or EEX	*1.765*	*00*
	4	*1.765*	*04*
1.241×10^{-7}	1.241	*1.241*	
	EXP, EE, or EEX	*1.241*	*00*
	7	*1.241*	*07*
	+/− or CHS	*1.241*	*−07*

Addition, Subtraction, Multiplication, and Division

To perform an arithmetic operation, you must press a series of keys. The two basic calculator types differ slightly as shown below:

"=" TYPE

On this type of calculator, you key in the operation as it is written. To determine 2 + 3 you would press "2," then "+," then "3," and finally "=." Examples of addition,

"ENTER" TYPE

On this type of calculator, you ENTER the first number, key in the second number, and *then* press the appropriate operation key. To determine 2 + 3, you would press the fol-

subtraction, multiplication, and division are given below.

lowing keys in order: "2," "ENTER," "3," and "+." Examples of addition, subtraction, multiplication and division are given below.

"=" TYPE

DESIRED OPERATION	PRESS	ON DISPLAY
176.5 + 12.41	176.5	*176.5*
	+	*176.5*
	12.41	*12.41*
	=	*188.91*
176.5 − 12.41	176.5	*176.5*
	−	*176.5*
	12.41	*12.41*
	=	*164.09*
176.5 × 12.41	176.5	*176.5*
	×	*176.5*
	12.41	*12.41*
	=	*2190.365*
176.5 ÷ 12.41	176.5	*176.5*
	÷	*176.5*
	12.41	*12.41*
	=	*14.22240129*

"ENTER" TYPE

DESIRED OPERATION	PRESS	ON DISPLAY
176.5 + 12.41	176.5	*176.5*
	ENTER	*176.5*
	12.41	*12.41*
	+	*188.91*
176.5 − 12.41	176.5	*176.5*
	ENTER	*176.5*
	12.41	*12.41*
	−	*164.09*
176.5 × 12.41	176.5	*176.5*
	ENTER	*176.5*
	12.41	*12.41*
	×	*2190.365*
176.5 ÷ 12.41	176.5	*176.5*
	ENTER	*176.5*
	12.41	*12.41*
	÷	*14.22240129*

Now try the following examples to check your techniques. The answers are given as they should be displayed on your calculator and without regard to significant digits.

$$39.21 + 9.57 = 48.78$$
$$1.021 + 0.24 = 1.261$$
$$39.21 \times 9.57 = 375.2397$$
$$1.021 \times 0.24 = 0.24504$$

$$39.21 - 9.57 = 29.64$$
$$1.021 - 0.24 = 0.781$$
$$39.21 \div 9.57 = 4.09717868$$
$$1.021 \div 0.24 = 4.25416667$$

Series or Chain Calculations

Many times, problems in this text require that you perform a series or chain of arithmetic operations. An example of this type of calculation would be summing up a list of numbers or performing a series of multiplications and divisions. You should generally perform these operations in the calculator *without* removing any intermediate answers. This approach minimizes round-off errors and copying errors.

"=" TYPE

DESIRED OPERATION	PRESS	ON DISPLAY
12.3 + 35.6 − 1.55	12.3	12.3
	+	12.3
	35.6	35.6
	=	47.9
	−	47.9
	1.55	1.55
	=	46.35
$\dfrac{12.3 \times 35.6}{1.55 \times 2.68}$	12.3	12.3
	×	12.3
	35.6	35.6
	=	437.88
	÷	437.88
	1.55	1.55
	=	282.5032258
	÷	282.5032258
	2.68	2.68
	=	105.4116514

"ENTER" TYPE

DESIRED OPERATION	PRESS	ON DISPLAY
12.3 + 35.6 − 1.55	12.3	12.3
	ENTER	12.3
	35.6	35.6
	+	47.9
	ENTER	47.9
	1.55	1.55
	−	46.35
$\dfrac{12.3 \times 35.6}{1.55 \times 2.68}$	12.3	12.3
	ENTER	12.3
	35.6	35.6
	×	437.88
	ENTER	437.88
	1.55	1.55
	÷	282.5032258
	ENTER	282.5032258
	2.68	2.68
	÷	105.4116514

Now try the following examples to check your technique. The answers are given as they should be displayed on your calculator and without regard to significant digits.

$$39.21 - 9.57 + 47.09 - 2.675 = 74.055$$

$$-39.21 + 9.57 - 47.09 - 2.675 = -79.405$$ (*Hint*: Key in 39.21 and use the CHS or +/− key to get −39.21.)

$$\frac{39.21 \times 9.57}{47.09 \times 2.675} = 2.97890260 \qquad \frac{39.21 \times 2.675}{47.09 \times 9.57} = 0.23274481$$

Square Roots, Squares, Logarithms, and Antilogarithms

Four other functions can be useful in working the problems in this book: taking a square root, squaring a number, taking a logarithm of a number, and taking an antilogarithm of a number. There is usually no difference in these operations from one calculator to the next, as long as you are able to identify the correct key for each operation. Some calculators designate two or more operations to a given key and require two keystrokes to accomplish these operations. In these cases, a shift or function key (often color coded with the label) should precede the desired operation. Check your manual if you are unsure. The typical key labels for these operations are as follows:

$$\text{square root key} = \sqrt{}$$

$$\text{squaring a number} = x^2$$

$$\text{taking a logarithm} = \log$$

$$\text{taking an antilogarithm} = 10^x$$

The following table illustrates examples of using these keys both with regular numbers and exponential numbers.

"=" OR "ENTER" TYPE

DESIRED OPERATION	PRESS	ON DISPLAY
$\sqrt{176.5}$	176.5 $\sqrt{}$	176.5 13.28533026
$(12.41)^2$	12.41 x^2	12.41 154.0081
log (176.5)	176.5 log	176.5 2.24674471
antilog (12.41)	12.41 10^x	12.41 2.570391 × 10¹²
$\sqrt{1.241 \times 10^{-7}}$	1.241 EXP, EE, or EEX 7 +/− or CHS $\sqrt{}$	1.241 1.241 00 1.241 07 1.241 −07 0.000352278
log (1.765 × 10⁴)	1.765 EXP, EE, or EEX 4 log	1.765 1.765 00 1.765 04 4.24674471

Now try the following examples to check your technique. The answers are given as they should be displayed on your calculator and without regard to significant digits.

$$\sqrt{72.471} = 8.51299007 \qquad (72.471)^2 = 5{,}252.045841$$

$$\log 72.471 = 1.86016425 \qquad \text{antilog } 72.471 = 2.958012 \times 10^{72}$$

Appendix 3

Quadratic Equations

Quadratic equations are equations in which the unknown quantity is raised to the *second* power, such as $x^2 = c$ or $ax^2 + bx + c = 0$ $(a \neq 0)$. We consider these two types of equations in this appendix. We need the solutions to quadratic equations for general chemistry problems involving chemical equilibria.

A3–1 Solution by Extraction of the Square Root

We solve the $x^2 = c$ form of a quadratic equation by extracting the square root, where **c** is equal to a positive number. To solve for x we must obtain the square root of both sides of the equation. The square root of the number is found by using your calculator. There are two roots to these equations, a *positive* root and a *negative* root. In chemistry the negative root is usually not considered in this type of equation. If the number is in exponential form, the power of 10 must be adjusted so as to give an *even*-numbered exponent.[1] This *even*-numbered exponent is divided by 2 (see Section 1–3).

EXAMPLE A3–1

Solve the following quadratic equations.

(a) $x^2 = 4.00$

SOLUTION Taking the square root of both sides of the equation gives

$$\sqrt{x^2} = \sqrt{4.00}$$

$$x = \pm 2.00 \quad \textit{Answer}$$

(b) $x^2 = 1.60 \times 10^{-3}$

---

[1] Calculators with the keys EXP, EE, or EEX automatically adjust the exponent, so you can read the square root directly from the display window without adjusting it. See Appendix 2 (Your Calculator).

SOLUTION Adjusting the *odd*-numbered exponent to give an *even*-numbered exponent (see Section 1–3) gives the equation

$$x^2 = 16.0 \times 10^{-4}$$

Next, taking the square root of both sides of the equation, and dividing the *even*-numbered exponent by 2, we obtain

$$\sqrt{x^2} = \sqrt{16.0 \times 10^{-4}}$$

$$x = \pm 4.00 \times 10^{-2} \qquad \textit{Answers}$$

A quadratic equation, as a linear equation (see Section 1–5), can be checked by substituting the values obtained for the unknown into the original equations.

(a) $(\pm 2.00)^2 = 4.00$

$4.00 = 4.00$

(b) $(4.00 \times 10^{-2})^2 = 16.0 \times 10^{-4}$

$= 1.60 \times 10^{-3}$

$1.60 \times 10^{-3} = 1.60 \times 10^{-3}$

Exercise A3–1

Solve the following quadratic equations.

(a) $2x^2 = 18.0$

(b) $x^2 = 6.40 \times 10^5$

A3–2   Solution by the Quadratic Formula

We can solve any quadratic equation by using the quadratic formula. The general form of a quadratic equation is $ax^2 + bx + c = 0 \ (a \neq 0)$. We use the quadratic formula to solve for x, yielding two roots, as follows:

$$x = \frac{-b \pm \sqrt{b^2 - 4ac}}{2a}$$

In chemistry problems, we must consider both roots in order to determine which root is reasonable. Before applying the equation given above, we must place the quadratic equation in the standard form $ax^2 + bx + c = 0 \ (a \neq 0)$. Thus, for the quadratic equation $2x^2 + x - 6 = 0$, $a = 2$, $b = 1$, and $c = -6$.

EXAMPLE A3–2

Solve the following quadratic equation.

(a) $2x^2 + x - 6 = 0$

SOLUTION As mentioned previously, $a = 2$, $b = 1$, and $c = -6$. Substituting these values into the quadratic formula, we have

$$x = \frac{-1 \pm \sqrt{(1)^2 - 4(2)(-6)}}{2(2)}$$

$$= \frac{-1 \pm \sqrt{1 + 48}}{4}$$

$$= \frac{-1 \pm \sqrt{49}}{4}$$

$$= \frac{-1 \pm 7}{4} \quad \text{or} \quad \frac{-1 + 7}{4}, \frac{-1 - 7}{4}$$

$$= \frac{3}{2}, \quad -2 \quad \textit{Answers}$$

(b) $x^2 - 8x = -15$

SOLUTION   Place this equation in the form $ax^2 + bx + c = 0$. Adding 15 to *both* sides of the equation gives us

$$x^2 - 8x + 15 = \cancel{-15} + \cancel{15} = 0$$

The values of a, b, and c are 1, -8, and $+15$, respectively. Substituting these values into the quadratic formula, we obtain

$$x = \frac{-(-8) \pm \sqrt{(-8)^2 - 4(1)(15)}}{2(1)}$$

$$= \frac{+8 \pm \sqrt{64 - 60}}{2}$$

$$= \frac{+8 \pm \sqrt{4}}{2}$$

$$= \frac{+8 \pm 2}{2} \quad \text{or} \quad \frac{+8 + 2}{2}, \frac{+8 - 2}{2}$$

$$= 5, \quad 3 \quad \textit{Answers}$$

Exercise A3–2

Solve the following quadratic equations.

(a) $x^2 + x - 2 = 0$ (b) $2x^2 - 6x = -4$

Problem

1. Solve the following quadratic equations.

(a) $x^2 = 25$
(b) $x^2 = 3.6 \times 10^{-3}$
(c) $x^2 = 20.0 \times 10^{-5}$
(d) $x^2 + 7x - 8 = 0$

(e) $6x^2 - x = 15$

(f) $x^2 + (1.76 \times 10^{-5})x - (1.80 \times 10^{-6}) = 0$

Answers to Exercises

A3–1. (a) ± 3.00; (b) $\pm 8.00 \times 10^2$

A3–2. (a) 1, -2; (b) 2, 1

Appendix 4

Vapor Pressure of Water at Various Temperatures

TEMPER-ATURE (°C)	PRESSURE[a]			TEMPER-ATURE (°C)	PRESSURE[a]		
	torr	atm	Pa		torr	atm	Pa
0	4.6	0.0061	610	32	35.7	0.0469	4755
5	6.5	0.0086	872	33	37.7	0.0496	5030
10	9.2	0.0121	1227	34	39.9	0.0525	5319
15	12.8	0.0168	1705	35	42.2	0.0555	5622
16	13.6	0.0179	1818	36	44.6	0.0586	5941
17	14.5	0.0191	1937	37	47.1	0.0619	6275
18	15.5	0.0204	2063	38	49.7	0.0654	6625
19	16.5	0.0217	2197	39	52.4	0.0690	6992
20	17.5	0.0231	2338	40	55.3	0.0728	7376
21	18.6	0.0245	2486	45	71.9	0.0946	9583
22	19.8	0.0261	2643	50	92.5	0.1217	12,333
23	21.1	0.0277	2809	55	118.0	0.1553	15,737
24	22.4	0.0294	2983	60	149.4	0.1965	19,915
25	23.8	0.0313	3167	65	187.5	0.2468	25,002
26	25.2	0.0332	3360	70	233.7	0.3075	31,157
27	26.7	0.0352	3564	75	289.1	0.3804	38,543
28	28.3	0.0373	3779	80	355.1	0.4672	47,342
29	30.0	0.0395	4005	85	433.6	0.5705	57,808
30	31.8	0.0419	4242	90	525.8	0.6918	70,094
31	33.7	0.0443	4492	95	633.9	0.8341	84,512
				100	760.0	1.0000	101,325

[a] Units in torr to nearest tenth; atmospheres to neartest ten-thousandth; pascals to nearest unit.

$\mathbf{A}$ppendix 5

Answers to Problems and Quizzes

Chapter 1

1. (a) 3; (b) 5; (c) 3; (d) 4

2. (a) 12.9; (b) 1.34; (c) 1.36; (d) 1.35

3. (a) 6.14; (b) 13.91; (c) 51; (d) 1.7; (e) 18

4. 6.25×10^5

5. 7.07×10^{-4}

6. (a) 5.74×10^3; (b) 5.53×10^2; (c) 5.01×10^3; (d) 3.72×10^5

7. (a) 9.05×10^5; (b) 4.74×10^2; (c) 2.41×10^4; (d) 2.21×10^{-3}

8. (a) 3.00×10^3; (b) 6.00×10^2; (c) 7.00×10^3; (d) 9.70×10^{-3}

9. (a) 3.03×10^6; (b) 1.10×10^{16}; (c) 8.49×10^6; (d) 9.26×10^9

10. (a) 6.72×10^6; (b) 3.42×10^{-3}; (c) 6.27×10^3; (d) 6.28×10^{-2}

11. (a) 4; (b) 3; (c) 5
   (d) $5(x + 2) = -20$
   $5x + 10 = -20$
   $5x + \cancel{10} - \cancel{10} = -20 - 10 = -30$
   $\qquad\qquad x = -6$

12. (a) 3.7×10^2;   (b) 9.5×10^1
   (c) 1.70×10^2;   (d) 3.45×10^{-2}

13. (a) $6.05 \times 10^3/18.7 \times 10^{10} = 0.324 \times 10^{-7} = 3.24 \times 10^{-8}$
   (b) 7.41×10^{-8}
   (c) $5.02 \times 10^6 + 6.5 \times 10^6 = 11.5 \times 10^6 = 1.15 \times 10^7$
   (d) 2.24×10^{10}

Chapter 1 Quiz

1. (a) 5; (b) 3; (c) 5; (d) 4

2. (a) 27.7; (b) 42.6; (c) 7350; (d) 18.3

3. (a) 2.43×10^3; (b) 7.52×10^{-3}; (c) 7.66×10^4; (d) 3.53×10^{-2}

4. (a) 1.25×10^5; (b) 7.74×10^{-15}; (c) 9.61×10^3; (d) 1.86×10^9

Chapter 2

1. (a) $2.75 \ \cancel{km} \times \dfrac{1000 \ \cancel{m}}{1 \ \cancel{km}} \times \dfrac{100 \ cm}{1 \ \cancel{m}} = 2.75 \times 10^5 \ cm$

   (b) 6.25×10^{-3} kg

   (c) $225 \ \cancel{mL} \times \dfrac{1 \ \cancel{cm^3}}{1 \ \cancel{mL}} \times \dfrac{(1 \ m)^3}{(100)^3 (\cancel{cm})^3} = 2.25 \times 10^{-4} \ m^3$

   (1 mL = 1 cm^3; see Table 2–2)

   (d) $7.55 \times 10^{-3} \ \mu$ (10 Å = 1 nm; 1000 nm = 1 μ; see Table 2–2).

2. 5.005 g

3. 2010.7505 m

4. (a) $8.00 \ \cancel{in.} \times \dfrac{2.54 \ \cancel{cm}}{1 \ \cancel{in.}} \times \dfrac{1 \ m}{100 \ \cancel{cm}} = 0.203 \ m$

   (b) 0.275 lb
   (c) 37.0 oz
   (d) 5660 mL or 5.66×10^3 mL

5. 6.2 mi

6. (a) 1.1 mi

   (b) $15 \ \cancel{laps} \times \dfrac{4\overline{0}0 \ \cancel{ft}}{1 \ \cancel{lap}} \times \dfrac{12 \ \cancel{in.}}{1 \ \cancel{ft}} \times \dfrac{2.54 \ \cancel{cm}}{1 \ \cancel{in.}} \times \dfrac{1 \ m}{100 \ \cancel{cm}} = 1800 \ m$ (two significant digits)

7. (a) $25°C = [(1.8 \times 25) + 32]°F = 77°F$
      $(25 + 273) = 298 \ K$
   (b) $-76°F, 213 \ K$

8. (a) $68°F = \dfrac{(68 - 32)}{1.8} = 2\overline{0}°C$

      $(2\overline{0} + 273) = 293 \ K$
   (b) $-87°C, 186 \ K$

9. $77 \ K = (77 - 273)°C = -196°C$
      $-196°C = [1.8(-196) + 32]°F = -321°F$

10. $-62.1°C$

11. $58.0°C$

12. 36.8°C, 36.7°C, 36.9°C

13. (a) $\dfrac{425 \text{ g}}{45.0 \text{ mL}} = 9.44 \text{ g/mL}$

 (b) 2.44 g/mL

14. (a) $335 \text{ g} \times \dfrac{1 \text{ mL}}{0.880 \text{ g}} = 381 \text{ mL}$

 (b) 297 mL

15. (a) $755 \text{ mL} \times \dfrac{1.05 \text{ g}}{1 \text{ mL}} = 793 \text{ g}$

 (b) $325 \text{ mL} \times \dfrac{1 \text{ cm}^3}{1 \text{ mL}} \times \dfrac{1 \text{ m}^3}{(100)^3 (\text{cm})^3} \times \dfrac{8.80 \times 10^2 \text{ kg}}{1 \text{ m}^3} \times \dfrac{1000 \text{ g}}{1 \text{ kg}} = 286 \text{ g}$

16. (a) $1.83 \times 1.00 \text{ g/mL} = 1.83 \text{ g/mL} = \text{density}$

 $35.0 \text{ mL} \times \dfrac{1.83 \text{ g}}{1 \text{ mL}} = 64.0 \text{ g}$

 (b) $28\overline{0}0 \text{ g}$

17. (a) $1.05 \times 1.00 \text{ g/mL} = 1.05 \text{ g/mL} = \text{density}$

 $285 \text{ g} \times \dfrac{1 \text{ mL}}{1.05 \text{ g}} \times \dfrac{1 \text{ L}}{1000 \text{ mL}} = 0.271 \text{ L}$

 (b) 0.940 L

18. (a) $2.09 \times 10^5 \text{ cm}^3$
 (b) $1.28 \times 10^4 \text{ in.}^3$

19. (a) 39.1 lb/gal

 (b) $\dfrac{5.00 \text{ g}}{1 \text{ cm}^3} \times \dfrac{1 \text{ lb}}{454 \text{ g}} \times \dfrac{(2.54)^3 (\text{cm})^3}{(1 \text{ in.})^3} \times \dfrac{(12)^3 (\text{in.})^3}{(1 \text{ ft})^3} = 312 \dfrac{\text{lb}}{\text{ft}^3}$

20. $15.0 \text{ cm} \times \dfrac{1 \text{ in.}}{254 \text{ cm}} \times \dfrac{1 \text{ ft}}{12 \text{ in.}} \times 30.0 \text{ in.} \times \dfrac{1 \text{ ft}}{12 \text{ in.}} \times 5.00 \text{ ft} = 6.15 \text{ ft}^3$

 $\dfrac{1340 \text{ kg}}{6.15 \text{ ft}^3} \times \dfrac{1000 \text{ g}}{1 \text{ kg}} \times \dfrac{1 \text{ lb}}{454 \text{ g}} = 48\overline{0} \text{ lb/ft}^3$

21. $\dfrac{32.0 \text{ lb}}{1 \text{ in.}^2} \times \dfrac{(1)^2 \text{ in.}^2}{(2.54)^2 \text{ cm}^2} \times \dfrac{(100)^2 \text{ cm}^2}{(1)^2 \text{ m}^2} \times \dfrac{454 \text{ g}}{1 \text{ lb}} \times \dfrac{1 \text{ kg}}{1000 \text{ g}}$

$= 2.25 \times 10^4 \text{ kg/m}^2$

22. $15 \text{ ft} \times \dfrac{12 \text{ in.}}{1 \text{ ft}} \times \dfrac{2.54 \text{ cm}}{1 \text{ in.}} \times 12 \text{ ft} \times \dfrac{12 \text{ in.}}{1 \text{ ft}}$

 $\times \dfrac{2.54 \text{ cm}}{1 \text{ in.}} \times 11 \text{ in.} \times \dfrac{2.54 \text{ cm}}{1 \text{ in.}} = 4.7 \times 10^6 \text{ cm}^3$

 $4.7 \times 10^6 \text{ cm}^3 \times \dfrac{0.915 \text{ g}}{\text{cm}^3} \times \dfrac{1 \text{ kg}}{1000 \text{ g}} = 4.3 \times 10^3 \text{ kg}$

Chapter 2 Quiz

1. 3.77×10^3 mL

2. 103.6°F

3. 1.12 kg

4. 0.109 L

5. 5.94 g/mL

Chapter 3

1. (a) and (c) compounds; (b) element; (d) mixture

2. (b), (c), (f), (g) physical; others chemical

3. (a), (b), (d), (e), (h), (j) physical; others chemical

4. (a) 12 atoms carbon, 22 atoms hydrogen, 11 atoms oxygen; total: 45 atoms
 (b) 1 atom nitrogen, 3 atoms hydrogen; total: 4 atoms
 (c) 1 atom carbon, 2 atoms chlorine, 2 atoms fluorine; total: 5 atoms
 (d) 1 atom tin, 2 atoms fluorine; total: 3 atoms
 (e) 3 atoms calcium, 2 atoms phosphorus, 8 atoms oxygen; total: 13 atoms
  [*Note*: Clear ( ) first.]

5. (a) $C_9H_8O_4$; (b) $C_8H_{10}N_4O_2$; (c) $C_{20}H_8Br_2HgNa_2O_6$;
 (d) $C_{10}H_{16}N_5O_{13}P_3$; (e) $C_7H_5N_3O_6$

6. density of nitric acid = 1.42×1.00 g/mL = 1.42 g/mL

$$165 \text{ g} \times \frac{1 \text{ mL}}{1.42 \text{ g}} \times \frac{1 \text{ L}}{1000 \text{ mL}} = 0.116 \text{ L}$$

7. 0.242 L

8. 12.4 lb

9. $V = 1.30$ cm $\times$ 1.30 cm $\times$ 1.30 cm = 2.20 cm^3

$$\frac{5.94 \text{ g}}{2.20 \text{ cm}^3} \times \frac{1 \text{ cm}^3}{1 \text{ mL}} = 2.70 \text{ g/mL}$$

10. $122\overline{0}$°F, 933 K

Chapter 3 Quiz

1. (a) Au; (b) Ra

2. (a) chemical; (b) and (c) physical

3. 5 atoms carbon, 12 atoms hydrogen, 1 atom sulfur; 18 atoms total

4. $C_9H_{11}NO_2$

5. 3.63 mL

Chapter 4

1. (a) $\boxed{\begin{array}{c}19p\\20n\end{array}}$ $19e^-$ (b) $\boxed{\begin{array}{c}22p\\24n\end{array}}$ $22e^-$

(c) $\boxed{\begin{array}{c}27p\\32n\end{array}}$ $27e^-$ (d) $\boxed{\begin{array}{c}31p\\38n\end{array}}$ $31e^-$

2. 68.9257 amu (0.6040) + 70.9249 amu (0.3960) = 69.72 amu

3. 121.8 amu

4. (a) $2 \times 1^2 = 2$ electrons   (b) $2 \times 2^2 = 8$ electrons
(c) $2 \times 6^2 = 72$ electrons   (d) $2 \times 7^2 = 98$ electrons

5. (a) $\boxed{\begin{array}{c}4p\\5n\end{array}}$ $2e^-$ $2e^-$ (b) $\boxed{\begin{array}{c}9p\\10n\end{array}}$ $2e^-$ $7e^-$
                         1   2   1   2

(c) $\boxed{\begin{array}{c}12p\\12n\end{array}}$ $2e^-$ $8e^-$ $2e^-$ (d) $\boxed{\begin{array}{c}16p\\16n\end{array}}$ $2e^-$ $8e^-$ $6e^-$
                              1   2   3   1   2   3

6. (a) He: or H̤e, etc,   (b) Li·
(c) ·N̈·   (d) :Ä̤r:

7. (a) $1s^2, 2s^1$ (1); (b) $1s^2, 2s^22p^3$ (5); (c) $1s^2, 2s^22p^6, 3s^23p^1$ (3); (d) $1s^2,$
$2s^22p^6, 3s^23p^63d^{10}, 4s^24p^6$ (8), or $1s^2, 2s^22p^6, 3s^23p^6, 4s^2, 3d^{10}, 4p^6$ (8); (e) $1s^2,$
$2s^22p^6, 3s^23p^63d^{10}, 4s^2$ (2), or $1s^2, 2s^22p^6, 3s^23p^6, 4s^2, 3d^{10}$ (2); (f) $1s^2, 2s^22p^6,$
$3s^23p^63d^{10}, 4s^24p^64d^{10}, 5s^25p^6, 6s^2$, (2), or $1s^2, 2s^22p^6, 3s^23p^6, 4s^2, 3d^{10}, 4p^6,$
$5s^2, 4d^{10}, 5p^6, 6s^2$ (2)

8. (a) 5513°F; (b) 2.257×10^4 kg/m³; (c) 1.408×10^3 lb/ft³

9. $1s^2, 2s^22p^6, 3s^23p^63d^{10}, 4s^24p^64d^{10}4f^{14}, 5s^25p^65d^6, 6s^2$ (Place one electron in
5d and then fill 4f to 14, returning again to the 5d to add 5 more electrons,
giving a total of 6 electrons in the 5d.) Or: $1s^2, 2s^22p^6, 3s^23p^6, 4s^2, 3d^{10}, 4p^6,$
$5s^2, 4d^{10}, 5p^6, 6s^2, 4f^{14}, 5d^6$

10. $^{12}_{6}$C: $1s^2, 2s^22p^2$; 4 valence electrons; $^{28}_{14}$Si: $1s^2, 2s^22p^6, 3s^23p^2$; 4 valence
electrons. Elements in the same column have similar valence electron config-
urations.

11. $1.00 \, \cancel{lb} \times \dfrac{454 \, \cancel{g}}{1 \, \cancel{lb}} \times \dfrac{1 \text{ proton}}{1.67 \times 10^{-24} \, \cancel{g}} = 2.72 \times 10^{26}$ protons

Chapter 4 Quiz

1. (a) $\boxed{\begin{array}{c}9p\\10n\end{array}}$ $2e^-$ $7e^-$ number of electrons
                         1   2   principal energy level number

(b) $\begin{pmatrix} 17p \\ 20n \end{pmatrix}$ $2e^-$ $8e^-$ $7e^-$ number of electrons

   1   2   3   principal energy level number

2. 63.55 amu

3. (a) 18 electrons; (b) 72 electrons

4. (a) $\cdot\dot{C}\cdot$; (b) $:\ddot{Ar}:$

5. (a) $1s^2$, $2s^22p^6$, $3s^23p^6$, $4s^1$ (1 valence electron)
   (b) $1s^2$, $2s^22p^6$, $3s^23p^63d^{10}$, $4s^24p^1$ (3 valence electrons).
   Or: $1s^2$, $2s^22p^6$, $3s^23p^6$, $4s^2$, $3d^{10}$, $4p^1$.

Chapter 5

1. (a), (b) metal; (c) metalloid; (d) nonmetal

2. (a) nonmetal; (b), (d) metal; (c) metalloid

3. (a) 1; (b) 4; (c) 6; (d) 8

4. (a) 8; (b) 7; (c) 3; (d) 5

5. (a) and (c); (b) and (d)

6. (a) and (d); (b) and (c)

7. (a) arsenic; (b) cesium; (c) aluminum; (d) lead

8. (a) barium; (b) magnesium; (c) lead (d) polonium

9. (a) barium; (b) silver; (c) phosphorus; (d) lead

10. (a) selenium; (b) arsenic; (c) indium; (d) rubidium

11. (a) metalloid; (b) 4; (c) Ge: $1s^2$, $2s^22p^6$, $3s^23p^63d^{10}$, $4s^24p^2$ or $1s^2$, $2s^22p^6$, $3s^23p^6$, $4s^2$, $3d^{10}$, $4p^2$; Si: $1s^2$, $2s^22p^6$, $3s^23p^2$; (d) more metallic; (e) 5.5×10^3 kg/m^3, 5.3×10^3 kg/m^3

12. 291°F, 417 K

13. 1.14×10^{-8} cm

Chapter 5 Quiz

1. (a) nonmetal; (b) metal; (c) metalloid; (d) metal

2. (a) 1; (b) 3; (c) 5; (d) 7

3. (a) and (d); (b) and (c)

4. (a) rubidium; (b) tellurium

5. (a) astatine; (b) tungsten

Chapter 6

1. (a) $+1 + $ ox no N $+ 3(-2) = 0$

   $+1 + $ ox no N $- 6 = 0$

   ox no N $- 5 = 0$

   ox no N $= +5$ or 5^+

(b) $+5$ or 5^+

(c) $+1 + $ ox no Cl $+ (-2) = 0$

   $+1 + $ ox no Cl $- 2 = 0$

   ox no Cl $- 1 = 0$

   ox no Cl $= +1$ or 1^+

(d) $+4$ or 4^+

(e) $+3$ or 3^+

(f) ox no I $+ 4(-2) = -1$

   ox no I $- 8 = -1$

   ox no $= 8 - 1 = +7$ or 7^+

(g) $+4$ or 4^+

(h) $+5$ or 5^+

(i) $+6$ or 6^+

(j) $+5$ or 5^+

2. (a) H $\overset{\cdot\cdot}{\underset{\cdot\cdot}{\text{F}}}{}^{\times}$, H—F;

(b)

$$\overset{\times\times}{\underset{\times\times}{\text{Cl}}}{}^{\times}_{\times}$$

$\overset{\times\times}{\underset{\times\times}{\text{Cl}}}{}^{\times}_{\times}\text{C}\,\overset{\times\times}{\underset{\times\times}{\text{Cl}}}{}^{\times}_{\times}$,

$$
\begin{array}{c}
\text{Cl} \\
| \\
\text{Cl}-\text{C}-\text{Cl} \\
| \\
\text{Cl}
\end{array}
$$

(c) $\overset{\times\times}{\text{N}}{}^{\times\times}_{\times\times}{\overset{\cdot\cdot}{\text{N}}}$, N≡N

(d) $\text{H}\overset{\times}{\underset{\times}{\text{C}}}{}^{\times}\overset{\cdot\cdot}{\underset{\cdot\cdot}{\text{O}}}\overset{\cdot}{\underset{\cdot}{\text{C}}}{}^{\cdot}\text{H}$,

$$
\begin{array}{c}
\text{H}\quad\text{H} \\
|\quad\ | \\
\text{H}-\text{C}=\text{C}-\text{H}
\end{array}
$$

3. (a) $-\overset{\cdot\cdot}{\underset{\otimes}{\text{S}}}{}^{\times}_{\times}\text{H}$, $[\text{S}-\text{H}]^{1-}$

(b) $-\overset{\cdot\cdot}{\underset{\otimes}{\text{O}}}{}^{\times}\overset{\times\times}{\underset{\times\times}{\text{S}}}\overset{\cdot\cdot}{\underset{\otimes}{\text{O}}}-$, $\begin{array}{c}:\overset{\cdot\cdot}{\text{O}}:\end{array}$

$$
\begin{bmatrix}
\text{O}-\text{S}-\text{O} \\
| \\
\text{O}
\end{bmatrix}^{2-}
$$

(c) $-\overset{\cdot\cdot}{\underset{\otimes}{\text{O}}}{}^{\times}\overset{\times\times}{\text{N}}{}^{\times}_{\times}\overset{\cdot\cdot}{\text{O}}$, $[\text{O}-\text{N}=\text{O}]^{1-}$

(d) $\overset{\times}{\underset{\times\times}{\text{O}}}{}^{\times}\overset{\cdot\cdot}{\underset{\times\times}{\text{Cl}}}{}^{\times}\overset{\times}{\underset{\times\times}{\text{O}}}{}^{\times}_{\otimes}-$, $\overset{\times}{\underset{\times\times}{\text{O}}}{}^{\times}$

$$
\begin{bmatrix}
\text{O}-\text{Cl}-\text{O} \\
| \\
\text{O}
\end{bmatrix}^{1-}
$$

4. (a) LiBr; (b) HgI_2; (c) Mg_3N_2; (d) $FeCl_3$; (e) CdO; (f) Ca_3P_2;

(g) LiH; (h) $Ba(NO_3)_2$; (i) $Al(ClO_4)_3$; (j) $Ba(MnO_4)_2$

5. (a) 2^+; (b) 6^+ and 2^-; (c) 4^+; (d) 7^+ and 1^-

6. (a) CaI_2, (b) Al_2O_3; (c) Mg_3As_2; (d) Na_2Te

7. (a) 2.16 Å + (2.16 − 2.02) Å = 2.30 Å (actual value is 2.35 Å)
(b) 2.60 g/mL + (2.60 − 1.54) g/mL = 3.66 g/mL (actual value is 3.74 g/mL)

8. (a) $Mg_3(AsO_4)_2$; (b) Al_2S_3; (c) Na_2SeO_4; (d) $Ba_3(AsO_4)_2$

9. (a), (c), and (d) ionic; (b) covalent

10. (a) IVA(14); (b) 4; (c) more metallic; (d) Pb

11. (a) IA(1); (b) 1; (c) 1^+; (d) Fr

12. H_3Q

Chapter 6 Quiz

1. (a) $+3$ or 3^+; (b) $+6$ or 6^+

2. (a)

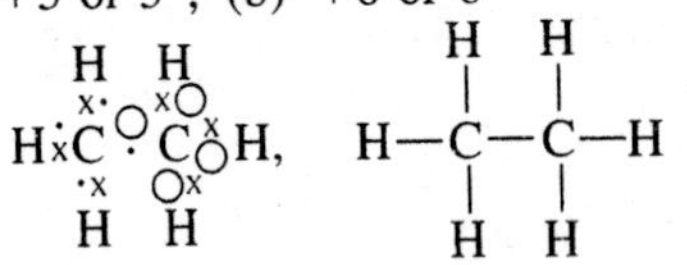

(b)

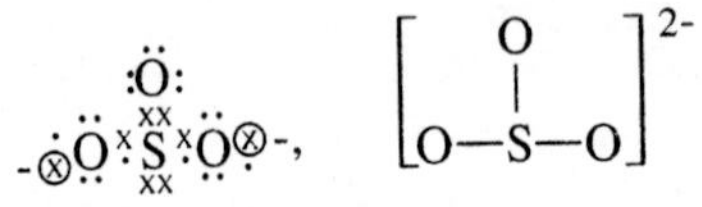

3. (a) K_3N; (b) FeI_3; (c) $Al_2(SO_4)_3$; (d) $Pb(CrO_4)_2$

4. (a) CaSe; (b) Al_2Te_3

5. (a) covalent; (b), (c), and (d) ionic

Chapter 7

1. (a) sulfur trioxide; (b) phosphorus tribromide; (c) silicon dioxide; (d) dinitrogen pentasulfide

2. (a) CO_2; (b) SO_2; (c) Cl_2O; (d) S_2F_{10}

3. (a) potassium iodide; (b) aluminum oxide; (c) strontium chloride; (d) silver bromide

4. (a) K_2O; (b) CaS; (c) $BiCl_3$; (d) Mg_3N_2

5. (a) tin(IV) iodide or stannic iodide; (b) lead(II) sulfide or plumbous sulfide; (c) iron(III) sulfide or ferric sulfide; (d) copper(I) bromide or cuprous bromide.

6. (a) $FeCl_2$; (b) SnS; (c) SnO_2; (d) Au_2O_3

7. (a) copper(I) cyanide or cuprous cyanide; (b) calcium hydrogen sulfate or calcium bisulfate; (c) potassium permanganate; (d) lithium sulfite

8. (a) Ag_3PO_4; (b) Cu_2CO_3; (c) $Zn(HCO_3)_2$; (d) $Al_2(SO_4)_3$

9. (a) sodium bromite; (b) potassium hypochlorite; (c) potassium chlorate; (d) calcium iodate

10. (a) $Cd(IO_2)_2$; (b) $Hg(ClO_3)_2$; (c) $Cu(ClO_4)_2$; (d) $Ca(ClO)_2$

11. (a) hydrogen bromide, hydrobromic acid; (b) hydrogen perchlorate, perchloric acid; (c) hydrogen permanganate, permanganic acid; (d) hydrogen cyanide, hydrocyanic acid

12. (a) HNO_3; (b) HIO_3; (c) H_2SO_4; (d) $H_2Cr_2O_7$

13. (a) bismuth hydroxide; (b) potassium hydroxide; (c) lead(II) hydroxide or plumbous hydroxide; (d) zinc hydroxide

14. (a) $Cd(OH)_2$; (b) $Ca(OH)_2$; (c) $Fe(OH)_3$; (d) $AgOH$

15. (a) (3), salt; (b) (1), acid; (c) (3), salt; (d) (2), base

16. (a) $Sn_3(PO_4)_2$; (b) $Ba(MnO_4)_2$; (c) $Ca(IO)_2$; (d) PCl_3; (e) $Mg(NO_2)_2$; (f) $Fe_3(PO_4)_2$

17. (a) iron(II) hydroxide or ferrous hydroxide; (b) diboron trioxide; (c) sodium oxalate; (d) chloric acid; (e) magnesium hydrogen carbonate or magnesium bicarbonate; (f) calcium cyanide

18. (a) CaF_2; (b) Ca^{2+}: $1s^2, 2s^22p^6, 3s^23p^6$; F^{1-}: $1s^2, 2s^22p^6$; (c) (3), salt

Chapter 7 Quiz

1. (a) lithium bromide
   (c) lead(IV) sulfide or plumbic sulfide
   (e) diphosphorus pentasulfide
   (g) tin(II) bromate or stannous bromate

   (b) aluminum iodide
   (d) tin(II) sulfate or stannous sulfate
   (f) calcium chlorite
   (h) ammonium nitrate

2. (a) BaO
   (c) Fe_2S_3
   (e) H_3PO_4
   (g) $SnCrO_4$

   (b) Ca_3P_2
   (d) SnF_2
   (f) $Al(OH)_3$
   (h) CaC_2O_4

3. (a) (3), salt
   (c) (1), acid

   (b) (1), acid
   (d) (2), base

Chapter 8

1. (a) 2×12.0 amu $= 24.0$ amu
       6×1.0 amu $= 6.0$ amu
       2×16.0 amu $= \underline{32.0 \text{ amu}}$
                           62.0 amu

   (b) 44.0 amu

(c) 1×40.1 amu $= 40.1$ amu   (d) 261.3 amu
    2×16.0 amu $= 32.0$ amu
    2×1.0 amu $= \underline{2.0 \text{ amu}}$
    $\phantom{2 \times 16.0 \text{ amu} = } 74.1$ amu

2. (a) $87.5 \text{ g } \cancel{C_2H_6O_2} \times \dfrac{1 \text{ mol } C_2H_6O_2}{62.0 \text{ g } \cancel{C_2H_6O_2}} = 1.41 \text{ mol } C_2H_6O_2$

(b) 0.827 mol

(c) $1.50 \times 10^{23} \text{ molecules } \cancel{C_2H_6O_2} \times \dfrac{1 \text{ mol } C_2H_6O_2}{6.02 \times 10^{23} \text{ molecules } \cancel{C_2H_6O_2}}$
$$= 0.249 \text{ mol } C_2H_6O_2$$

(d) 16.3 mol

3. (a) $1.25 \text{ mol } \cancel{CO_2} \times \dfrac{44.0 \text{ g } CO_2}{1 \text{ mol } \cancel{CO_2}} = 55.0 \text{ g } CO_2$

(b) $1.25 \text{ mol } \cancel{CO_2} \times \dfrac{6.02 \times 10^{23} \text{ molecules } CO_2}{1 \text{ mol } \cancel{CO_2}}$
$$= 7.52 \times 10^{23} \text{ molecules } CO_2$$

4. (a) $75.0 \text{ g } \cancel{C_2H_6O_2} \times \dfrac{1 \text{ mol } \cancel{C_2H_6O_2}}{62.0 \text{ g } \cancel{C_2H_6O_2}} \times \dfrac{6.02 \times 10^{23} \text{ molecules}}{1 \text{ mol } \cancel{C_2H_6O_2}}$
$$= 7.28 \times 10^{23} \text{ molecules}$$

(b) 1.14×10^{24} molecules

5. (a) $2.45 \text{ mol } \cancel{C_2H_6O_2} \times \dfrac{2 \text{ mol O atoms}}{1 \text{ mol } \cancel{C_2H_6O_2}} = 4.90 \text{ mol O atoms}$

(b) $2.45 \text{ mol } \cancel{C_2H_6O_2} \times \dfrac{2 \text{ mol O atoms}}{1 \text{ mol } \cancel{C_2H_6O_2}} \times \dfrac{16.0 \text{ g O}}{1 \text{ mol O atoms}} = 78.4 \text{ g O}$

6. (a) $12.6 \text{ L STP} \times \dfrac{1 \text{ mol } CO_2}{22.4 \text{ L STP}} = 0.562 \text{ mol } CO_2$

(b) $12.6 \text{ L STP} \times \dfrac{1 \text{ mol } \cancel{CO_2}}{22.4 \text{ L STP}} \times \dfrac{44.0 \text{ g } CO_2}{1 \text{ mol } \cancel{CO_2}} = 24.8 \text{ g } CO_2$

7. (a) $7.00 \text{ g } \cancel{N_2} \times \dfrac{1 \text{ mol } \cancel{N_2}}{28.0 \text{ g } \cancel{N_2}} \times \dfrac{22.4 \text{ L } N_2 \text{ STP}}{1 \text{ mol } \cancel{N_2}} = 5.60 \text{ L } N_2 \text{ STP}$

(b) 0.112 L STP

8. (a) $\dfrac{0.572 \text{ g}}{3.20 \text{ L STP}} \times \dfrac{22.4 \text{ L STP}}{1 \text{ mol}} = 4.00 \text{ g/mol, 4.00 amu}$

(b) 28.0 amu

9. (a) $\dfrac{44.0 \text{ g } CO_2}{1 \text{ mol } \cancel{CO_2}} \times \dfrac{1 \text{ mol } \cancel{CO_2}}{22.4 \text{ L STP}} = 1.96 \text{ g/L STP}$

(b) 0.714 g/L STP

10. (a) $1 \times 40.1 = 40.1$ amu $\dfrac{40.1 \text{ amu}}{200 \text{ amu}} \times 100 = 20.0\% \text{ Ca}$

$2 \times 79.9 = 16\overline{0}$ amu $\dfrac{16\overline{0} \text{ amu}}{200 \text{ amu}} \times 100 = 80.0\% \text{ Br}$
FM $CaBr_2 = 200$ amu

(b) 19.4% Ca, 34.3% Cl, 46.4% O

11. (a) $0.500 \text{ g} + 0.400 \text{ g} = 0.900 \text{ g oxide}$

$$\frac{0.500 \text{ g metal}}{0.900 \text{ g oxide}} \times 100 = 55.6\% \text{ metal}$$

(b) 57.9%

12. (a) $48.0 \text{ g Zn} \times \dfrac{1 \text{ mol Zn}}{65.4 \text{ g Zn}} = 0.734 \text{ mol Zn}$

$52.0 \text{ g Cl} \times \dfrac{1 \text{ mol Cl}}{35.5 \text{ g Cl}} = 1.46 \text{ mol Cl}$

$\dfrac{0.734}{0.734} = 1 \text{ Zn}; \dfrac{1.46}{0.734} = \text{approximately 2 Cl: } ZnCl_2$

(b) $FeBr_3$

13. (a) $80.0 \text{ g C} \times \dfrac{1 \text{ mol C atoms}}{12.0 \text{ g C}} = 6.67 \text{ mol C atoms}$

$20.0 \text{ g H} \times \dfrac{1 \text{ mol H atoms}}{1.0 \text{ g H}} = 2\bar{0} \text{ mol H atoms}$

$\dfrac{6.67}{6.67} = 1 \text{ C}; \dfrac{2\bar{0}}{6.67} = 3 \text{ H: } CH_3\text{—empirical formula}$

Molecular formula:

$$\begin{array}{l} 1 \times 12.0 = \\ 3 \times\ \ 1.0 = \end{array} \dfrac{\begin{array}{l} 12.0 \text{ amu} \\ 3.0 \text{ amu} \end{array}}{15.0 \text{ amu}} \quad \dfrac{30.0 \text{ amu}}{15.0 \text{ amu}} = 2$$

$$(CH_3)_2 = C_2H_6$$

(b) C_6H_{14} (*Hint:* In the calculations for the empirical formula, the ratio of C to H is calculated as 1 to 2.33, respectively.)

14. C_3H_6

15. $\dfrac{10\bar{0} \text{ mg}}{10\bar{0} \text{ mL blood}} \times \dfrac{1 \text{ g}}{1000 \text{ mg}} \times \dfrac{1 \text{ mol}}{46.0 \text{ g}} \times \dfrac{6.02 \times 10^{23} \text{ molecules}}{1 \text{ mol}} =$

$1.31 \times 10^{19} \text{ molecules/mL blood}$

16. 259 mL

17. 8.40×10^{19} molecules

18. 6.52 g/L (STP)

Chapter 8 Quiz

1. 3.27×10^{23} molecules

2. 20.4 amu

3. 12.1% C; 16.2% O; 71.7% Cl

4. $C_{10}H_{12}NO$

5. $C_3H_6O_3$

Chapter 9

(The numbers represent the coefficients in front of the formulas in the balanced equation.)

1. (a) $1 + 1 \rightarrow 1 + 2$
   (b) $2 \rightarrow 2 + 3$
   (c) $1 + 1 \rightarrow 1 + 2$
   (d) $2 + 3 \rightarrow 1 + 6$
   (e) $2 + 2 \rightarrow 2 + 1$
   (f) $3 + 1 \rightarrow 1$
   (g) $4 + 3 \rightarrow 2$
   (h) $2 + 1 \rightarrow 2 + 1$

2. (a) $2Fe + 3Cl_2 \rightarrow 2FeCl_3$
   (b) $2KNO_3 \xrightarrow{\Delta} 2KNO_2 + O_2$
   (c) $Ca + 2H_2O \rightarrow Ca(OH)_2 + H_2$
   (d) $NaOH + H_2SO_4 \rightarrow NaHSO_4 + H_2O$

3. (a) $2 + 1 \rightarrow 2\ CaO$
   (b) $2 + 3 \rightarrow 2SO_3$
   (c) $1 + 1 \rightarrow Ca(OH)_2$
   (d) $2 + 3 \rightarrow 2AlCl_3$

4. (a) $2 \rightarrow 2Hg + O_2$
   (b) $2 \rightarrow 2H_2 + O_2$
   (c) $1 \rightarrow SrO + CO_2$
   (d) $1 \rightarrow CaSO_4 + 2H_2O$

5. (a) $1 + 1 \rightarrow ZnCl_2 + Ni$
   (b) $1 + 2 \rightarrow PbCl_2 + H_2$
   (c) $2 + 2 \rightarrow 2NaOH + H_2$
   (d) $1 + 2 \rightarrow 2NaCl + Br_2$

6. (a) $1 + 2 \rightarrow PbCl_{2(s)} + 2HNO_3$
   (b) $1 + 3 \rightarrow Bi(OH)_{3(s)} + 3NaNO_3$
   (c) $1 + 1 \rightarrow PbSO_{4(s)} + 2KC_2H_3O_2$
   (d) $1 + 2 \rightarrow CaCl_2 + H_2O + CO_{2(g)}$

7. (a) $1 + 2 \rightarrow Ca(C_2H_3O_2)_2 + 2H_2O$
   (b) $1 + 2 \rightarrow 2FePO_{4(s)} + 3H_2O$
   (c) $1 + 2 \rightarrow K_2CO_3 + H_2O$ also $1 + 1 \rightarrow KHCO_3$
   (d) $1 + 2 \rightarrow BaCl_2 + H_2O$

8. (a) $FeS + 2HCl \rightarrow FeCl_2 + H_2S_{(g)}$
   (b) $CaCO_3 + H_2SO_4 \rightarrow CaSO_{4(s)} + H_2O + CO_{2(g)}$
                       or slightly soluble
   (c) $CdO + 2HCl \rightarrow CdCl_2 + H_2O$
   (d) $2Al + 3PbCl_2 \rightarrow 2AlCl_3 + 3Pb_{(s)}$

   9. $CaCO_3 + 2HCl \rightarrow CaCl_2 + CO_2 + H_2O$

   10. $TiCl_4 + 2BCl_3 + 5H_2 \rightarrow TiB_2 + 10HCl$

   11. $Mg_3N_2 + 6H_2O \rightarrow 2NH_3 + 3Mg(OH)_2$

Chapter 9 Quiz

1. (a) $3 + 4 \rightarrow 2 + 3$
   (b) $2 + 13 \rightarrow 8 + 10$

2. $HC_2H_3O_2 + NaHCO_3 \rightarrow NaC_2H_3O_2 + CO_2 + H_2O$

3. (a) $2 + 1 \rightarrow 2BaO$
   (b) $2 \rightarrow 2NaNO_2 + O_2$
   (c) $1 \rightarrow BaCl_2 + 2H_2O$
   (d) $1 + 2 \rightarrow I_2 + 2NaCl$
   (e) $2 + 3 \rightarrow 2AlCl_3 + 3Sn$
   (f) $2 + 1 \rightarrow Ag_2S_{(s)} \rightarrow 2HNO_3$

Chapter 10

1. (a) $HgCl_{2(aq)} + H_2S_{(aq)} \rightarrow HgS_{(s)} + 2H^{1+} + 2Cl^{1-}$
   Net: same as above
   (b) $Mg^{2+}_{(aq)} + SO_4^{2-}_{(aq)} + 2Na^{1+}_{(aq)} + 2OH^{1-}_{(aq)} \rightarrow$
   $$Mg(OH)_{2(s)} + 2Na^{1+} + SO_4^{2-}$$
   Net: $Mg^{2+}_{(aq)} + 2OH^{1-}_{(aq)} \rightarrow Mg(OH)_{2(s)}$
   (c) $CaO_{(s)} + 2H^{1+}_{(aq)} + 2Cl^{1-}_{(aq)} \rightarrow Ca^{2+} + 2Cl^{1-} + H_2O$
   Net: $CaO_{(s)} + 2H^{1+}_{(aq)} \rightarrow Ca^{2+}_{(aq)} + H_2O$
   (d) $Al(OH)_{3(s)} + 3H^{1+}_{(aq)} + 3Cl^{1-}_{(aq)} \rightarrow Al^{3+} + 3Cl^{1-} + 3H_2O$
   Net: $Al(OH)_{3(s)} + 3H^{1+}_{(aq)} \rightarrow Al^{3+} + 3H_2O$
   (e) $Fe^{2+}_{(aq)} + SO_4^{2-}_{(aq)} + 2NH_4^{1+}_{(aq)} + S^{2-}_{(aq)} \rightarrow FeS_{(s)} + 2NH_4^{1+} + SO_4^{2-}$
   Net: $Fe^{2+}_{(aq)} + S^{2-}_{(aq)} \rightarrow FeS_{(s)}$

2. (a) $Mg^{2+}_{(aq)} + 2Cl^{1-}_{(aq)} + 2Na^{1+}_{(aq)} + CO_3^{2-}_{(aq)} \rightarrow$
   $$MgCO_{3(s)} + 2Na^{1+} + 2Cl^{1-}$$
   Net: $Mg^{2+}_{(aq)} + CO_3^{2-}_{(aq)} \rightarrow MgCO_{3(s)}$
   (b) $2H^{1+}_{(aq)} + 2Cl^{1-}_{(aq)} + Ba^{2+}_{(aq)} + 2OH^{1-}_{(aq)} \rightarrow 2H_2O + Ba^{2+} + 2Cl^{1-}$
   Net: $H^{1+}_{(aq)} + OH^{1-}_{(aq)} \rightarrow H_2O$
   (c) $Pb^{2+}_{(aq)} + 2NO_3^{1-}_{(aq)} + 2K^{1+}_{(aq)} + 2OH^{1-}_{(aq)} \rightarrow$
   $$Pb(OH)_{2(s)} + 2K^{1+} + 2NO_3^{1-}$$
   Net: $Pb^{2+}_{(aq)} + 2OH^{1-}_{(aq)} \rightarrow Ph(OH)_{2(s)}$
   (d) $Ca_{(s)} + 2H^{1+}_{(aq)} + 2Br^{1-}_{(aq)} \rightarrow Ca^{2+} + 2Br^{1-} + H_{2(g)}$
   Net: $Ca_{(s)} + 2H^{1+}_{(aq)} \rightarrow Ca^{2+} + H_{2(g)}$
   (e) $2Ag^{1+}_{(aq)} + 2NO_3^{1-}_{(aq)} + 2Na^{1+}_{(aq)} + CO_3^{2-}_{(aq)} \rightarrow$
   $$Ag_2CO_{3(s)} + 2Na^{1+} + 2NO_3^{1-}$$
   Net: $2Ag^{1+}_{(aq)} + CO_3^{2-}_{(aq)} \rightarrow Ag_2CO_{3(s)}$

3. (1) Molecular: $CaCO_3 + H_2SO_4 \rightarrow CaSO_4 + H_2O + CO_2$

Total: $CaCO_3 + 2H^{1+} + SO_4^{2-} \rightarrow Ca^{2+} + SO_4^{2-} + H_2O + CO_2$

Net: $CaCO_3 + 2H^{1+} \rightarrow Ca^{2+} + H_2O + CO_2$

(2) Molecular: $SO_3 + H_2O \rightarrow H_2SO_4$

Total and net: $SO_3 + H_2O \rightarrow 2H^{1+} + SO_4^{2-}$

Chapter 10 Quiz

1. (a) $Cd^{2+}_{(aq)} + 2Cl^{1-}_{(aq)} + H_2S_{(aq)} \rightarrow CdS_{(s)} + 2H^{1+} + 2Cl^{1-}$

Net: $Cd^{2+}_{(aq)} + H_2S_{(aq)} \rightarrow CdS_{(s)} + 2H^{1+}$

(b) $Fe^{3+}_{(aq)} + 3NO_3^{1-}_{(aq)} + 3Na^{1+}_{(aq)} + 3OH^{1-}_{(aq)} \rightarrow$
$$Fe(OH)_{3(s)} + 3Na^{1+} + 3NO_3^{1-}$$

Net: $Fe^{3+}_{(aq)} + 3OH^{1-}_{(aq)} \rightarrow Fe(OH)_{3(s)}$

(c) $Mg_{(s)} + 2H^{1+}_{(aq)} + 2Cl^{1-}_{(aq)} \rightarrow Mg^{2+} + 2Cl^{1-} + H_{2(g)}$

Net: $Mg_{(s)} + 2H^{1+}_{(aq)} \rightarrow Mg^{2+} + H_{2(g)}$

2. $S^{2-}_{(aq)} + 2H^{1+}_{(aq)} \rightarrow H_2S_{(g)}$

Chapter 11

1. $30.0 \text{ g Zn} \times \dfrac{1 \text{ mol Zn}}{65.4 \text{ g Zn}} \times \dfrac{2 \text{ mol HCl}}{1 \text{ mol Zn}} \times \dfrac{36.5 \text{ g HCl}}{1 \text{ mol HCl}} = 33.5 \text{ g HCl}$

2. 0.933 g

3. 3.09 g

4. $94.0 \text{ g Ca}_3(PO_4)_2 \times \dfrac{1 \text{ mol Ca}_3(PO_4)_2}{310 \text{ g Ca}_3(PO_4)_2} \times \dfrac{3 \text{ mol CaCl}_2}{1 \text{ mol Ca}_3(PO_4)_2}$
$$= 0.910 \text{ mol CaCl}_2$$

5. 0.500 mol

6. 27.8 g

7. 1.206 g

8. Step I: $1.65 \text{ g N}_2 \times \dfrac{1 \text{ mol N}_2}{28.0 \text{ g N}_2} = 0.0589 \text{ mol N}_2$

$1.40 \text{ g H}_2 \times \dfrac{1 \text{ mol H}_2}{2.0 \text{ g H}_2} = 0.700 \text{ mol H}_2$

Step II: $0.0589 \text{ mol N}_2 \times \dfrac{2 \text{ mol NH}_3}{1 \text{ mol N}_2} = 0.118 \text{ mol NH}_3$

$0.700 \text{ mol H}_2 \times \dfrac{2 \text{ mol NH}_3}{3 \text{ mol H}_2} = 0.467 \text{ mol NH}_3$

Step III: $0.118 \text{ mol NH}_3 \times \dfrac{17.0 \text{ g NH}_3}{1 \text{ mol NH}_3} = 2.01 \text{ g NH}_3$

9. $\dfrac{13.7 \text{ g} = \text{actual yield}}{18.7 \text{ g} = \text{theoretical yield}} \times 100 = 73.3\%$

10. $42.0 \; \text{g FeS} \times \dfrac{1 \; \text{mol FeS}}{87.9 \; \text{g FeS}} \times \dfrac{1 \; \text{mol H}_2\text{S}}{1 \; \text{mol FeS}} \times \dfrac{22.4 \; \text{L H}_2\text{S(STP)}}{1 \; \text{mol H}_2\text{S}}$

$$= 10.7 \; \text{L H}_2\text{S(STP)}$$

11. 19.5 L

12. 7.95 L

13. $4.00 \; \text{L NH}_{3(g)} \times \dfrac{1 \; \text{L N}_{2(g)}}{2 \; \text{L NH}_{3(g)}} = 2.00 \; \text{L N}_2$

14. 32.7 L

15. 152 L

16. (a) exothermic; (b) 1.25×10^2 kcal (*Hint*: kilo = 1000 units; see Table 2–1.)

17. (a) endothermic; (b) $2.09 \; \text{kcal} \times \dfrac{2 \; \text{mol F}_2}{11.0 \; \text{kcal}} \times \dfrac{38.0 \; \text{g F}_2}{1 \; \text{mol F}_2} = 14.4 \; \text{g F}_2$

18. (a) exothermic; (b) 1.01×10^5 J

19. (a) $CH_{4(g)} + 2O_{2(g)} \xrightarrow{\Delta} CO_{2(g)} + 2H_2O_{(g)}$

(b) $8.00 \; \text{g CH}_4 \times \dfrac{1 \; \text{mol CH}_4}{16.0 \; \text{g CH}_4} \times \dfrac{4 \; \text{mol H atoms}}{1 \; \text{mol CH}_4}$

$$= 2.00 \; \text{mol H atoms}$$

(c) 12.5 mol O_2

(d) 32.0 g O_2

(e) 16.8 L CO_2 (STP)

(f) 11.2 L O_2

(g) 48.4 g CO_2

(h) 84.1%

20. $30.0 \; \text{g Fe} \times \dfrac{1 \; \text{mol Fe}}{55.8 \; \text{g Fe}} \times \dfrac{2 \; \text{mol HCl}}{1 \; \text{mol Fe}} \times \dfrac{36.5 \; \text{g HCl}}{1 \; \text{mol HCl}}$

$\times \dfrac{100 \; \text{g HCl soln}}{35.0 \; \text{g HCl}} \times \dfrac{1 \; \text{mL HCl soln}}{1.18 \; \text{g HCl soln}}$

$$= 95.0 \; \text{mL of 35.0\% HCl soln}$$

21. (a) 38.2 g; (b) 89.0%

Chapter 11 Quiz

1. 78.6 g

2. 1.15 g

3. 0.411 L

4. 3.33 L

5. 450 cal

Chapter 12

1. $V_{new} = 4\overline{00} \text{ mL} \times \dfrac{3\overline{00} \text{ torr}}{195 \text{ torr}} = 615 \text{ mL}$

2. $25\overline{0}$ torr

3. $V_{new} = 1\overline{00} \text{ mL} \times \dfrac{278 \text{ K}}{3\overline{00} \text{ K}} = 92.7 \text{ mL}$

4. $-33°C$

5. $P_{new} = 2\overline{00} \text{ torr} \times \dfrac{4\overline{00} \text{ K}}{3\overline{00} \text{ K}} = 267 \text{ torr}$

6. $89°C$

7. $V_{new} = 395 \text{ mL} \times \dfrac{60.0 \text{ torr}}{76\overline{0} \text{ torr}} \times \dfrac{273 \text{ K}}{3\overline{00} \text{ K}} = 28.4 \text{ mL}$

8. 1970 torr

9. $P_{dry\ gas} = 600.0 \text{ torr} - 23.8 \text{ torr} = 576.2 \text{ torr}$ (see Appendix 4)

$V_{new} = 155 \text{ mL} \times \dfrac{576.2 \text{ torr}}{76\overline{0} \text{ torr}} \times \dfrac{273 \text{ K}}{298 \text{ K}} = 108 \text{ mL}$

10. 232 mL

11. $V = \dfrac{0.0300 \text{ mol} \times 0.0821 \dfrac{\text{atm} \cdot \text{L}}{\text{mol} \cdot \text{K}} \times 303 \text{ K}}{1.20 \text{ atm}} = 0.622 \text{ L}$

$0.622 \text{ L} \times \dfrac{1000 \text{ mL}}{1 \text{ L}} = 622 \text{ mL}$

12. 0.213 mol

13. (a) 0.330 L

(b) $\dfrac{0.443 \text{ g}}{0.330 \text{ L}} \times \dfrac{22.4 \text{ L}}{1 \text{ mol}} = 30.1 \text{ g/mol, } 30.1 \text{ amu}$

(c) $80.0 \text{ g C} \times \dfrac{1 \text{ mol C atoms}}{12.0 \text{ g C}} = 6.67 \text{ mol C atoms}$

$20.0 \text{ g H} \times \dfrac{1 \text{ mol H atoms}}{1.0 \text{ g H}} = 2\overline{0} \text{ mol H atoms}$

for C: $\dfrac{6.67}{6.67} = 1$; for H: $\dfrac{2\overline{0}}{6.67} = 3$

CH_3 = empirical formula

$\begin{aligned} 1 \times 12.0 &= 12.0 \text{ amu} \\ 3 \times 1.0 &= \underline{3.0 \text{ amu}} \\ &\ \ \ 15.0 \text{ amu} \end{aligned}$ $\dfrac{30.0 \text{ amu}}{15.0 \text{ amu}} = 2$

$(CH_3)_2 = C_2H_6$ = molecular formula

14. (a) 60.2 mL

(b) 197 amu

(c) $C_2HBrClF_3$

15. (a) $2KClO_3 \xrightarrow{\Delta} 2KCl + 3O_2$

(b) $10.0 \text{ g } \cancel{KClO_3} \times \dfrac{1 \text{ mol } \cancel{KClO_3}}{122.6 \text{ g } \cancel{KClO_3}} \times \dfrac{3 \text{ mol } \cancel{O_2}}{2 \text{ mol } \cancel{KClO_3}} \times \dfrac{22.4 \text{ L } O_2 \text{STP}}{1 \text{ mol } \cancel{O_2}}$

$$= 2.74 \text{ L } O_2 \text{STP}$$

(c) $V_{\text{new}} = 2.74 \text{ L} \times \dfrac{76\bar{0} \text{ torr}}{63\bar{0} \text{ torr}} \times \dfrac{308 \text{ K}}{273 \text{ K}} = 3.73 \text{ L } O_2 \text{ at } 35°C \text{ and } 63\bar{0} \text{ torr}$

16. (a) $Mg + 2HCl \rightarrow H_{2(g)} + MgCl_2$

(b) 0.479 L

(c) 0.621 L

17. $n = \dfrac{1.10 \text{ atm} \times 10.0 \text{ L}}{0.0821 \dfrac{\text{atm} \cdot \text{L}}{\text{mol} \cdot \text{K}} \times 303 \text{ K}} = 0.422 \text{ mol}$

$\left(\text{Note the units. In division of fraction, you invert and multiply; therefore,}\right.$

$\left. \dfrac{1}{\dfrac{1}{\text{mol}}} = 1 \times \dfrac{\text{mol}}{1} = \text{mol}\right)$

$0.442 \text{ mol} \times 28.0 \dfrac{\text{g}}{\text{mol}} = 12.4 \text{ g } N_2$

Chapter 12 Quiz

1. 103 mL

2. 173 mL

3. 183 mL

4. 1.26 atm

5. 22.1 L

Chapter 13

1. (a) $\dfrac{8.60 \text{ g NaCl}}{98.0 \text{ g solution}} \times 100 = 8.78\%$

(b) 20.0%

2. (a) $35\bar{0} \text{ g } \cancel{H_2O} \times \dfrac{15.0 \text{ g } K_2SO_4}{85.0 \text{ g } \cancel{H_2O}} = 61.8 \text{ g } K_2SO_4$

(b) 2.22 g

3. (a) $\dfrac{175 \text{ g } \cancel{C_2H_6O}}{65\bar{0} \text{ g } \cancel{H_2O}} \times \dfrac{1 \text{ mol } C_2H_6O}{46.0 \text{ g } \cancel{C_2H_6O}} \times \dfrac{1000 \text{ g } \cancel{H_2O}}{1 \text{ kg } H_2O} = 5.85 \text{ } m$

(b) 3.57 m

4. (a) $0.500 \text{ mol } C_2H_6O \times \dfrac{46.0 \text{ g } C_2H_6O}{1 \text{ mol } C_2H_6O} = 23.0 \text{ g } C_2H_6O$

$100\overline{0} \text{ g } H_2O + 23 \text{ g } C_2H_6O = 1023 \text{ g solution}$

$35\overline{0} \text{ g solution} \times \dfrac{23.0 \text{ g } C_2H_6O}{1023 \text{ g solution}} = 7.87 \text{ g } C_2H_6O$

(b) 38.9 g

5. (a) $60.0 \text{ g } C_6H_{12}O_6 \times \dfrac{1 \text{ mol } C_6H_{12}O_6}{18\overline{0}.0 \text{ g } C_6H_{12}O_6} \times \dfrac{1.00 \text{ kg } H_2O}{2.00 \text{ mol } C_6H_{12}O_6}$

$\times \dfrac{1000 \text{ g } H_2O}{1 \text{ kg } H_2O} = 167 \text{ g } H_2O$

(b) 41.4 g

6. (a) $\dfrac{75.0 \text{ g } C_2H_6O}{45\overline{0} \text{ mL solution}} \times \dfrac{1 \text{ mol } C_2H_6O}{46.0 \text{ g } C_2H_6O} \times \dfrac{1000 \text{ mL solution}}{1 \text{ L solution}} = 3.62 \ M$

(b) 0.889 M

7. (a) $50\overline{0} \text{ mL solution} \times \dfrac{1 \text{ L solution}}{1000 \text{ mL solution}} \times \dfrac{0.100 \text{ mol NaOH}}{1 \text{ L solution}}$

$\times \dfrac{40.0 \text{ g NaOH}}{1 \text{ mol NaOH}} = 2.00 \text{ g NaOH}$

The sodium hydroxide (2.00 g) is dissolved in sufficient water to make the total volume of the solution equal to $50\overline{0}$ mL.

(b) 0.556 g

The calcium chloride (0.556 g) is dissolved in sufficient water to make the total volume of the solution equal to $25\overline{0}$ mL.

8. (a) $5.00 \text{ g NaBr} \times \dfrac{1 \text{ mol NaBr}}{102.9 \text{ g NaBr}} \times \dfrac{1 \text{ L solution}}{0.100 \text{ mol NaBr}}$

$\times \dfrac{1000 \text{ mL solution}}{1 \text{ L solution}} = 486 \text{ mL solution}$

(b) 49.2 mL

9. (a) $\dfrac{8.75 \text{ g NaOH}}{45\overline{0} \text{ mL solution}} \times \dfrac{1 \text{ eq NaOH}}{40.0 \text{ g NaOH}} \times \dfrac{1000 \text{ mL solution}}{1 \text{ L solution}} = 0.486 \ N$

(b) 0.0467 N

10. (a) $25\overline{0} \text{ mL solution} \times \dfrac{1 \text{ L solution}}{1000 \text{ mL solution}} \times \dfrac{0.0100 \text{ eq } H_2SO_4}{1 \text{ L solution}}$

$\times \dfrac{49.0 \text{ g } H_2SO_4}{1 \text{ eq } H_2SO_4} = 0.122 \text{ g } H_2SO_4$

(b) 3.53 g

11. (a) $60.0 \text{ g } H_2SO_4 \times \dfrac{1 \text{ eq } H_2SO_4}{49.0 \text{ g } H_2SO_4} \times \dfrac{1 \text{ L solution}}{4.00 \text{ eq } H_2SO_4}$

$\times \dfrac{1000 \text{ mL solution}}{1 \text{ L solution}} = 306 \text{ mL solution}$

(b) 674 mL

12. (a) $\dfrac{0.00570 \text{ mol } Ca(OH)_2}{1 \text{ L solution}} \times \dfrac{2 \text{ eq } Ca(OH)_2}{1 \text{ mol } Ca(OH)_2} = 0.0114 \ N$

(b) $1.89 \ M$

13. (a) $\dfrac{98.0 \text{ g } H_2SO_4}{2.0 \text{ g } H_2O} \times \dfrac{1 \text{ mol } H_2SO_4}{98.1 \text{ g } H_2SO_4} \times \dfrac{1000 \text{ g } H_2O}{1 \text{ kg } H_2O} = 5\overline{0}0 \ m$

(b) $\dfrac{98.0 \text{ g } H_2SO_4}{100 \text{ g soln}} \times \dfrac{1 \text{ mol } H_2SO_4}{98.1 \text{ g } H_2SO_4} \times \dfrac{1.83 \text{ g soln}}{1 \text{ mL soln}} \times \dfrac{1000 \text{ mL soln}}{1 \text{ L soln}}$

$$= 18.3 \ M$$

(c) $\dfrac{18.3 \text{ mol } H_2SO_4}{1 \text{ L soln}} \times \dfrac{2 \text{ eq } H_2SO_4}{1 \text{ mol } H_2SO_4} = 36.6 \ N$

14. (a) $16.1 \ m$; (b) $12.0 \ M$; (c) $12.0 \ N$

15. $5.11 \ m$

Chapter 13 Quiz

1. 0.924%

2. $28.9 \ g$

3. $0.140 \ M$

4. $3.10 \ N$

5. $29.2 \ N$

Chapter 14

1. (a) $8 + 1 + 10H^{1+} \rightarrow 1 + 4 + 4H_2O$
   SO_4^{2-}: reduced, oxidizing agent
   I^{1-}: oxidized, reducing agent

(b) $3 + 2 + 8H^{1+} \rightarrow 3 + 2 + 4H_2O$
   NO_3^{1-}: reduced, oxidizing agent
   Cu: oxidized, reducing agent

(c) $1 + 1 + 4H^{1+} \rightarrow 1 + 1 + 2H_2O$
   SO_4^{2-}: reduced, oxidizing agent
   Cu: oxidized, reducing agent

(d) $6 + 12OH^{1-} \rightarrow 2 + 10 + 6H_2O$
   Cl_2: reduced, oxidizing agent
   Cl_2: oxidized, reducing agent

2. $Ca(OH)_2 + 2HCl \rightarrow CaCl_2 + 2H_2O$

$$12.4 \text{ mL soln} \times \dfrac{1 \text{ L}}{1000 \text{ mL}} \times \dfrac{0.100 \text{ mol HCl}}{1 \text{ L soln}} \times \dfrac{1 \text{ mol } Ca(OH)_2}{2 \text{ mol HCl}}$$

$$= 0.000620 \text{ mol } Ca(OH)_2$$

$$\dfrac{0.000620 \text{ mol } Ca(OH)_2}{32.0 \text{ mL soln}} \times \dfrac{1000 \text{ mL}}{1 \text{ L}} = 0.0194 \ M$$

3. $Na_2CO_3 + 2HCl \rightarrow 2NaCl + CO_2 + H_2O$

$$0.625 \text{ g } Na_2CO_3 \times \frac{1 \text{ mol } Na_2CO_3}{106.0 \text{ g } Na_2CO_3} \times \frac{2 \text{ mol HCl}}{1 \text{ mol } Na_2CO_3}$$

$$= 0.0118 \text{ mol HCl}$$

$$\frac{0.0118 \text{ mol HCl}}{30.8 \text{ mL soln}} \times \frac{1000 \text{ mL}}{1 \text{ L}} = 0.383 \text{ } M$$

4. (a) $pH = -[(-8.0000)] = 8.0000, 8.00$
$pOH = 14.00 - 8.00 = 6.00$

(b) $11.70, 2.30$

(c) $pH = -[(-6.6990)] = 6.6990, 6.70$
$pOH = 14.00 - 6.70 = 7.30$

(d) $6.20, 7.80$

(e) $pH = -[(-8.2757)] = 8.2757, 8.28$
$pOH = 14.00 - 8.28 = 5.72$

5. (a) $3C_2H_6O + 2Cr_2O_7^{2-} + 16H^{1+} \rightarrow 3C_2H_4O_2 + 4Cr^{3+} + 11H_2O$

(b) $1.11 \text{ mg } Cr^{3+} \times \frac{1 \text{ g}}{1000 \text{ mg}} \times \frac{1 \text{ mol } Cr^{3+}}{52.0 \text{ g}} \times \frac{3 \text{ mol } C_2H_6O}{4 \text{ mol } Cr^{3+}}$

$$\times \frac{46.0 \text{ g } C_2H_6O}{1 \text{ mol } C_2H_6O} \times \frac{1000 \text{ mg}}{1 \text{ g}} = 0.736 \text{ mg } C_2H_6O$$

6. (a) $2NaOH + H_2SO_4 \rightarrow Na_2SO_4 + 2H_2O$

$$0.200 \text{ g } NaOH \times \frac{1 \text{ mol } NaOH}{40.0 \text{ g } NaOH} \times \frac{1 \text{ mol } H_2SO_4}{2 \text{ mol } NaOH}$$

$$= 0.00250 \text{ mol } H_2SO_4$$

$$\frac{0.00250 \text{ mol } H_2SO_4}{5.70 \text{ mL soln}} \times \frac{1000 \text{ mL}}{1 \text{ L}} = 0.439 \text{ } M$$

(b) $\dfrac{0.439 \text{ mol } H_2SO_4}{1 \text{ L soln}} \times \dfrac{2 \text{ eq. } H_2SO_4}{1 \text{ mol } H_2SO_4} = 0.878 \text{ } N$

7. (a) $2IO_3^{1-} + 5HSO_3^{1-} \rightarrow 5SO_4^{2-} + I_2 + 3H^{1+} + H_2O$

(b) $577 \text{ g } I_2$

Chapter 14 Quiz

1. $4 + 1 + 11H^{1+} \rightarrow 4 + 1 + 4H_2O$
AsO_4^{3-}: reduced, oxidizing agent
Zn: oxidized, reducing agent

2. $7.45 \text{ } M$

3. $1.70 \text{ to } 1.92$

Appendix 3

1. (a) ± 5.0 (b) $\pm 6.0 \times 10^{-2}$
(c) $\pm 1.41 \times 10^{-2}$ (d) $1, -8$

(e) $x = \dfrac{-(-1) \pm \sqrt{(-1)^2 - 4(6)(-15)}}{12}$

$x = \dfrac{+1 \pm \sqrt{1 + 360}}{12} = \dfrac{+1 \pm 19}{12}$ or $\dfrac{+1 + 19}{12}, \dfrac{+1 - 19}{12} = \dfrac{5}{3}, -\dfrac{3}{2}$

(f) $1.33 \times 10^{-3}, -1.35 \times 10^{-3}$

GLOSSARY

acid   A substance that yields hydrogen ions [$H^{1+}_{(aq)}$] when dissolved in water.

acid oxide   A nonmetal oxide.

actual yield   Amount of product that is actually obtained in a given reaction.

anion   Ion with a negative charge.

atom   The smallest particle of an element that can undergo chemical changes in a reaction.

atomic mass scale   Relative scale of atomic masses, based on an arbitrarily assigned value of exactly 12 atomic mass units (amu) for the *mass* of carbon-12.

atomic number   Number of protons found in the nucleus of an atom of a given element.

Avogadro's (a'vō·ga'drō) **number** (N)   The number of atoms in exactly 12 grams of carbon-12 (approximately 6.02×10^{23}); it is equivalent to 1 mol of a substance.

base   A substance that yields hydroxide ions (OH^{1-}) when dissolved in water.

basic oxide   A metal oxide.

bond angle   The angle defined by two bonds that connect three atoms in a molecule.

bond length   The distance between the nuclei of covalently bonded atoms.

Boyle's law   Principle that, at constant temperature, the volume of a fixed mass of a given gas is *inversely* proportional to the pressure it exerts.

calorie (cal)   A unit of measurement of heat energy. It is defined as the amount of heat required to raise the temperature of 1.00 g of water from 14.5°C to 15.5°C. A *kilocalorie* is equal to 1000 calories.

catalyst   A substance that speeds up a chemical reaction but is recovered relatively unchanged at the end of the reaction.

cation   Ion with a positive charge.

Charles' law   Principle that, at constant pressure, the volume of a fixed mass of a given gas is *directly* proportional to the temperature in *kelvins*.

chemical bonds   The attractive forces that hold atoms together as compounds

chemical change   Change in the substance that can be observed only when a change in the composition of the substance is occurring. *New substances are formed.*

chemical equation   A shorthand way of expressing a chemical change (reaction in terms of symbols and formulas.

chemical property   Property of a substance that can be observed when the substance undergoes a change in composition.

combination reaction   A process in which two or more substances (either elements or compounds) react to produce one substance:

$$A + Z \longrightarrow AZ, \qquad \text{where } A \text{ and } Z \text{ are elements or compounds}$$

compound   A pure substance that can be broken down by various chemical means into two or more different substances.

conservation of mass, law of   Mass can be neither created nor destroyed, and the total mass of the substance involved in a physical or chemical change remains constant.

coordinate covalent bond   A bond formed when *both* the electrons in an electron-pair bond are supplied by *one* atom.

covalent bond   A bond formed by the sharing of electrons between atoms.

Dalton's law of partial pressure   Principle that each gas in a mixture of gases exerts a partial pressure equal to the pressure it would exert if it were the only gas present in the same volume. The total pressure of the mixture is the sum of the partial pressures of all the gases present.

decomposition reaction   A process in which one substance undergoes a reaction to form two or more substances:

$$AZ \longrightarrow A + Z, \qquad \text{where } A \text{ and } Z \text{ are elements or compounds}$$

density   Mass of a substance occupying a unit volume:

$$\text{Density} = \frac{\text{mass}}{\text{volume}}$$

double bond   Chemical bond in which two atoms share two pairs of electrons (four electrons).

double-replacement reaction   A process in which two compounds undergo a reaction where the positive ion (cation) of one compound exchanges with the positive ion (cation) of the second compound:

$$AX + BZ \longrightarrow AZ + BX$$

electrolytes   Substances whose aqueous solution can conduct an electric current.

electrolytes, strong   Substances whose aqueous solutions conduct sufficient electric current to produce a *bright glow* in a standard light bulb because there are many ions present in the solution. These substances are soluble salts and strong acids and bases.

electrolytes, weak   Substances whose aqueous solutions conduct sufficient electric current to produce a *dull glow* in a standard light bulb because only a few ions exist in solution. These substances are slighly soluble salts and weak acids and bases.

electron   A particle that has a relative charge of -1 (actual charge $= -1.602 \times 10^{-19}$ coulomb) and a mass of 9.109×10^{-28} g or 5.486×10^{-4} amu (relatively negligible).

electronegativity   A measure of the tendency of an atom to attract a pair of electrons in a covalent bond.

element   A pure substance that cannot be broken down into simpler substances by ordinary chemical means. All of its atoms have the *same* atomic number.

empirical formula   The formula of a compound that contains the smallest whole number ratio of atoms present in a molecule or formula unit of a compound.

endothermic reaction   A reaction in which heat energy is absorbed.

equation, chemical   A shorthand way of expressing a chemical change (reaction) in terms of symbols and formulas.

equation, word   Expresses a chemical equation in words instead of symbols and formulas.

equivalence point   The point at which an acid or base is exactly neutralized in the titration process; it is also called the *end point*.

excess reagent   Reactant in a chemical reaction that is *not* completely used up in the reaction, so called because, when the last amount of the new compound is formed, some of this reactant will be left over.

exothermic reaction   A reaction in which heat energy is evolved.

exponent   A number or symbol written as a superscript above another number or symbol, the base, denoting the number of times the base is to be multiplied by itself.

exponential notation   A form for expressing a number using a product of two numbers; one of the numbers is a decimal and the other is a power of 10.

formula unit   Generally, the smallest combination of *charged* particles (ions) in which the opposite charges present balance each other so that the overall compound has a net charge of zero, such as NaCl (Na^{1+}, Cl^{1-}).

Gay-Lussac's (gā'lü·sȧk') **law**   Principle that, at constant volume, the pressure of a fixed mass of a given gas is *directly* proportional to the temperature in *kelvins*.

Gay-Lussac's law of combining volumes   Principle that, at the same temperature and pressure, whenever gases react or gases are formed, they do so in the ratio of small whole numbers by volume.

groups   The 18 vertical columns in the Periodic Table.

heat of reaction   The number of calories or joules of heat energy evolved or absorbed in a given chemical reaction per given amount of reactants and/or products.

heterogeneous matter   Matter not uniform in composition and properties and consisting of two or more physically distinct portions or phases unevenly distributed.

homogeneous matter   Matter uniform in composition and properties throughout.

homogeneous mixture   Matter homogeneous throughout but composed of two or more pure substances whose proportions may be varied in some cases *without limit*.

hydrate   A crystalline substance that contains chemically bound water molecules in definite proportions. An example is Epsom salts, magnesium sulfate *hepta*hydrate ($MgSO_4 \cdot 7H_2O$).

ideal-gas equation   Formula that allows scientists to vary not only the temperature, pressure, and volume of a gas, but also its mass; it is expressed mathematically as $PV = nRT$, where P is pressure, V is volume, n is the amount of gas in moles, T is temperature, and R is universal gas constant.

ion   Atom or group of atoms in which the number of electrons do not equal the number of protons; it carries either a positive or negative charge.

ionic bond   A bond formed by the transfer of one or more electrons from one atom to another. The resulting positively and negatively charged species attract each other and form a bond.

ionic charge   The charge on an ion. The ion may consist of a single atom or a group of atoms bonded together.

ionic equation   A method of expressing a chemical reaction involving compounds that exist mostly in ionic form in aqueous solution.

ionic equation, net   An ionic equation that shows only those *ions* that have undergone a chemical change.

isotopes   Atoms having different atomic masses or mass numbers but the same atomic number.

joule (J)   The standard unit for the measurement of heat energy in the International System of units (SI units).

Lewis structures   A method of expressing the covalent bonds among atoms in a molecule using the rule of eight or octet rule and dots (:) to represent bonds.

limiting reagent   Reactant in a chemical reaction that is completely used up in the reaction, so called because the amount of the reactant limits the amount of new compounds that can be formed.

mass   The quantity of matter in a particular body.

mass number   Sum of the number of protons and neutrons in the nucleus of an atom of an element.

matter   Anything that has mass and occupies space.

metalloid   Elements (except aluminum) that lie on the heavy *black* stair step line in the Periodic Table; they have both metallic and nonmetallic properties.

metals   Elements that generally have a high luster, conduct electricity and heat well, are malleable and ductile, have high densities and melting points, are hard, and do not readily combine with one another.

mixture   Matter composed of two or more substances, each of which retains its identity and specific properties.

molality (*m*)   The measure of the concentration of a solution expressed as the number of moles of solute per *kilogram* of *solvent*:

$$m = \text{molality} = \frac{\text{moles of solute}}{\text{kilogram of } \textbf{\textit{solvent}}}$$

molarity (*M*)   The measure of the concentration of a solution expressed as the number of moles of solute per *liter* of *solution*:

$$M = \text{molarity} = \frac{\text{moles of solute}}{\text{liter of } \textit{solution}}$$

molar mass   The mass in grams of one mole of a substance.

molar volume of a gas   The volume occupied by 1 mol of any gas, 22.4 L of gas molecules at 0°C and 760 torr.

mole (mol)   The amount of a substance that contains the same number of atoms, formula units, molecules, or ions as there are atoms in exactly 12 g of carbon-12, (approximately 6.02×10^{23} atoms).

molecular formula   A formula composed of an appropriate number of symbols of elements representing *one* molecule of the given compound. Also defined as the formula containing the *actual* number of atoms of each element in *one* molecule of the compound.

molecule   The smallest particle of a pure substance (element or compound) that can exist and still retain the physical and chemical properties of the substance, such as O_2 and H_2O.

neutralization reaction   A process in which an acid (**H*X***) or an acid oxide reacts with a base (***M*OH**) or basic oxide. In most of these reactions, water is one of the products:

$$\text{H}X + M\text{OH} \longrightarrow MX + \text{HOH}$$

neutron   A particle that has no charge and a mass of 1.6748×10^{-24} g or 1.0087 amu (approximately 1 amu).

nonelectrolytes   Substances whose aqueous solution *not* conduct an electric current.

nonmetals   Elements that generally are dull, do not conduct electricity and heat well, are brittle, have low densities and melting points, are soft, and combine readily with metals and with other nonmetals.

normality (*N*)   The measure of the concentration of solution expressed as the number of equivalents of solute per *liter* of *solution*:

$$N = \text{normality} = \frac{\text{equivalents of solute}}{\text{liter of } \textit{solution}}$$

octet rule   In the formation of molecules from atoms most molecules attempt to obtain this stable configuration of 8 electrons in the valence energy level of each atom.

orbital   A region in space about the nucleus where a particular electron might be expected to be found. A 1*s* orbital is that region about the nucleus where a 1*s* electron is most likely to be found.

oxidation   A chemical change in which a substance loses electrons, or one or more elements in it increase in oxidation number.

oxidation number   Usually, a positive or negative whole number used to describe the combining capacity of an element in a compound.

oxidizing agent   The substance reduced in an oxidation–reduction reaction.

percent by mass   The measure of the concentrations of a solution expressed as parts by mass of solute per 100 parts by mass of *solution*:

$$\text{Percent by mass} = \frac{\text{mass of solute}}{\text{mass of } \textit{solution}} \times 100$$

percent yield Percent of the theoretical yield that is actually obtained in a chemical reaction.

$$\text{Percent yield} = \frac{\text{actual yield}}{\text{theoretical yield}} \times 100$$

Periodic Law Elements with similar chemical properties appear at regular intervals in the Periodic Table when they are listed in order of increasing atomic number.

periods The seven horizontal rows in the Periodic Table.

pH Quantitative way of expressing the acidic or basic nature of solutions using the negative logarithmic value of their hydrogen ion (H^{1+}) concentration, **pH** = $-\log[H^{1+}]$; a solution with pH < 7 is acidic; one with pH > 7 is basic; one with pH = 7 is neutral.

physical change Change in a substance that occurs with no change in the composition of the substance taking place.

physical property Property of a substance that can be observed without changing the composition of the substance.

pOH Quantitative way of expressing the acidic or basic nature of solutions using the negative logarithmic value of their hydroxide ion (OH^{1-}) concentration, **pOH** = $-\log[OH^{1-}]$; a solution with pOH < 7 is basic; one with pOH > 7 is acidic; one with pOH = 7 is neutral.

$$\textbf{pH + pOH = 14.00}$$

polyatomic ion An ion consisting of two or more covalently bonded atoms that possess a net negative or positive charge on the ion.

pressure Force per unit area.

proton A particle that has a relative charge of +1 (actual charge = +1.602 $\times 10^{-19}$ coulomb) and a mass of 1.6725×10^{-24} g or 1.0073 amu (approximately 1 amu).

reaction, combination A process in which two or more substances (either elements or compounds) react to produce one substance.

$$A + Z \longrightarrow AZ, \qquad \text{where } A \text{ and } Z \text{ are elements or compounds}$$

reaction, decomposition A process in which one substance undergoes a reaction to form two or more substances:

$$AZ \longrightarrow A + Z, \qquad \text{where } A \text{ and } Z \text{ are elements or compounds}$$

reaction, double-replacement A process in which two compounds undergo a reaction where the positive ion (cation) of one compound exchanges with the positive ion (cation) of the second compound:

$$AX + BZ \longrightarrow AZ + BX$$

reaction, neutralization   A process in which an acid (**H**X) or an acid oxide reacts with a base (**M**OH) or a basic oxide. In most of these reactions, water is one of the products:

$$H\!X + M\!OH \longrightarrow MX + HOH$$

reaction, single-replacement   A process in which an element (**A** or **X**) replaces another element (**B** or **Z**) in a compound:

1. A metal replacing a metal ion in its salt or hydrogen ion in an acid:

$$A + B\!Z \longrightarrow AZ + B$$

2. A nonmetal replacing a nonmetal ion in its salt or acid:

$$X + B\!Z \longrightarrow BX + Z$$

reducing agent   The substance oxidized in an oxidation-reduction reaction.

reduction   A chemical change in which a substance gains electrons, or one or more elements in it decrease in oxidation number.

representative elements   The elements in the A group elements (1, 2, 13, 14, 15, 16, 17, 18).

rule of eight or octet rule   In the formation of molecules from atoms, most molecules attempt to obtain this stable configuration of 8 electrons in the valence energy level of each atom.

rule of two   An exception to the rule of eight (octet rule). A completed first principal energy level is also a stable configuration. Helium atoms and hydrogen atoms in the combined state obey this rule.

salt   Ionic compound made up of a positively charged ion (cation) and a negatively charged ion (anion).

scientific notation   A more systematic form of exponential notation in which the decimal part must have *exactly one* nonzero digit to the left of the decimal point.

significant digits (figures)   The number of digits in a measurement that are known to be precise, along with a final digit about which there is some uncertainty.

single-replacement reaction   A process in which an element (**A** or **X**) replaces another element (**B** or **Z**) in a compound:

1. A metal replacing a metal ion in its salt or hydrogen ion in an acid:

$$A + B\!Z \longrightarrow AZ + B$$

2. A nonmetal replacing a nonmetal ion in its salt or acid:

$$X + B\!Z \longrightarrow BX + Z$$

solute   Substance dissolved in a solution; it is usually present in lesser quantity than the solvent.

solution   Homogeneous mixture composed of two or more pure substances; its composition can be varied usually *within certain limits.*

solvent   Dissolving substance in a solution; it is usually present in greater quantity than the solute.

specific gravity   Density of a substance divided by the density of some substance taken as a standard, usually water at 4°C:

$$\text{Specific gravity} = \frac{\text{density of substance}}{\text{density of water at 4°C}}$$

It is the ratio of the density of the substance to the density of the standard.

stoichiometry   The measurement of the relative quantities of chemical reactants and products in a reaction.

structural formula   Formula showing the arrangement of atoms within a molecule, using a dash (—) for each pair of electrons shared between atoms.

substance, pure   Homogeneous matter characterized by definite and constant composition, and definite and constant properties under a given set of conditions.

temperature   Degree of hotness of matter.

theoretical yield   The amount of products obtained when we assume that all of the limiting reagent (reactant completely used up) forms products and that none of the products are lost during isolation and purification.

titration   A procedure for determining the *concentration* of an acid or base by adding a base or an acid of *known concentration.*

transition elements   The elements in the B group elements and the group VIII elements (3, 4, 5, 6, 7, 8, 9, 10, 11, 12).

triple bond   Chemical bond in which two atoms share three pairs of electrons (six electrons).

valence electrons   Electrons occupying the highest principal energy level in an atom.

volume   Cubic space taken up by matter.

weight   Measure of the gravitational force of attraction between the body's mass and the mass of the planet or satellite on which it is weighed.

word equation   Expresses a chemical equation in words instead of symbols and formulas.

Index

Rules for the Solubility of Inorganic Substances in Water

1. Nearly all *nitrates* (NO_3^{1-}) and *acetates* ($C_2H_3O_2^{1-}$) are *soluble*.

2. All *chlorides* (Cl^{1-}) are *soluble*, except AgCl, Hg_2Cl_2, and $PbCl_2$. ($PbCl_2$ is soluble in hot water.)

3. All *sulfates* (SO_4^{2-}) are *soluble*, except $BaSO_4$, $SrSO_4$, and $PbSO_4$. ($CaSO_4$ and $AgSO_4$ are only slightly soluble.)

4. Most of the *alkali metal* [Group IA(1), Li, Na, K, etc.] salts and *ammonium* (NH_4^{1+}) salts are *soluble*.

5. All the common *acids* are *soluble*.

6. All *oxides* (O^{2-}) and *hydroxides* (OH^{1-}) are **insoluble**, except those of the alkali metals and certain alkaline earth metals [Group IIA(2), Ca, Sr, Ba, Ra]. [$Ca(OH)_2$ is only moderately soluble.]

7. All *sulfides* (S^{2-}) are **insoluble**, except those of the alkali metals, alkaline earth metals, and ammonium sulfide.

8. All *phosphates* (PO_4^{3-}) and *carbonates* (CO_3^{2-}) are **insoluble**, except those of the alkali metals and ammonium salts.

Electromotive Series or Activity Series

Li
K
Ba
Ca
Na
Mg
Al
Zn
Fe
Cd
Ni
Sn
Pb
(**H**)
Cu
Hg
Ag
Au